Merry Xmas to

Ruth

[illegible]

1986

INTENSIVE GRAZING MANAGEMENT: FORAGE, ANIMALS, MEN, PROFITS

Burt Smith

PingSun Leung

George Love

The Graziers Hui
Hawaii

Cover graphics by Ms. Marlene Honma.
Graphs by Mr. Kulavit Wanitprapha.

Library of Congress Catalog Number 86-081674

The Graziers Hui, P.O. Box 1944, Kamuela, Hawaii 96743

PREFACE

The management involved in the process of obtaining meat varies from hunter-gather tribes on one end, to intensively managed farms on the other. A modern U.S. poultry operations or a New Zealand strip grazing bull farm would be examples of intensively managed operations. The typical modern ranch utilizing continuous grazing as its major grazing policy is not all that far removed in management effort from the tribal hunter-gather societies of the not too distant past. If that sounds a bit heavy, consider that the U.S. national average of percent calves weaned is 70% (60% in Australia), most animals in the wild do that good, many better. Other than providing, water, some supplements and a little health care; just what does the average rancher do in the way of management? Don't confuse white picket fences, expensive bulls, a fleet of trucks, a lot of equipment (heavy metal), computers and all the other trappings of "modern technology", with management. Management is defining goals and objectives, setting production targets, then making it happen.

The open range worked reasonably well under a certain set of social and environmental conditions, as did the philosophy of: "All you have to do is produce it, the rest will take care of itself." The advent of barb wire radically changed those conditions and physically eliminated the open range; the consumer has re-defined the market. While the environment and the market has changed the perception of both has not change at the same pace. Thanks in part to the "Cowboy Myth" the tradition of the "Wild West" (open range), is very much alive, but not so well. As long as cattle are raised, simply as away to keep the land occupied, reduce local property taxes or for various tax reasons, the open range mentality works admirably. However, in the modern scheme of things, competition from other meats and foods, rising land and material cost, etc., as a way to make a living the open range, "Wild West" way of thinking just doesn't work anymore.

As anyone that has ever changed religions or nationalities can attest, changing the way you think about a subject is difficult at best. It is not that all the little bits and pieces that make up the sum total of a ranchers knowledge are wrong, they're not. It is simply a different way of putting those bits and pieces together to make a whole. However, one does not change ones lifestyle or way of doing business unless one is dissatisfied. If you are satisfied with your operation, then there is little point in continuing, unless you wish to know what others are up to. However, if you are dissatisfied, this book is a starting point.

ACKNOWLEDGEMENTS

Very likely this book would not have been written had we not ended up in Hawaii. Hawaii, with its dazzling array of environments, all within short distances of each other, has allowed observations to be made on the effects of different grazing practices and how they relate to the many varied environments. Then there are the intrepid Hawaiian Ranchers, who keep plugging away even when the price of calves falls below 32 cents a pound (65 cents a pound for the same calf on the mainland U.S.). Without their continued faith in the industry, which is mirrored by ranchers everywhere, there would be little point in writing a book on grazing.

Other than in a few instances we have not attempted to give credit for every word or piece of work that is not original. To attempt to do so would make this book look like a scientific paper and as unreadable. At the end of each chapter is provided a suggested reading section which will allow the reader to pursue different subjects in greater detail. The section consists mostly of books, which can be ordered from book stores, or if the book or pamphlet is not in general circulation, we have attempted to provide an address where you can write. Scientific papers, which are not easily available to the average rancher, have been, for the most part, omitted.

We would like to mention a few individuals that have shaped our thinking: The late Dr. Gerald O. Mott, Professor of Agronomy at the University of Florida; Dr. Howard T. Odum, Professor of Environmental Ecology, University of Florida; Mr. Allan Savory, Center for Holistic Resource Management, Albuquerque, New Mexico; Mr. Keith Milligan, District Farm Advisory Officer, Ministry of Agriculture and Fisheries, New Zealand.

We would especially like to thank Dr. An Peischel, of the University of Hawaii at Hilo and Mr. Gordon Cran, a Kau rancher, for their critical review of earlier drafts of this book and in particular Mr. Kulavit Wanitprapha for the long hours spent in final preparation of this endeavor.

We are indebted to the College of Tropical Agriculture and Human Resource, University of Hawaii at Manoa for their assistance and cooperation during the many months that this book was in preparation.

The senior author would like to acknowledge Dr. Larry White, Extension Range Specialist, Texas A&M, who, when at the Univ. of Florida, conned him into staying with agriculture; Mr. Paul Penrod, a trader from Idaho, who taught him what being a cattleman and friendship was all about and finally, his wife Martha, who thought she was marrying an engineer and ended up having to cope with a great deal of dirt and manure, but then, with his initials (B.S.), what else should she expect.

ABOUT THE AUTHORS

BURT SMITH: Born in Douglas Wyoming, way back when. Married with three children. Spent 13 years as owner-operator of a diversified farm and dairy in Northwestern Utah, a cattle ranch in Central Nevada and a diversified farm in Central Kentucky. Spent five years as an engineer for General Dynamics Astronautics, San Diego California and Thiokol Chemical Co. Brigham City Utah. Received Bachelors degree from San Diego State College in Mathematics, Masters of Science degree from Western Kentucky University in Biology and PhD from the University of Florida in Agronomy, Range Management and Systems Ecology. Presently is the Extension Specialist in Livestock and Pasture Management, University of Hawaii at Manoa. In 1985, nominated as Livestock Man of the Year at the annual meeting of the Hawaii Cattlemen Association.

PINGSUN LEUNG: Assistant Professor of Agricultural and Resource Economics at the University of Hawaii at Manoa. Research and teaching in applied quantitative methods, production economics and computer applications. Has worked closely with the senior author on various intensive grazing projects and workshops. Received B.B.A. in business economics and statistics, M.S. in information and computer science, and Ph.D. in agricultural systems analysis.

GEORGE LOVE: Resource Conservationist, Soil Conservation Service, Honolulu, Hawaii. Family has been involved in ranching for four generations in West Texas. Received a B.S. in Animal Science and an M.S. in Range Science from Sul Ross State University. Has worked with the senior author for the last six years in conducting trials and workships on intensive grazing. This training has also been presented on the islands of Guam, Tinian, Rota and Saipan.

CONTRIBUTORS

Gordon Cran: Owner-Manager, Kapapapla Ranch, Big Island, Hawaii.

Steve Damron: Assistant Professor Animal Science, University of Hawaii at Hilo; presently at Univ. of Wisconsin (PhD).

Ron Guyton: Researcher, Department of Agronomy and Soils, University of Hawaii Experiment Station (PhD).

Steve Hughes: Assistant Professor, Department of Animal Science, University of Hawaii, Hilo (PhD).

Wade McCall: Soil Specialist, Department of Agronomy and Soils, University of Hawaii at Manoa (PhD).

John Morgan: Owner-Manager, Kualoa Ranch, Oahu, Hawaii.

Phlip Motooka: Extension Specialist in Weed Science, University of Hawaii at Manoa (PhD).

An Peischel: Assistant Professor, Department of Animal Science, University of Hawaii, Hilo (PhD).

Peter Rotar: Professor, Department of Agronomy and Soils, University of Hawaii at Manoa (PhD)

Earl Spence: Representative KM Seed Co. Kamuela; Former Agronomist, Parker Ranch, Kamuela, Hawaii.

Jack Sprague: District Conservationist (retired), Soil Conservation Service, Kamuela, Hawaii.

Sheldon Whitney: Researcher (retired), Department of Agronomy and Soils, University of Hawaii at Manoa (PhD).

TABLE OF CONTENTS

PREFACE iii
ACKNOWLEDGEMENTS iv
ABOUT THE AUTHORS v
CONTRIBUTORS vi
LIST OF TABLES xiii
LIST OF FIGURES xv
GLOSSARY xvii
METRIC CONVERSIONS xxvi

CHAPTER

1. AN INTRODUCTION TO INTENSIVE GRAZING: A slightly different way of looking at an old problem. 1
   - Energy in Pastoral Agriculture 2
   - Control 6

2. GETTING STARTED 9
   - First Steps 10
   - Inventories 11
   - Objectives 13
     - Production Targets 14
     - Performance Targets 16
   - Initial Wrap-up 16
     - Summary 17
   - Some Thoughts on Organization 17
     - The 80-20 Rule 17
     - People 18
       - The Family Outfit 18
       - Employees 19
       - Middle Management Control 20
     - Motivation 21
   - Other Considerations 23
     - Thoughts on Management of the Larger Ranch; by Earl Spence 23
     - The Leaseholder Point of View; by Gordon Cran 26
     - Contract Agreements 27
     - Diversification; by John Morgan 28

3. FINANCIAL ANALYSIS 30
   - Unit of Production 31
   - Enterprise Analysis 31
   - Yields and Associated Cost 33
     - Turnover Rate 37
   - Stock Flows 37
     - Percent Calf Crop 37
       - What is the Percent Calf Crop and Weaning Weight Worth? 38
     - Herd Structure 38
     - Herd Replacement 40
     - The Stock Flow Chart 40
   - Cash Flows 42
     - Cash Flow Budgets 42

4. SOIL BASICS 46
Characteristics 48
Cations, Anions and Other Creatures 50
Organic Matter 50
Other Soil Factors 51
Soil Organisms 52
PH or Soil Reaction 53
Irrigation 55
Soil Fertility 55
Fertilization 58
Manure 60

5. PLANT BASICS 63
Metabolic Pathways 64
The C3 Cycle 65
The C4 Cycle 65
The CAM Cycle 66
Leaf Area Index and Canopy Cover 67
Grass 68
Classification 68
Tropical and Temperate Grasses 69
Growth and Yield 72
Effects of Grazing 74
Residual Dry Matter 76
Estimates for Minimum Grazing 78
Estimating Forage Quantity 78
Forage Quality 79
The Plant 79
The Pasture and/or Range 81
Leaves and Stems 84
Forage Legumes 84
Management Considerations 87
Toxic Forages 88

6. ANIMAL BASICS 93
Behavior 94
General 95
Social 96
Sexual 98
Reproduction and Maternal 99
Herding 100
Environmental 101
Stress 102
Foraging 103
Nutrition 106
Minerals 108
Salt 109
Vitamins 109
Oxygen 110
Water 111
Quantity 112
Quality 112
Water Supply 114

Energy 116
Protein 118
Animal Unit 119
Reproduction and Genetics by An Peischel and Burt Smith 120
Reproduction 121
Genetics 123
Selection 124
Bulls 126
Sheep 128

7. ANIMAL HEALTH 131
The Healthy Animal 132
Giving Injections 133
Other Methods of Administrating Drugs 135
Drugs 136
Pharmaceuticals 136
Biologicals 137
Parasites 138
Worming 139
Other Parasites 140
Diseases 140
Prevention 140
Treatment 142
Care of Newborn 143
Colostrum 144
Grafting a Young to a Different Mother 145
Raising the Calf in Confinement 146
Care at Branding 146
Castration 147
Dehorning 149
Other 149

8. BASIC ECOLOGY 151
More on Energy 153
The Concept of Ecosystem Climax 154
Pulsing 156
Energy in Genetics 159
Cycles 160
Water Cycle 160
Mineral Cycles 161
Production and Other Things 162
Limiting Factor 163
The Role of the Animal 164
Rangeland 166
Ecological Trends 167

9. ELECTRIC FENCING AND DESIGN 169
Some Electrical Basics 171
Components 172
The Energizer 173
Purchasing an Energizer 174
Safety 175
Grounding 175

	Wire	176
	Temporary Wire	178
	Some No-No's	179
	Post	180
	Others	181
	The Animal	183
	Training	185
	Livestock	185
	Wildlife	187
	The Fence	188
	Livestock Fences	190
	Game Fences	190
	Miscellaneous Fence Designs	193
	Rejuvenating Non-Electric Fences	193
	Gates	193
	Floodgates	195
	Testing	197
10.	**GRAZING UNIT DESIGN**	199
	Planning the Unit - Wheel Design; by Jack Sprague	201
	Laying out the Hub	204
	Block Designs	208
	Texas Block	210
	New Zealand Block	211
	Strip and Break Grazing	212
	In Actual Fact	214
	Corridors and Alleyways	215
	Working Facilities	215
	Sheep	216
	Cattle	218
11.	**BASIC GRAZING**	221
	Essential Requirements	222
	Tether Grazing	223
	Natural, Continuous Grazing and Set Stocking	224
	Sub-division	226
	Rotational Grazing	228
	A Little History	228
	Principals of Rotational Grazing, or the "Hang Loose" Approach to Grazing	230
	Forage Allocation	231
	Feed Budgets	234
	Determinations of Forage Quality	234
	Sample Feed Budget for a Small Cow/Calf Outfit	235
	The All Purpose Chart - APC	237
	Shifting	238
	Extensive vs Intensive	241
	Management	241
	Extensive Management	241
	Intensive Management	242
	Environments	242
	Extensive Environments	242
	Intensive Environments	244

12.	**FORAGE CONSERVATION**	**246**
	Hay and Silage	247
	Silage; by Ron Guyton & Burt Smith	248
	Hay	252
	Other Forage Conservation Techniques	254
	Standing Hay	254
	Feed Banking	255
	Living Hay	255
	Intercropping	256
	Supplementation	257
13.	**HI-TECH**	**261**
	Computers	262
	What a Computer Does	262
	Terms	264
	Getting Started	266
	Controlling Pasture Weeds with Herbicides; by Philip S. Motooka	267
	Foliar Application	267
	Calibration	268
	Managing Herbicides for Efficacy and Economy	270
	Surfactants	272
	Drift and Drift Control	272
	Spot Treatments	273
	Japanese Quarter Horses	275
	Electronics	276
	Pasture Meters	277
	Scales	278
	Forage Fed and Bull Beef	280
	Forage Finishing	280
	Bull Grazing Management	281
	Cull Cows	283
	Marketing	284
	Summary	285
14.	**INTENSIVE GRAZING MANAGEMENT**	**286**
	Basics	287
	Planning	288
	Flexibility	288
	Knowledge	289
	Resource Tools	289
	Forage	289
	Animal	290
	Supplemental	291
	Sacrifice	293
	Meeting the Production Targets	293
	Some Uses of Resource Tools	294
	Staying on Track	298
	Standing Crop	299
	Drought	299
	Drought Forecasting	300
	Growth Rate	301

Yield 301
Soil Moisture 302
Growth Potential 302
Prediction of Yield 302
Quality 303
Handling the Information 303
And Finally 305

THE HAWAIIAN ADDENDUM 307
Physical 309
Climate 310
Biological 311
Hawaiian Forages 312
Pasture and Animal Feeds; by Peter Rotar 312
Recommended Forages for Hawaii; by Sheldon Whitney 315
Intensive Grazing in Hawaii 318

APPENDIXES

1. Forage Nomenclature 324
2. Suggested Intake levels for the Major Nutrients 330
3. Suggested Intake levels for the Trace Minerals 334
4. Energy and Protein Requirements for Steers 337
5. Feed Information for the Newborn; by Steve Damron 339
6. Additional of Fence and Unit Designs 342

INDEX 348

TABLES:

3.1 Required price per pound to break even at different percent calf crop, weaning weights, and cost per mother cow. 39
3.2 Stock flow of cow/calf operation for one year. 41
3.3 Cash flow chart for first quarter. 44

4.1 Essential minerals for plant growth. 48
4.2 Affect of slope on actual land area, rainfall and aerial fertilization application. 52
4.3 Classification of irrigation water by toxic constituents. 56
4.4 Components of various fertilizers. 58
4.5 Chemical composition of different animal manures. 60
4.6 Distribution of various minerals contained in dung and urine. 61

5.1 Crude protein content of three common Hawaiian sub-tropical legumes and three grasses at different stages of re-growth. 86

6.1 Diet selection of four classes of range livestock 106
6.2 Heritability of different economic traits in cattle. 107
6.3 The essential minerals. 108
6.4 Estimated daily water intake of cattle in gallons, at different temperatures. 113
6.5 Total water requirements of sheep in gallons, at different temperatures. 113
6.6 Recommended limits of concentrations of some possible toxic substances found in drinking water for livestock. 114

7.1 Vital signs of common livestock. 133
7.2 First aid kit for ruminants. 144

8.1 Annual productivity and respiration of growth type and steady state ecosystems. 162
8.2 Relative biomass of different consumers. 166

9.1 Cost in man and vehicle hours of construction of various types of livestock fences in Hawaii. 171
9.2 Electrical resistance of hi-tensile fence wire as a function of diameter (gauge). 177
9.3 Methods preferred by different animals in attempting to get past a fence. 184
9.4 Distance cattle react to different types of electric fencing material. 186

10.1 Cell hub planning guide. 206
10.2 Comparison of block vs wagon wheel designs. 209

11.1 Effect of the number of paddocks on a grazing unit. 226
11.2 Feed budget for breeding bulls. 236
11.3 Feed budget for mature cows. 236
11.4 Feed budget for replacement heifers. 237

13.1 Increasing weed control cost with increasing volumes of herbicide spray per acre (volume-rates). 268
13.2 Efficacy and properties of herbicides for pasture-range weed control. 271
13.3 Farm bike size according to various factors. 275
13.4 Sample format and equations for determining average daily gain (ADG). 279

14.1 Efficiency of utilization with various harvesting methods. 298
14.2 Spreadsheet format for determinations of yield and quality functions. 304

A1 Common and scientific names of grasses and probable date of introduction to Hawaii. 313
A2 Recommendations of various grasses for Hawaii. 316
A3 Recommendations of various legumes for Hawaii. 317

A1.1 Common and scientific names of some temperate (C_3 - cool season) grasses. 324
A1.2 Common and scientific names of some tropical (C_4 - warm season) grasses. 326
A1.3 Common and scientific names of some temperate, sub-tropical and tropical legumes. 328

A2.1 Calcium (Ca): requirements for cattle and sheep. 330
A2.2 Phosphorus (P): requirements for cattle and sheep. 331
A2.3 Sodium (Na): requirements for cattle and sheep. 332
A2.4 Magnesium (MG): requirements for cattle and sheep. 333

A3.1 Suggested levels of trace minerals in the diet of mature beef cattle. 334

A4.1 Daily requirements of feed yard animals. 337
A4.2 Daily feed requirements, in Mcals per day, for fattening steers on pasture. 338

A5.1 Suggested whole milk feeding schedule for calves. 339
A5.2 Feeding schedule - other species. 340
A5.3 Composition of different mammals milk. 341
A5.4 Milk replacers for different livestock. 341

FIGURES:

1.1 Energy flow diagram of livestock grazing operation. 4
1.2 Think Big. 7

2.1 Any road will get you there. 14
2.2 Diversification, wind farm. 27

3.1 Additive effect of direct and variable cost. 32
3.2 Expected yield from a single input item. 34
3.3 Cost of obtaining successive increments of yield. 35
3.4 Managerial requirements to obtain high yields and low per unit cost. 35
3.5 Effect of yield and price on profit or loss. 36
3.6 Scheduling budget sessions. 43

4.1 Soil erosion. 47
4.2 Available soil water in different textures of soil. 49
4.3 Soil colloids attract cations but not anions. 51
4.4 Soil formation in action. 57

5.1 Some plants will grow anywhere. 64
5.2 Equation for photosynthesis showing energy stored. 65
5.3 Percent photosynthesis rate (P.S.) as a function of temperature for three different types of plants. 67
5.4 Dependence of productivity as a function of latitude. 68
5.5 Fast growth of tropical grasses. 70
5.6 Basic growth curve. 72
5.7 Decline of net pasture production with increasing pasture mass. 73
5.8 Senescence. 75
5.9 Decline of crude protein content in kikuyu grass as a function of regrowth time. 81
5.10 Relationship of residual dry matter and average daily gains for late spring lambs. 82
5.11 Pre-grazing pasture mass, feed on offer, intakes and daily gains of ewes. 83
5.12 Comparison of relative yields of temperate and tropical grasses and legumes. 86

6.1 Druthers. 94
6.2 Relationship between tiller length, number of bites, size of bites and the resultant intake of grazing sheep. 105
6.3 Water trough for larger paddocks. 115
6.4 Water trough for smaller paddocks hidensity. 116
6.5 Energy flow through an animal and associated energy losses. 117
6.6 Bull libido testing corral. 127

7.1 Cell center. 141

8.1 Depreciation is an energy cost. 153
8.2 Climax as a function of rainfall. 156
8.3 Energy pulsed systems. 157
8.4 Standard yield curve. 163
8.5 Management makes a difference. 165

9.1 The electric pulse from an energizer. 173
9.2 One type of splices for Hi-tensile wire. 177
9.3 Quick connect for temporary fence. 179
9.4 Homemade anchor from PVC pipe. 181

9.5 Nutrient transfer. 189
9.6 Multi-wire fence designs. 191
9.7 Predator fence for feral pigs. 192
9.8 Low cost gate made from discarded elevator cable. 194
9.9 A well constructed floodgate. 195

10.1 An elaborate hub design. 200
10.2 Three basic layouts utilizing the wheel design. 202
10.3 The use of a "dog-leg" to bring a radial into the hub and two methods of further sub-dividing a radial paddock. 203
10.4 A water center for 16 paddocks. 205
10.5 A well designed and solidly constructed circular raceway around a water hub. 207
10.6 A water center that has been divided into two sections. 207
10.7 Center hub with outside raceway fence constructed with electrified wire and gates. 208
10.8 Three major block designs for grazing units. 209
10.9 Strip grazing in Texas. 213
10.10 Cattle grazing on the long acre in New Zealand. 214
10.11 Outdoor sorting den for red deer. 216
10.12 Diagram of a rectangular and circular sheep corral. 217
10.13 Diagram of two circular cattle corrals. 219

11.1 Animal control. 222
11.2 Effect of graze period on the quality of daily forage intake. 228
11.3 All purpose chart (APC). 238
11.4 One-way gate. 240

12.1 Packing a pit silo. 248

13.1 Basic components of a microcomputer. 264

14.1 Basic format. 287
14.2 Animal power. 295
14.3 A high performance machine. 305

A1 Wind. 308
A2 Reluctant surfer. 318
A3 Water wings. 319
A4 Swimming lessons. 320
A5 Up and over. 321

A6.1 "N" brance. 342
A6.2 Dead man post and brace. 342
A6.3 Australian gate. 343
A6.4 Hub design. 343
A6.5 Paddock expansion for hub. 344
A6.6 Proper grounding. 344
A6.7 Underground gate transmission. 345
A6.8 Overhead gate transmission. 345
A6.9 Water gap. 346
A6.10 Two wire visual barrier fence. 346
A6.11 Three wire visual barrier fence. 347

Used with permission of NEA.

They don't plan to fail. They just fail to plan.

GLOSSARY

GLOSSARY

ACUTE: Having a sharp point or angle. In medicine it means coming to a crises quickly; violent.

ALLOCATION: The process of parceling out, over a period of time, a specified amount of forage, or supplement, that is required to meet predetermined animal needs.

AMINE: An organic compound derived from ammonia. A substance that is dissolvable in water. Certain herbicide formulations can only be dissolved in water.

AMPS: A unit of electrical rate of flow. One amp is the rate of electron flow that results when a pressure of one volt is applied across a resistance of one ohm.

ANIMAL UNIT EQUIVALENT: An attempt to express the impact of different species of grazing animals in terms of the AU. This is accomplished by dividing the metabolic weight of the animal into 98.5 (which is the metabolic weight of 1000 lb, or 455 kg cow). The metabolic weight is found by first converting the weight into kilograms then raising this number to the 0.75 power.

ANIMAL UNIT: The animal unit (AU) is the amount of forage (dry matter) required to feed one mature, 1000 lb cow (with or without a calf) for one day. Normally given as 2.5% of the live weight of the animal. A rough unit in that it does not consider the quality of the forage or the physiological state of the animal.

BASE SATURATION: Term used by soil scientist to describe the percent of a soil that is saturated with cations other than hydrogen and aluminum.

BLOCK DESIGN: A grazing unit laid out in rectangular, square or irregular shaped paddocks. The area in each paddock is seldom the same and corridors may or may not be present.

BREAK GRAZING: The daily or twice daily allocation of forage required to satisfy stated performance targets. Normally used in conjunction with temporary electric fence and within the confines of a permanent boundary fence.

CALORIE: The amount of energy (heat) required to raise one gram of pure water one degree Centigrade.

CALVES: Young cattle of either sex that are still nursing their mothers. At weaning they become weaners.

CARRYING CAPACITY (CC): The total number of animals, all species, usually expressed in AU's, that the area is capable of supporting year after year.

CATION: A positively charged ion or molecule. In soils an ion of calcium, phosphorus or other basic ion that is attracted to a negatively charged soil particle.

CELL: An area of land that is handled as an individual grazing unit. Has been closely associated with the wagon wheel design.

CHRONIC: Prolonged or lingering; continuing for a long period.

CLASS: Animals grouped as to same species and/or breed and of similar sex and/or weight and/or physiological state; eg. 900 to 1100 lb dry cows, or 500 lb stocker steers.

COME AWAY: Refers to the length of time that a paddock takes to recover from grazing. A relative term and will vary from season to season even though the paddock is grazed to the same residual level.

CONTINUOUS GRAZING: Maintaining a group or herd of grazing animals in one paddock or pasture for extended periods of time, usually a year or more. Not to be confused with open range grazing or set stocking.

CONTROLLED GRAZING: A New Zealand term describing a livestock grazing program based on maintaining control of the animal, animal numbers, area to be grazed and the length of time the area is to be grazed or rested.

CORRIDOR: A fenced lane for movement of livestock, usually to water or to a fresh paddock. Can be a used as a regular paddock, or under certain conditions may be used as a sacrifice area.

COW DAY: The amount of feed required to feed one cow for one day. Generally denotes the amount of feed required for maintenance only. Similar to animal day or stock day.

CULL: To remove animals from a herd or ranch, based upon failure to meet pre-selected performance targets.

CYCLE: The length of time it takes to graze all of the paddocks in a grazing unit.

DAYS GRAZED (G): The number of days that stock are physically in one paddock during one grazing cycle.

DAYS REST (R): The number of days that a paddock is to be rested, no stock present, during one cycle.

DRY MATTER: Weight of forage after moisture content has been removed through oven drying.

DYSTOCIA: Difficulty in the process of giving birth; the offspring is too large or the dam's pelvis is too small, or both.

ENERGY: A measure of everything. Usually thought of as the ability to do work, but is more. There are different qualities of energy based on how much of a low grade energy was required to make the higher grade. Measured in heat units (kilocalorie).

ENVIRONMENT: The sum total of all climatic, soil, biological and human factors that affect a given operation or ranch. In energy terms, the total energy flows into and out of a system.

ESTER: Any organic compound that is formed from an acid and alcohol. A substance that is dissolved in oil. Certain herbicide formulations are dissolvable only in oil (such as diesel oil).

EXTENSIVE: An environment and/or type of operation that requires a large area of land to supply the necessary yearly forage requirements for grazing animals; generally more than 5 acres per cow. Also applies to a low level of managerial skills applied to a particular ranch or farm enterprise.

FEED ON OFFER (FOO): The total amount of feed or forage that is available to the grazing animal, including supplements.

FLIP-FLOP: In grazing, refers to a 2 paddock system where the herd grazes one paddock for a period of time then is moved to the next paddock for a similar period, then back to the first paddock. In electrical terms, a system that alternates between on or off.

FLOGGING: To remove all of the forage material capable of being removed by the animal or machine used. To flog a paddock is to graze it until bare ground shows. Used to remove old or rank material build-up, such as prior to stock seeding, mechanical re-seeding or other soil preparation.

FOLLOW THROUGH: Where one herd of animals immediately follows another herd through a grazing cycle; eg. running weaners ahead of the main cow herd.

FORAGE ALLOCATION: The allotment or budgeting of forage (FOO) to obtain desired production targets.

FORAGE CONSERVATION: Forage that is produced but whose use is deferred until some later point in time; hay and silage are the two prime examples. Also used to mean the planting of a crop (usually an annual) to be used for grazing during a period of time when normal forage production is low or has stopped, eg. planting annual rye grass or barley to graze during the winter.

GOMER: An animal used to detect cows in heat. Usually bulls that have been made sterile, or unable to penetrate. However, cows that have been given hormone shots also work well and in some cases better than males.

GRAM ATOMIC WEIGHT: The weight in grams, equal to the atomic weight of the element or compound under discussion.

GRAZING DAY: The day as viewed by the grazing animal. How the day is broken up into different times and lengths of grazing. Necessary when less than daily moves are to be used. For example, a cow will typically obtain 50% of its daily intake between sunup and 2:00 PM; therefore shifts should be made prior to sunup and/or after 2:00 PM.

GRAZING DENSITY: The number of animals per unit of land at a given point in time, for example 50 cows/acre on 10 January.

GRAZING FORMULAS:
One herd: $G = R/(P-1)$.
More than one herd in a follow through: $G = R(P-H)$.
More than one herd, not in a follow through: $G = R/[(P/H)-1]$
Where G = days grazed; R = days rested; P = number of paddocks;
H = number of herds.

GRAZING PRESSURE: The physical pressure placed on a given area of land. The grazing density times the length of stay, per unit area, determines the grazing pressure, eg. 20 cow days per acre.

GRAZING UNIT: An area of land which is planned for grazing management as one unit; with one or more herds.

GROUND: In electric fencing, the return pathway back to the energizer, usually carrying a negative charge (-). In a one wire system, the earth serves as the ground, in drier areas there is a ground wire associated with the hot, or positive (+) wire. Also what cowboys hit when they get buck off.

HEAVY METAL: Machinery items that are bought more for convenience rather than as a way to make money, or at least save money.

HERD (H): A collection of grazing animals grouped together and handled as a distinct unit. A herd implies some social structure to the group.

HOGGET: A sheep between weaning and 14-16 months of age; there are ram, ewe and wether hoggets. In the meat trade a hogget carcass is from a sheep between the ages of 12 to 24 months.

HOLISTIC: A term first used in 1922 by A. J. Lotka, incorporated by Jan Smuts in 1926 in his philosophy "Holism". Recently adopted by ecologist. The current meaning is, in essence, the whole is greater than the sum of all the parts. In ranching situations it implies consideration and awareness of all aspects (biological, environmental, economic, cultural and social) and their manipulation to obtain stated objectives.

INDUCTION: Electrical power transfer without physical contact, such as used in transformers. Often occurs in electric fencing when a neutral or poorly grounded earth wire is run too close to a live or hot wire.

INTENSIVE: An environment that meets the yearly forage requirements of a grazing animal on a small area of land; for example, less than 5 acres per cow. Also, an operation that is not intensive by virtue of the environment alone, but can be made so through various management techniques, eg. sub-division, irrigation, fertilization, etc. Also applies to the level of management brought to bear in a grazing or business operation.

JOINING: Putting the rams with the ewes.

JOULE (J): An energy term used in most countries, except the U.S. One Joule equals one watt for one second. One calorie = 4.187 J.

KILOCALORIE (Kcal): A unit used to measure the heat equivalent of energy. The amount of heat required to raise one liter of water one degree Centigrade. One Kcal equals 3.97 Btu's, or 4187 Joules, or 1.17 watt-hours.

LAMB: Young sheep between birth and weaning.

LAX GRAZING: Light grazing, little forage removed, creaming the paddock, grazing only the highest quality material. Used when superior animal performance is desired.

METABOLIC WEIGHT: The weight of the animal in kilograms, raised to the 0.75 power. Used to estimate the maintenance energy requirements of various mammals.

MOB STOCKING: Placing a mob of animals on a unit for a short period of time, usually, but not always for maintenance or flogging.

MOB: A large number of animals similar to a herd, but managed and/or grazed at high densities.

OHM: A unit of electrical resistance to the flow of electrons. One ohm is the resistance that will limit the flow rate to one amp when a pressure of one volt is applied. Ohms = volts/amps.

OPEN RANGE: The "natural" pre-settlement condition of the land. Large areas of grazing land measured in millions of acres; unfenced. Characterized by large herds moving, under their own initiative, over the area during the course of the year. A situation that existed in the western U.S. with bison, prior to settlers with their introduction of cattle and barb wire. Never truly existed with domestic cattle as their movements were always restricted in some manner by humans. The closest it came with cattle was in the days before the big trail drives out of Texas. Remnants may still be found in some parts of Africa.

PADDOCK (P): A subdivision of a grazing unit. Used to in conjunction with an intensive grazing program. Is more than a pasture, as implies an associated level of managerial effort.

PASTURE: Simply refers to a fenced area (any size) where animals are kept, or grazed for varying lengths of time. May also refer to the forage eaten by grazing animals.

PERFORMANCE TARGETS: The pre-selected level of performance that a given class of animals is expected to attain within a given time.

PERFORMANCE: Normally used in reference to animals and is a measure as to how well or poorly an individual or group of similar animals are doing; eg. daily rate of gain, percent calf crop. A time period is either stated or implied. Expressed on an individual animal basis; eg. lbs/animal/day.

POSTING: The dissection of a dead animal in order to determine cause of death, similar to an autopsy.

PRODUCTION TARGETS: The pre-selected production that is expected to be attained on a unit of land within a given time period.

PRODUCTION: Amount of material, plant or animal, produced per unit area of land during a specified time period, for example, pounds of beef/acre/day.

PUGGING: The effects of hoof action on wet or soggy paddocks.

PUT AND TAKE: An attempt to adjust grazing pressure by adding to, or taking animals from a pasture, depending upon present forage conditions.

RATIONALE GRAZING: Term used by Andre Voisin in the late 1930's and early 40's to describe a form of rotational grazing based on growth rate of plants.

RESIDUAL: The amount of forage remaining in a paddock when animals are removed or the level that the paddock is grazed down to before moving the animals; usually expressed on a dry matter basis, however forage height is often used.

RHIZOME: A underground stem, different from a root in that it has nodes; thicken and contains food storage.

ROTATIONAL GRAZING: Any program or system of pasture utilization involving regular shifts of animals from paddock to paddock.

SACRIFICE: The philosophy of sacrificing the few to save the many. Used under adverse conditions where additional time is required for some desired event to occur. An example would be keeping animals on poor feed for a short period of time to allow the rest of the paddocks to attain a certain level of growth.

SAVORY GRAZING METHOD (SGM): Also termed Holistic Resource Method. A holistic approach to resource management advocated by Mr. Allan Savory. An approach to environmental modification through the application of various grazing techniques.

SELECTION: Picking one item over another based on some criteria. Given sufficient mass of forage material animals will select (graze) certain species over others, or even certain parts of a plant over other parts, eg. leaves over stems. Used in the culling process to select animals that can survive and produce under a given set of environmental conditions.

SET STOCKED: Similar to continuous grazed, except that an effort is made on a regular basis to adjust animal numbers to forage conditions by "put and take".

SHIFTS: The movement of animals from one paddock to another.

SIGMOID GROWTH CURVE: The rate at which an organism or group of organisms grow over time. A pasture after hard grazing, starts growing slowly, picks up speed, then tapers off at maturity. If this growth is plotted against time it produces an "S" shaped curve. If the overall growth is fast, the "S" stands upright. In slow growth the "S" sort of falls over on it's side, (lazy S).

SOIL PROFILE: The layering (horizons) of soil, noted in ditch banks and excavations. Starting at the surface, there is usually an organic layer (mulch or O layer), below this is a layer of mineral material (top soil or A layer), below is the subsoil (B layer or horizon), followed by the C horizon which is composed of unconsolidated material (rocks, stones, subsoil, etc.), finally bedrock or parent material.

STANDING CROP: The amount of forage that is physically present at a particular point in time; expressed as dry matter weight per area of land, lbs/acre or kg/ha. Does not consider future growth, only what is available at present. It is measured to the grazing level of the animal, or the cutting height of the machine, for which the pasture is intended. Not to be confused with yield.

STEER: A young castrated male bovine. When mature called ox or bullock.

STOCK DAY: The amount of forage a stated species, breed or class of livestock will consume in one day; similar to cow day.

STOCK SEEDING: The use of livestock to establish or improve vegetative cover in a paddock. Paddock is first flogged, seed applied, then animals used to tramp in seed.

STOCKER: Young, growing cattle of either sex; may be weaners, yearlings or even two year olds. Often refers to operations that grow out weaners until ready for the feedyard.

STOCKING RATE (SR): The number of livestock, or animal units that can be grazed on an area of land over a given period of time and meet their predetermined performance targets. Traditionally has been expressed as one cow (sheep, goat, deer) to however many acres (hectares) required; for example 1:10 (one cow to 10 acres).

STOLON: A horizontal stem, runner, on the surface of the ground, capable of taking root.

STOMATA: Small, microscopic openings found on the leaves and some stems of plants. Various gases, such as oxygen, carbon dioxide and water vapor, that are required or formed as a result of the plants metabolism, flow into or out of the plant through these openings. Stomata are able to vary their size, depending on the water content in the plant. During drought stomata reduce the size of the opening, thus cutting down on the amount of water lost through respiration.

STRIP GRAZING: Similar to break grazing, but differs in that it takes place in an existing small paddock; consequently is not as efficient as some forage will be either over or under utilized. Typical example is in a small rectangular paddock where the temporary fence is moved forward allowing the animals fresh forage, and a back fence is moved up behind the animals to keep them from grazing the old area.

SUBDIVISION: The dividing up of an area of land or pasture into smaller grazing areas termed paddocks.

SWARD: A vague term that refers to the plant density of a paddock. Also means, to a lesser extent, the forage composition of a paddock. Sometimes used in place of pasture.

TOOLS: Any of a number of management and/or biological techniques that may be used to manipulate the energy flows into or out of a ranch or farm system.

TOXIC: Due to, or caused by, a poison; poisonous.

TUCKER: An Australian and/or New Zealand term meaning feed or food.

VOLTS: A unit of electrical pressure which causes electrical current to flow.

WAGON WHEEL: A design of a grazing unit that from the air looks like a wheel with spokes. The center (hub) normally contains water, minerals and often stock handling facilities; the spokes (radials) are the fence lines radiating out to the perimeter. Useful in environments where effective soil moisture and/or stock handling or livestock water are problems. Generally not recommended in rainfall zones above 20 to 30 inches a year unless to overcome a particular problem.

WATTS: Units of electrical rate of doing work (power). Watts = volts times amps. One horsepower equals 746 watts.

WEANERS: Young growing animals of either sex that have just been weaned from their mothers. Cattle are generally called weaners until they're about a year old, then they're called yearlings.

WETHER: A castrated male sheep of any age.

YEARLINGS: Young growing animals of either sex that are about one year old; usually with two permanent incisors. Less than a year - short yearlings, over a year - long yearlings.

YIELD: The amount of forage that an area has or is expected to produce over a stated period of time; expressed as dry matter weight per unit of land (lbs/acre or kg/ha). Should be measured to the height of intended utilization. Not to be confused with standing crop.

METRIC CONVERSIONS

A ROUGH APPROACH TO METRICS:

A gram weighs about as much as a paper clip.
A pound is a half a kilogram, just a tad less.
1 liter is about 1 quart, just a tad more.
1 meter is about a yard, just a tad more.
3 meters is just about 10 feet, just a tad less.
1 hectare is 2.5 acres, just a tad less.

To convert Centigrade (C) degrees to Fahrenheit (F) degrees, double the Centigrade and add 32 (this works pretty good to about 30 degrees Centigrade; at 30 deg. C you will be 6 deg. F higher than it really is). To go from degrees F to degrees C it's just the reverse, subtract 32 and then divide by 2.

METRICS AND CONVERSIONS - FOR THOSE WHO LIKE TO BE ACCURATE:

Kilo in metric language means 1000, so anything with kilo in front of it means 1000 of whatever it was in front of, for example 1 kilograms = 1000 grams. All units in metrics go up and down by shifting the decimal point around, there are no conversions as in 3 feet to a yard.

1 metric ton (MT) = 1000 kilograms (kg) = 1,000,000 grams (g)
1 kilometer (km) = 1000 meters (m) = 100,000 centimeters (cm)
1 meter (m) = 100 centimeters (cm) = 1000 millimeters (mm)
1 centimeter (cm) = 10 millimeters (mm)
1 cubic meter = 1000 liters (l)
1 hectare (ha) = 10,000 square meters

1 gram (g) = 0.0352 ounces (oz); 1 oz = 28.35 g
1 kilogram (kg) = 2.205 pounds (lbs); 1 lb = 454 g = 0.454 kg
1 metric ton (MT) = 1.1 English tons (T); 1 T = 0.907 MT

1 centimeter (cm) = 0.394 inches (in); 1 in = 2.54 cm
1 meter (m) = 3.281 feet (ft); 1 ft = 0.305 m
1 kilometer (km) = 0.621 miles (mi); 1 mi = 1.609 km

1 liter (l) = 1.057 quarts (qt); 1 qt = 0.946 l
1 liter (l) = 0.264 gallons (gal); 1 gal = 3.785 l
1 cubic meter = 264.17 gallons (gal); 1 gal = 0.004 cubic meters

1 square m = 10.764 square ft; 1 sq ft = 0.0929 sq m
1 square km = 0.386 square miles; 1 sq mi = 2.59 sq km
1 hectare (ha) = 2.471 acres (ac); 1 ac = 0.405 ha
Note: 1 acre = 43,560 square feet

1 kg/ha = 0.981 lbs/ac; 1 lb/ac = 1.12 kg/ha
1 Metric Ton/ha = 0.446 Ton/ac; 1 T/ac = 2.24 MT/ha

Centigrade = (F - 32)(5/9); Fahrenheit = C(9/5) + 32

Used with permission of NEA.

An attitude re-adjustment is often made under stress.

1 AN INTRODUCTION TO INTENSIVE GRAZING: A slightly different way of looking at an old problem.

A man went into a doctor's office for a check-up and was told to strip to the waist, so he took off his shoes, socks and pants.

Now this may not seem to have much to do with grazing and it doesn't, but it does make a point. Stripping to the waist, from the bottom up, is a bit out of the ordinary; but the job was done, as per instructions. The point is this: There are many ways to do a job, and some ways are better than others, but rare indeed are the occasions when there is only one way. Many in the livestock business have either conditioned themselves, or been conditioned to believe that there is only one way to graze livestock; there are other ways. All it takes is an open mind and a little imagination.

That intensive grazing management (IGM) can work, is no longer in question. It has been demonstrated on the mainland USA and in New Zealand where virtually the entire livestock sector is practicing it in one form or another. It was introduced to Hawaii in 1980, and by 1985 two ranches have joined the 700 club; over 700 lbs of beef (liveweight) produced per acre per year, without irrigation or fertilization. Thus far every unit that has been placed in operation in Hawaii has achieved not only improved animal performance, but also significant increases in stocking rate (see the Hawaii Addendum for more on IGM in Hawaii). The same is occurring on the mainland U.S. with greater and greater frequency.

Yet failures have occurred and will occur in the future. When they happen we say that it was management, and while true, it is not a very satisfying answer.

As you will see as you journey through this book, IGM requires a knowledge of many things. Yet simply knowing is not enough, one must be able to select the appropriate tool (knowledge) to fit whatever conditions arise. More, one must be able to predict those conditions before they arise; in short, manage. The more intensely one manages the land, the greater the probability of survival and success. But the flip side has to do with the greater penalty one pays for mismanagement. The higher the production, the greater the flow of energy, materials and goods into and out of the system and the greater the control required. Like airports, if it's a dirt strip and only the local crop duster uses it, not much control needed, but if it's a large international airport with jets landing every minute, the level of control required is awesome in comparison.

ENERGY IN PASTORAL AGRICULTURE:

Energy is universal and flows everywhere. It can be stored for later use or used immediately. It can be concentrated as in an explosion, or spread around so thin as to have little noticeable effect, such as trying to raise the water level by spitting in the ocean. We normally think of energy in the form of gasoline, electricity, oil, coal, etc; stuff that makes things go. Yet, everything can be represented by energy: fertilizer, plants, animals, even humans. The important thing to remember is that there is only one primary source of energy, the sun. With the exception of nuclear fission and fusion, every other thing that we think of as energy is a secondary source; one way or another it all came from the sun. That bag of fertilizer or gallon of gas was at one time sunlight. All other inputs into a ranching operation, other than sunlight, are energy subsidies or supplements.

Energy flows through a system, it does not cycle back around again, such as minerals do. However, just as there are different chemical compounds using the same elements, there are different states or qualities of energy. The energy contained in a plant is of higher quality than that of the sun, the energy of an animal higher than that of a plant, and the energy contained by the manager, higher than all. It has to do with how much primary energy went into the making of the secondary source. For example, it takes about 1000 kilocalories (Kcals) of sunlight to make 1 Kcal of plant, 2 Kcals of plant to make 1 Kcal of coal, and 3.4 Kcal's of coal to make 1 Kcal of electricity.[1]

Just because humans have a higher quality energy than plants, doesn't mean much, unless that high quality energy is used as high quality energy. Wood is low quality energy and electricity is high quality energy, both can warm a house, but only one can run a computer. The thing about high quality energy is that it can be used as an amplification factor, that is, it can be used to increase the amount of usable (net) low quality energy for a particular function. Knowledge is high quality energy, for instance knowing which plants are weeds and which are desired food plants. If this knowledge is put to use by weeding, a greater amount of sunlight (low quality energy) will go to the food plants; this is energy amplification. Without this high quality energy (knowledge of which plants are weeds) weeding a garden could be a total disaster. Using high quality energy as a control, that is to amplify low energy sources, presents an apparent exception to the statement that energy is not cycled.

Figure 1.1 is a simplified energy diagram of a typical livestock grazing operation. The sun provides the low quality energy that the plant utilizes for production. The climate (a direct product of the sun) provides a higher quality energy in the form of warmth and rainfall to the plant (control actions). Livestock graze the plants and in turn are marketed and eaten by humans. Man acts as the high quality energy source controlling fertilizer inputs to the plants, timing of livestock entry into paddocks and marketing of the animals. The animals grazing on the plants are another control function, as are the decomposer organisms that influence soil properties that in turn affect the plants. Some energy is lost as heat at every process.[2] There are many more such control functions taking place every day on any ranch or farm.

The amount of energy striking one square meter (1.2 sq. yds), of the upper atmosphere is 5110 kilocalories (Kcal) per day. By the time the sun's energy gets down to the surface of the earth, only 3400 Kcal per square meter are left. This energy drives several physical processes, such as expansion and cooling of rocks, creating thermals and evaporation of water which drives the earth's weather; it also is the prime driver in making plants grow. Only plants are capable of using this energy and turning it into complex organic compounds, such as carbohydrates, proteins and fats.

However, the plant can only use the energy in the visible part of the spectrum, about 50% of the total, or 1700 Kcals per square meter per day. The best laboratory conversion recorded under high light intensities is 10% or 170 Kcals. The efficiency of conversion in the real world is low, only about 3.5% at best for a tropical rain forest and 1.8-2.0% for sugar cane and most other high yielding commercial crops. Assuming that we could somehow or other match the natural production of a tropical rain forest, the amount of potential forage dry matter (DM) production on a per acre basis would be 21.6 tons/acre/year. In actual fact, only a fraction of this, 13.4 tons/acre have actually been recorded.

[1] One Kcal is the same as one calorie in human nutrition.
[2] There is no free lunch.

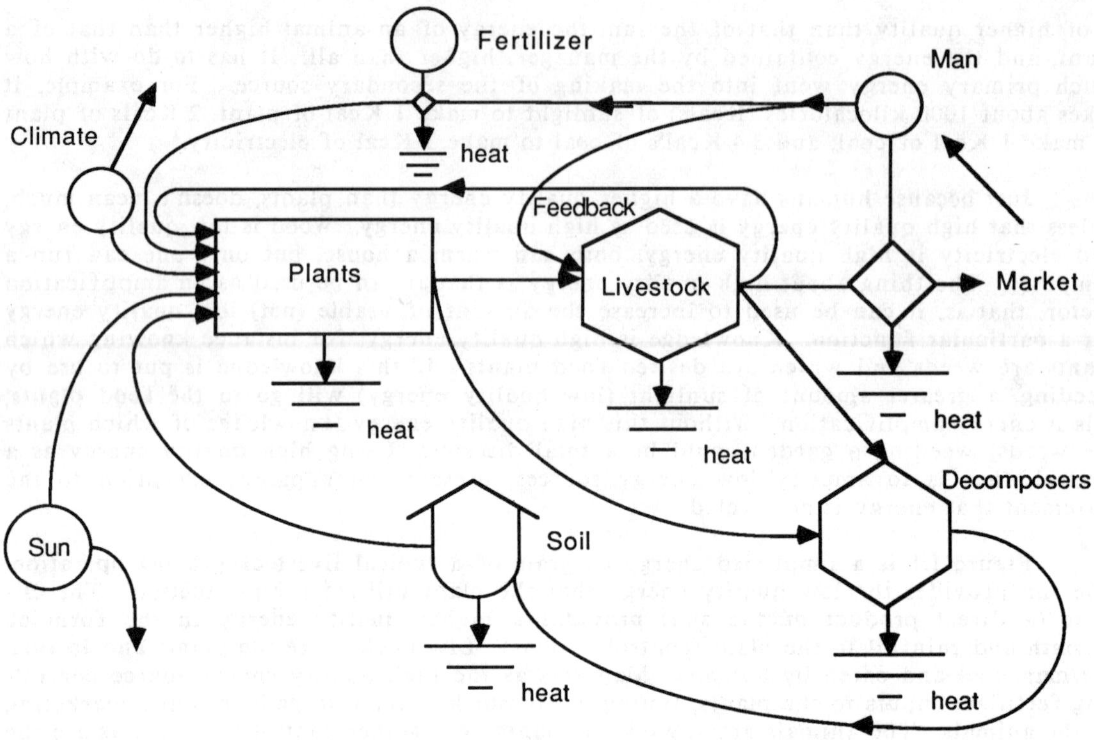

FIGURE 1.1. ENERGY FLOW DIAGRAM OF LIVESTOCK GRAZING OPERATION.

If the highest recorded production, in what is supposed to be the best the natural world has to offer, is only 13.4 tons/acre, how is it possible to have recorded in the scientific literature yields of 37.9 tons/acre/year of dry matter (DM) for elephant grass (Napier grass)? Forages grown by scientists are usually on small plots, or even in pots, and are liberally supplied with a great deal of energy. This energy is in the form of fertilizer, control of weeds, pests, often irrigated and, if done in the laboratory all the other environmental factors plus the very high quality energy input of the scientist himself.[3] In Hawaii yields of 19 tons DM/acre/year of alfalfa (Florida variety 77) have been produced on plot trials. Again considerable care in the form of fossil fuel subsidies were administered.

Agricultural systems, using fertilizers, tractors, cultivators, herbicides, etc., can outproduce those of the natural world, but only if outside energy is supplied. When the amount of outside energy subsidies are calculated some interesting results show up. For example, the total energy cost of growing, harvesting, transporting and storing of grain usually exceeds the energy content of the grain itself, often by several times. When grain is fed to animals, as in feedyards, the total energy input exceeds, by an order of magnitude, the energy contained in the finished product. Grain feeding livestock, as presently practiced, will be around only as long as there are grain surpluses and cheap energy.

[3] While it may be argued that a scientist represents high quality energy, the fact remains that it took a lot of high quality energy to educate him/her.

There are more efficient methods of utilizing grain. Several studies have demonstrated that small amounts of grain, fed in conjunction with a quality forage program, can obtain almost a one to one conversion.

There are two points to be considered a little closer. The first is that the majority of the steak consumed in the U.S. is a product of a great deal of oil (fossil fuel); a situation that will eventually change. The second, and most important, is that by pulsing a solar system with energy, grain in this case, a larger, more effective yield is possible.

A tropical rain forest is the most efficient converter of sunlight in the natural world. But what it makes from all that sunlight is gross production, the costs of production (respiration) have not been taken out; the net production of a tropical rain forest is almost zero (Chapter 8 gets into this in greater detail). Like it or not, humans live on net energy, as do all animals.[4] In nature the needs of the plant community come first, what's left over is available for animal consumption. For example: In a solar based tropical rain forest it takes about one square mile to support one human for one year; however on a solar based paddy rice culture in a similar environment, it only takes about one acre to support one human for one year.

The energy flows in a tropical rain forest are fairly predictable, energy moves thru without much fuss. On paddy rice, or for that matter any farming enterprise, energy flows are pulsed. There is a big pulse in field preparation, in planting, irrigation, harvesting, etc. It is this pulse of controlled energy that shifts the production from a low to a high net return. One of the highest net producers in the natural world are tidal marshes. They are pulsed twice daily by tides bringing in nutrients and removing waste products and surplus (net) production. However, if the pulse comes too irregularly and/or too often and strong, the organisms simply do not have time to build the necessary structures that will allow them to get through the next energy pulse and still maintain a high level of production. A wave pounded beach is an example of where the energy pulse is too large, too often and too irregular.

A barrel of crude oil contains 1600 Megacalories (Mcal). This energy, in the form of crude, is not very effective. It must be further processed (raised to a higher energy state) before it is usable. This is done by refining into gasoline, diesel oil or fertilizer. The amount of energy in a barrel of crude oil is capable of making 415 pounds of urea at 45% nitrogen (N), or 187 pounds of N. If this nitrogen is spread over 100 acres of good pasture land (1.87 lbs N/acre) not much is going to happen, the various microbes will get it before the plants do. If all of it is put on one acre, quite a bit can happen; forage production can better than double. However, if it is put on a tenth of an acre (a rate of 1870 lbs N/acre) very likely everything will turn brown.

How much energy can be added to a unit of land by way of energy subsidies? That depends on the natural energy flows, frequencies and the physical features that already exist there. In other words, that's the natural environment which includes soils, fertility, rainfall, temperature, seasons, sunlight, wind, etc. If adequate controls are in place the total of all these natural energy flows multiplied by 2.5 is about the maximum amount of energy subsidies that can be added (in increments) to a system and still maintain an overall viable ecosystem. It will, of course, not be the same system as the original. In fact, it will be a very different system, one totally dependent upon energy subsidies.

---

[4] For a better understanding of net and gross consider your pay check; what the company pays you (salary) is gross, what you take home is net, the difference is taxes and all those other deductions.

An average 1000 lb cow requires about 16 Mcal of energy in the form of food per day. The reason that she requires this amount is that she expends this amount of energy during the course of the day, just doing the things a cow does. Interestingly, 100 cows expend, during the course of a day, about the same amount of energy that's in a barrel of crude oil. However, like crude oil, cow energy must first be processed before it becomes effective. If 100 cows are allowed access to 100 acres for one day, cow energy is spread over too large an area to be really effective. However, if these 100 cows are placed on one acre for a day, the energy is sufficiently concentrated to be effective. But if they are placed on a tenth of an acre, the energy is too concentrated and everything turns brown.

Most cow energy is lost as heat, just as most of the energy in a barrel of crude oil is lost as heat during the manufacturing process of urea. However, some of that cow energy will be returned in the form of fertilizer. Those 100 cows grazing on one acre will return, depending upon the quality of the pasture, about 43 lbs of N, 10 lbs of phosphate (P_2O_5) and 35 lbs of potash (K_2O) per acre per day.[5] Because it is concentrated and is of sufficient size, this fertilization acts in a positive manner rather than being spread out all over the range, too thin to do much good. An additional amount of this pulse of cow energy will go into pruning the forage which stimulates the plant to increased production, trampling action, and all the other things that happen when a herd of cattle graze an area for a short time period.

Because it is a controlled pulse it can be timed to increase natural energy flows (positive feedback loops) or if poorly timed, waste the natural energy in unproductive efforts. If this pulse is managed, it can be used to stimulate different rates of re-growth by moving the herd when selected levels of pasture residuals are reached. It can be used to break up soil capping (crust) and plant seed or to remove old unproductive growth. If the cattle are under strict control the energy input is under control and many desirable things can be made to happen. On the other hand if the cattle are not under control, their energy is also out of control and overgrazing and other undesirable things result. Which happens depends on you, the manager.

CONTROL:

There are three "laws" that affect energy in natural systems. 1. The familiar law of conservation of energy; energy is neither created nor destroyed. Energy entering a system must either be stored in that system and eventually used or must flow out. 2. The second law is that in any process some of the energy is degraded, loses quality (second law of thermodynamics). Usually this means turning into heat, which is the lowest form of energy. Most of the energy contained in forages is converted into heat by the grazing animal. 3. The third law is; the system which uses energy best, survives. This is often called "the maximum-power principal". This law explains why some systems survive and others fail. The system that captures the most energy and uses it most effectively will out-compete others.

Well that's fine for natural systems, but what about the systems of man? Those that study this field (energetics, or systems ecology) believe that these same laws are just as valid in the various systems of man as they are in nature. In fact, the same symbols, flow diagrams and equations are used interchangeably.

[5] An average cow on pasture voids about 20 lbs of urine and 56 lbs of manure daily.

FIGURE 1.2. THINK BIG. The larger the energy forces that are being used, the greater the amount of energy needed for control. Used with permission of The Gallagher Export Co.

In nature the energy flows are captured by various organisms; plants capturing primary energy (sunlight), animals capturing secondary energy sources (plants or other animals). Much is lost as heat (basal metabolism), but the surplus is used to build high quality structures (reproduction) which are used in turn to capture more energy, which is then used to extend the range of the organism and thus to increase the chance of species survival. Consider a colony of bees just introduced to an island where no bees have been before. After finding a suitable place to live, energy from nectar and pollen is used to build a hive and the surplus stored as honey, the hive is enlarged and finally a new swarm is formed and the hive splits. The new swarm finds a home, some distance away and the process is repeated. Each new hive reduces the possibility that any one single catastrophic event will eliminate the species from their new island home. Likewise in the human world, the individual that is successful will be copied by others; consider all books, seminars and classes put on by people that have made it big.

The key in both natural and human systems is control. Energy must be controlled in order to be used effectively.[6] The more energy that flows into a system, the tighter must be the control. Uncontrolled energy is almost always destructive. Consider the amount of time and effort that has gone into the control of fossil fuels. Consider also the resulting waste products, toxic material and other environmental problems that the nations of the world are having just with fossil fuels. It is a wonder that the attempts to control nuclear energy, which represents energies several orders of magnitude greater than any fossil fuel, have gotten as far as it has. The greater the magnitude of energy, the greater the amount of energy that must be used, just to control it.

There is another way to obtain control in grazing programs rather than simply overpowering it with fossil fuel inputs such as grain, fertilizers and the like. It requires an understanding of the basic forces that act on an agricultural enterprise. It requires knowledge of a great many tools and the ability to use those tools, either separately or in combination, to obtain desired goals or objectives; and for this it requires control of the animal. It requires the ability to pinpoint the various places in the overall operation where a small amount of energy input will act as an amplification factor, thereby increasing the total amount of sunlight harvested and marketed. It requires thought and the ability to manage; we call it Intensive Grazing Management (IGM).

SUGGESTED READING:

Energy Basis for Man and Nature. 1976. H. T. & E. C. Odum. McGraw-Hill Book Co.

Perspectives in Grassland Ecology. 1979. Norman R. French, Ed. Springer-Verlag.

Systems Ecology: An Introduction. 1983. H.T. Odum. John Wiley & Sons.

[6] Effective and efficient are not the same. To be efficient is to do things well; to be effective is to do the right things.

Used with permission of NEA.

Learning the basics can be a difficult period.

2 GETTING STARTED

Contrary to popular belief the first step in going into intensive grazing management (IGM) is not to build a lot of fence nor start rotating animals. The first basic step required of all business operations that are contemplating a major change in operation, such as going intensive, is to re-examine the goals. Having a well defined set of goals is the most important aspect of any business; if you don't know where you're going, then it doesn't matter much which road you take.

THE FIRST STEPS:

Contact by letter, all owners and designated heirs; this group may just be your wife and family or, in a larger operation, it could involve stockholders, board members, etc. In the letter state where and when the meeting will take place and what will be discussed and state the reason for calling the meeting. For example: A meeting has been called at 10:00 AM, Tuesday 11 April 1985, at the offices of the Wikiwiki Cattle Co. for all owners and heirs of the above named company. The purpose of this meeting is to discuss and formulate long range goals of the Wikiwiki Cattle Co. This should be followed by a paragraph that sets out the major reasons why you feel a re-examination of the goals of the organization are necessary.

Even if the only persons involved are you and your spouse it is important that this be set up as a formal meeting, held during the day, in a business like setting and not something briefly discussed during a TV commercial. By setting a time and place, the objective and what is to be discussed at this meeting, you have let everyone concerned know how much importance you place on it. You have also given each member time to think about the goals of the business; it may have been a long time since anyone thought much about the why's of being involved in this particular type of business.

The meeting should be formal but friendly. You are not out to skin anybody. Coffee and other light refreshments should be available. Call the meeting to order at the appointed time. Explain briefly your reasons for holding the meeting then throw it open for discussion. Maintain control, don't let people get to rambling on about what they did last summer, but keep the discussion on the goals of the operation. Make sure that everyone there says something; ask Timid Terry what his thoughts are if he doesn't voice them on his own. Take notes on what is said, ask questions if need be; in other words get everything out on the table. The last part of the meeting should be devoted to formulating a goal or goals for the organization. This is done by obtaining a consensus, rather than majority vote. When you take a vote you force a position, for or against. This is not what you are trying to accomplish. You are attempting to arrive at a set of statements that give meaning to the total organization, not trying to find out how many line up on which side of the fence.

In order for a statement to be considered a "Goal", it must contain two parts: 1. A quantity and\or quality statement. 2. A time or date in the future when this quantity and/or quality statement is to be completed. A statement such as: "The Wikiwiki Cattle Co. will always be in the cattle business" is not a goal, it's a wish. Goals have to state something that is definable, they must be "future time" oriented (where should the organization be 5, 10 years down the road and they need to be personalized so that you and the other owners can identify with them. Not "Make a lot of money", but "Reduce the mortgage by $50,000 during the next 5 years". Not "Have the best purebred herd in the state", but "Win the Grand Championship at the state livestock show within 4 years".

The goals of a corporate ranching organization will tend to be more abstract than that of a family held ranch. But either way, goals are the reasons why you're busting your butt doing what you're doing. At first you may think that they are no more than to put food on the table, get the kids through college, or prepare for retirement. But they should be much more, something to shoot for, something for you, like taking a trip around the world. Think BIG.

It may not be possible to get a meaningful consensus as to the goals of the organization in the first meeting. After all, this just might be the first time anybody ever thought about what the ranch ought to be doing, let alone where it should be at some future point in time. Get the best you can, with the understanding that there will be a meeting in a few weeks to update these goals and that even they will be subject to revision from time to time. Goals are not something that are cast in concrete, putty perhaps, but not concrete. They should be re-examined every year as personal and/or business environments are changing rapidly in today's world.

Generally speaking, only owners or their representatives set goals. The different levels of management may be called in to advise, but rarely for consent. As soon as a set of goals have been formulated they should be communicated to the work force; it's nice for people to know which direction the company is trying to go.

INVENTORIES:

The next step is to find out just exactly what resources your ranch or operation has on hand and/or is available to it. Call a meeting of management. In a family run ranch it may only be you and your spouse, and possibly the hired man. On a larger operation it would be the head, or heads of all major departments. At this meeting the goals (they should have been distributed to all concerned earlier) should be discussed in general terms, also possible ways to obtain those goals. At some point stress the problem of trying to attain goals if you don't know what you have to work with. Prior to closing the meeting have different persons prepare the following prior to the next meeting:

1. BASIC RESOURCE INVENTORY: Number of paddocks and acreage in each, kinds of forage found in each (what percent and the quality), soils (fertile, unfertile, special problems), topography, (steep, rolling, flat), water (streams, lakes, springs, ponds, dams, windmills, easy or hard access), unique features (cliffs, caves, historical sites, wildlife, forestry and recreation potential), etc.

2. PHYSICAL INVENTORY: This should include all equipment, major supply items, buildings, sheds, watering facilities, fences, and especially all "Heavy Metal" items.[1] In addition to noting how many of each, the condition and the expected life, as well as the estimated market worth of each major item should be recorded.

3. LIVESTOCK AND GAME INVENTORY: List kinds, breeds, sex and age. Also include a stock flow, length of calving season, average weaning weight, percent calves weaned, and the major livestock programs now going on, such as, AI, crossbreeding, replacement selection criteria, etc. The kinds, location and estimated numbers of game animals (both mammals and birds) should also be included.

[1] "Heavy Metals" are those machinery items that spend most of their useful life sitting in a shed, depreciating.

4. FINANCIAL INVENTORY: Record the cash on hand, loans, debts, borrowing ability, cash flow, past balance sheets, any financial or enterprise analyses that have been made, etc.

5. PERSONNEL INVENTORY: This is perhaps the most important inventory of all. List the personal skills and abilities that each member can contribute to the overall operation, also include personal job or work preferences if possible. This list should also include hobbies and outside interest as well. Prepare a brief description of any consultants or financial and/or tax accountants that are being used in connection with your operation,and note their good and bad points, as well as any possible alternates.

If possible, have different people conduct the inventory rather than the ones most familiar with a given area. There are a number of advantages to this which outweigh the obvious disadvantage. For one thing, it is not necessary to record every nut and bolt, or every obscure plant or animal, which may happen if you put the resident expert in charge of the inventory for his department. Secondly, you will be letting others find out about different functions of the overall ranch operation, something that will be increasingly helpful in the time to come. Third, a new person will have a different viewpoint, and is more likely to question why something is done, or why that fat cow, the one that has never had a calf, is still on inventory. Admittedly this approach must be handled with some care to avoid ruffling too many feathers.

There is an interesting story that occurred in a large packing plant that was killing 1200 animals a day. A young financial manager was hired who knew absolutely nothing about the slaughtering operation. During his introductory tour, he saw a worker making a cut between the hide and the carcass at about mid-rib and placing a stick to hold the hide away from the ribs. The next step in the operation was to remove the hide by a mechanical hide stripper which usually broke the stick. The new man could see no reason for it; come to find out, neither could anyone else. As it turned out, the process was started when the plant was killing just a few head per day. One man was doing the whole butchering operation and he used the stick to separate the hide so that the carcass would continue to cool while he was busy doing other things. One worker, using 1200 sticks a day, doing a job, for no other reason than simply because that was the way it had always been done. The young man earned his entire salary the first day on the job because he wasn't afraid to ask about something he didn't understand while everyone took it for granted.

With the above in mind, you may want to add an outside person to your management team. Someone that doesn't know very much about ranching but does know a great deal about something you don't know much about, such as a banker, lawyer, accountant, etc. The idea is to have someone that can contribute and also ask those embarrassing questions. There are a number of ways that this person can function, but as a member of the board is perhaps the best. The important thing is to use him.

Make sure that all have a copy of the above inventories before calling the group together again. At this meeting, discuss the goals in light of the various inventories just completed. Begin the process of formulating specific objectives to meet the ranch goals by assigning each person to come up with a set of objectives to be presented at the next meeting, in one week.

OBJECTIVES:

Objectives may be thought of as a series of enterprise analyses (Chapter 3). They are the broad blueprints of how you plan to meet the stated goals. Goals are the overall strategy, objectives are the tactics used to implement that strategy. For example, if a goal was to make an economic return from all land not suited for livestock production by a certain date, then the first step would be to look at the inventories. Assume [2] there is an area of land that is very unproductive but has a superb view and some real trail ride possibilities, also there is a high resident population of game birds near that area. The personnel inventory shows that you have a person whose hobby is community theater (not afraid to stand up in front of a group). Maybe, just maybe, that combination can be turned into a lucrative recreational enterprise. Of course there are a lot of things that have to be done before setting up the Wikiwiki Trail Ride or Dude Ranch, but would that opportunity have presented itself for consideration if you had not had your goals and inventories in front of you?

Another example might be a family operation employing part time help. You have set your goals: Pay $50,000 on the mortgage principal; take a round-the-world trip for the family (estimated cost, $10,000) and do all this within 5 years. The completed inventories are in front of you. The time has now arrived to determine just how you are going to go about obtaining these goals.

You are presently running a cow/calf operation, 80% weaned calf crop, calves average 450 lbs at an average weaning age of 8 months, calving season is 3 months. Many might be tempted to think of an objective as: 1. Improve the cow/calf operation by: a. Reduce the calving season to 2 months; b. Increase the weaned calf crop to 90%; c. Increase weaning weights to 475 lbs; d. Reduce the average weaning age to 6 months. As noble as these might be, they are not objectives. They are "performance targets". Forget what you are doing with the operation at this point in time and let your mind run free; think BIG.

Since your goals did not state that you had to raise cattle, or even livestock for that matter, seriously look at other enterprises or mix of enterprises.[3] Look at the recreational possibilities, dude ranch, cabin lots, trail rides, tours, hunting, etc. Check it out, talk to travel agents, local hotel or motel operators, realtors, bankers; there's a lot of money being spent for recreation.

Have you ever thought about running sheep, or goats? How about farming deer? Just because you have always been a cow man, doesn't mean that you always have to be just a cow man. If your goals are to make money from livestock, then all livestock enterprises should be considered. If you're not sure that you can stomach goats, run a trial. But make it a fair trial, don't throw a bunch of goats out there with wild dogs or coyotes and expect them to survive, like those old mossy-back cows do. If you're running cattle in sheep country, maybe you had better take a hard look at the why of it. Above all, be honest with yourself. If you're doing all this just to maintain a life style, admit it to yourself; you don't have to tell your banker or the IRS (they probably already know). However, you need to know the why, so you can take steps to see that you can enjoy that

---

[2] I dislike using the word assume, ever since a fellow pointed out that it breaks down to ass-u-me, then proceeded to say that when you assume something you invariably make an ass out of either you or me.

[3] If your goal was simply to make a lot of money, sell the ranch and put the money away at interest.

life style for as long as possible. Chances are your heirs won't be able to enjoy it but what the hell.

FIGURE 2.1. ANY ROAD. If you don't know where you're going, any road will get you there.

IMPORTANT: Hang the goals and objectives up in the office where you and everybody else can see them; you're not the only one that needs to be reminded from time to time, just where the organization is supposed to be going. Remember, goals and objectives are not cast in concrete, they need to be updated and changed as conditions dictate.

You are now at the point where you can begin planning how your outfit is going to reach these objectives.

PRODUCTION TARGETS:

A production target is the amount, of whatever is being produced, that is required to make the enterprise economical. When production targets are considered along with objectives they become, in effect, an enterprise analysis (Chapter 3).

Production targets are determined for each objective. Initially only consider one objective at a time. When you have completed all the production targets for each objective and are reasonably satisfied as to their accuracy, then re-work them so they dove-tail into an overall management program.

For example, the ranch consists of 300 acres of good grass in a semi-tropical environment. Eighty acres could be cropped and 20 of these could be irrigated. Under traditional management the average stocking rate (SR) is one mature cow or animal unit (AU) to 3 acres. Rainfall is 30 inches per year and is generally dry during August and September. Low temperatures limit growth (one half that of spring and fall) beginning in mid-December to mid-March. Overhead and fixed costs are $35 per acre per year with interest at 12% per year. Calves are worth at weaning, $0.40 per pound liveweight and 850 lb stockers are worth $0.35 per pound liveweight. Mature cows are worth $300 each and bulls $1200 each. Two objectives will be considered: 1. Cow/calf operation. 2. Stocker, utilizing raised calves.[4]

COW/CALF: Total variable cost per cow, includes vet. cost, minerals, necessary handling, transportation, water, etc., is $65 per head, including interest, but does not include any return to labor, management or capital. If 270 acres are designated for the cow/calf enterprise to be run under traditional management, 90 mature cows would be carried year round. Total cost of this enterprise is: 90 head X $65 = $5850 plus 270Ac X $35 = $9450 for a total cost of $15,300. At $0.40/lb this means that 38,250 lbs of calf would have to be produced, or 85 calves weighing 450 lbs each. To accomplish this a weaned calf crop of 95% would have to be obtained each year. The production target is 142 lbs of calf per acre per year, or $56.80/acre.

If intensive grazing was used, one AU to 1.5 acres could be run. The costs of doing this would be an additional $15 per head (cost of additional fencing amortized over a 5 year period, interest included). The overhead cost would remain the same, $9450; the variable cost would now be: 180 head X $80 = $14,400. Adding the overhead cost gives a total cost of $23,850. At $0.40/lb this requires 59625 lbs of calf, or 133 calves weighing 450 lb each. To get this number requires a weaned calf crop of 74%. If a calf crop of 90% is taken as a reasonable average, this means that 29 calves, or $5,220 would be available to pay for labor, management and capital. The production target for this scenario is 221 lbs of calf/acre, or $88.40 per acre per year. The stocking rate could be increased further by use of other tools, such as fertilization and forage conservation (Chapter 12), but each additional tool has a cost associated with it and the same procedure of comparing added benefits and added costs would be required to analyze it's cost effectiveness.

STOCKER: If 30 acres are set up as an intensive stocker operation, with fertilization and feeding 4 lbs of molasses per day, the total variable cost is $100 per animal, including interest. The overhead cost is $35/Ac X 30Ac = $1050/year. The stocking rate is two animals to the acre, total cost is $7,050 per year; if it takes one year to put on the required 400 pounds. The expected return is 400 lbs X 60 head X $0.35/lb = $8,400, for a net return to labor, management and capital of $45.00 per acre per year, or $1350 total. The production target in this instance is 800 lbs of beef per acre per year, or $280/acre/year.

---

[4] The above values are based on 1985 Hawaiian conditions. The reason Hawaiian data is used is that the value of the animal is very low compared to the mainland U.S. (more in line with world price), yet the cost of operation is generally higher than the mainland.

The cow/calf yielded a net return of $19.33 per acre and the stocker a return of $45.00 per acre, or over $25 more per acre for the stocker operation. If all the land was suitable for a stocker operation then a stocker operation would make more economic sense. However, in this example only 80 acres are suitable. If all 80 acres were used, 160 calves would be required, which is 28 more than the reduced cow/calf enterprise could produce at a 90% calf crop.

At this point a technique called iteration or looping is employed. All this means is that the procedure is repeated, but with different numbers, in an attempt to obtain the best (most economical) mix. In other words, try 80 acres for the stocker, with the additional calves purchased. Irrigation of the 20 acres and forage conservation could also be included, but should not be part of the initial sizing of the enterprise mix. Why?

The above example has been streamlined for sake of clarity, in actual fact there are other considerations that would have to be reckoned with. Half of the calves will be heifers, of which a certain number would have to be kept for replacements for the cow/calf operation. There would also be a minimum of five classes of animals; bulls, mature cows, cull cows, steers and heifers. This would present no great problem, but would require that certain allowances be made in the handling procedures adopted, such as multiple herds. The cull cows might also be considered as another enterprise, if some extra weight was to be put on them before selling.

PERFORMANCE TARGETS:

Performance targets refer to what is expected out of the animals, eg. pounds per day gain. As might be expected they are closely tied to production targets (production per acre or hectare). In crops performance targets are not normally used, performance being figured on the basis of a unit of land, so many bushels per acre (production targets). However, under today's economics the land is becoming the standard unit of measure. So, while individual animal performance and thus herd performance remains an important measuring tool, the ultimate measure of success is production per acre; unless you are leasing land on a animal unit basis.

In the above example the stockers were figured to gain 400 lbs per year, or 1.1 lbs/day gain. Because this is an average number, it is possible to have the animal gain higher during parts of the year than during others and still average 1.1 lbs/day over the long term. This has important implications when dealing with the seasonal growth patterns of forage plants, or when grazing more than one group of animals that have different performance targets (multiple herds, leader-followers). The major point is that once production targets have been set, performance targets can be adjusted, through the use of various management tools, to take advantage of seasonal or other effects, increasing or decreasing the level of herd performance, and still meet the pre-set production target for the year.

INITIAL WRAP-UP:

When all production targets have been completed for each objective and copies given to all parties, call the next meeting. Meanwhile you, as head honcho, need to work all of the production targets into a overall management program and present it to the group, prior to the next meeting, for their consideration, discussion and assignment of priorities. At the next meeting do not push for acceptance, but rather ask for suggestions as to ways to make it better. After the meeting incorporate any changes into the final version. Be sure to take the various suggestions and/or criticisms seriously, and if you do not use them make sure you jot down why you did not and let the group know why at the next meeting. Remember, you as a manager are managing people as well as things. You are going to need all the help you can get, so pay attention to all of your resources.

If it seems that you are having an awful lot of meetings, you are, and you should continue to have them; one formal meeting a year to discuss strategy (Goals), one formal meeting a month to discuss tactics (Objectives and Production Targets), and a formal weekly operational meeting (Performance Targets). What about the morning get-togethers to discuss today's work; if you're doing your job at the other meetings you should be able to cancel those morning meetings as everybody will know what's expected of them and what they have to do. As much fun as those early morning plantation style get togethers are, they are basically a waste of time for all concerned. If you truly have people that can not work without constant supervision, find something they can do. Remember that's your job as manager. If in the final analysis they are truly hopeless, or un-trainable, get rid of them. But don't be in a hurry to use the ax. People when given a little responsibility and authority, have a way of rising to the occasion.

SUMMARY:

A. Set Goals.

B. Determine inventories.

C. Formulate specific objectives to meet goals.

D. Estimate production target for each objective {enterprise analysis}.

E. Maintain communication with all aspects of operation.)

SOME THOUGHTS ON ORGANIZATION:

There are many ways to organize a ranch or farm business, as there are in any business. Numerous books deal in great detail with how to do this. Almost any book by Peter Drucker will give one some excellent ideas. It is not the purpose of this book to stray very far into this area. There are a few items that will be mentioned.

THE 80-20 RULE:

There seems to be a similar condition that exists in almost every organization and that is; the lions share of the time and effort that a organization spends, goes towards producing the least amount of production. This is the so-called "80% - 20%" rule. In other words, 80% of the total effort produces 20% of the production; conversely 20% of the effort produces 80% of the production.[5] Think about it for a bit, before you junk the concept.

A really successful business identifies those tasks that produce, and then spends most of its time and effort developing them. Those task's that require a lot of effort, but produce relatively little are identified. Maybe they still have to be done to keep the operation running smoothly, maybe not.

---

[5] In ranch context this would be spending one hour a day planning cattle marketing strategy and the rest of the day fussing with a colt.

One way to identify those task or operations, if they are large enough, is to do an enterprise analysis of them to find out just how much they are making, or costing. On the smaller jobs, select some criteria such as: If this job is not done today or this week, will it make me (or cost me if it doesn't get done) money? If it will make money, then how much? Select some dollar amount, say $200. Those jobs that are in excess of $200 (either make or cost), give them an "A" priority, those under the $200 give a "B" priority, and those jobs that you can not put a dollar value on, but would be nice it they got done, a "C" priority. Then give your attention to the "A" jobs first. Use a little horse-sense though, if you drive 5 miles to do an "A" job and there is a "C" job that also needs to be done that won't take too long, then do that while you're there. With many "C" type jobs the major problem is simply getting there. You may not be able to justify spending all that time getting there just for a "C" job, but if you have a higher priority job to do in the same area then there is nothing wrong in doing both. However, if you seem to having a large number of "C" jobs, then better check the 80-20 rule.

In the typical ranch operation the above will be hard to do. However, after you have formulated your goals, objectives and production targets, the important jobs will be a lot easier to identify.

PEOPLE:

This wouldn't be such a bad place if there weren't any people in it. How often have you thought that? You may not believe it now, but the biology of intensive grazing is child's play compared to the managing of the people that are going to be involved with it. There are a number of excellent books listed in the suggested reading section that go into this in far greater detail than this section will. In dealings with people, often the head and the heart are pulling in opposite directions; this is especially true of those nearest and dearest to you. In most cases, for the long term the head should win 80% of the time, in actual fact it's probably the other way around.

THE FAMILY OUTFIT:

If you're single, own your own place, do your own work and live like a hermit, you don't have to worry too much about other people.[6] If you're a family rancher it's a different story altogether. You are worrying about each member of the family and they are worrying about you. Generally, motivation is not a major item, you know why you're doing it: for your kids. You may not have a decent set of goals written down anywhere, but somewhere in the recesses of your mind, you're thinking: "Some day this will all be theirs." Meanwhile you work 'em like slave labor; then get all up tight when they turn 18, leave home and get a job selling shoes in the big city. You don't put a 10 year old kid on a horse for 12 hours a day, 7 days a week and expect him to hang around after he's legal, unless he's got less smarts than the horse.

If you truly want to turn the outfit over to your children, then start by giving them some responsibility and authority, maybe even a whole enterprise to run on their own. You may not like some of their decisions, that's too bad. How else can they learn, unless they make their own mistakes?

What usually happens is that you keep them in school as long as you can, so they can learn all those smarts you never had a chance to. Then soon as they get home, all fired up with enthusiasm and new ideas, and announce that you should sell the horses

[6] We've known a few like this, mostly sheepherders.

and buy a motorcycle,[7] or worse yet get rid of the registered herd and raise sheep and goats. Right away, CONFLICT.

Lets face it, you don't really want to give up this place you've been busting your butt on these last 20 or so years to some wise ass college kids, even if they are your own. There's nothing wrong with that attitude. The problem is that in our farm and ranch culture turning the place over to the kids is the way it's supposed to happen, and if it doesn't happen that way, you end up feeling guilty. So you make vague promises just to keep them around. Meanwhile, you run the outfit just like you always did, at first listening patiently to all the new ideas, and just as patiently explaining why they won't work. Soon the ideas come farther and farther apart, and finally stop. You're thinking the kid finally has developed some sense. Around this point in time, one of two things have happened, the ones with get up and go, got up and left, or they sort of turn into a ball of wax, essentially a hired man with tenure. When you finally do retire or die, your child is suddenly faced with all sorts of management decisions that he has never had any experience with. He will be extremely lucky if he can hang onto the place for even a few years.

The ability to manage is not a genetic trait, it must be learned. Book learning can give some ideas and help to get one started. But as every operation is different, books don't have all the answers. That comes from experience, making decisions, good and bad and having to live with the outcome.

If you really don't want to quit or retire, then don't; but be honest with your kids. Explain that since you read somewhere that most people die five or so years after they retire, and since you're only 40, you just wouldn't feel right about cashing in just yet. There are other ways to handle the problem; expand the operation, add another enterprise, diversify, etc. Whatever you do, see to it that they have some training for it, then put them in charge.

Draw up a formal contract, detailing how this is going to work, build in a time frame for these things to happen. Give them responsibility and the authority that goes with that responsibility. Initially the final say should be mostly yours, but very quickly should become joint, then as time passes, or goals are achieved, should pass totally to them, according to the contract. Your function should be that of providing counsel and guidance, not command. Hard? You bet.[8]

EMPLOYEES:

The industrial revolution was successful because the major organizations operated on the assumptions that people hate to work; they have to be driven and threatened with punishment to get them to work; they like security, aren't ambitious, have to be told what to do and dislike responsibility. As long as the average worker was uneducated, and the punishment was death, instant dismissal or the equivalent, this policy worked reasonably well. Well, it's outdated by at lease 60 years, yet many organizations still cling to the image of the Roman legions as their role model.

The Japanese have been able to achieve great things by incorporating the worker, in varying degrees, into the overall management of the company.[9] Several large corporations in this country have adopted a more enlightened policy, with various de-

[7] It makes it easier if you think of them as Japanese quarter horses.
[8] But, like old age, not so bad when you consider the alternative.
[9] I'm sure there're a few bugs in the system that they aren't telling us about.

grees of success. Top people don't respond well to threats. They don't have to. Today, people are very mobile and with all the other jobs opening up, it is increasingly difficult to hold them in agriculture. Employers are slowly giving up the old plantation mentality, and are beginning to give employees authority as well as responsibility. With intensive grazing this is a must, as the romance of being there on weekends wears thin real quick.

The only time man lives on bread alone is when he doesn't have any bread. Human needs form a hierarchy, after the first is satisfied, the second is important, when that has been met, the third, etc. In general, they can be listed as follows: 1. Bodily; food, clothing and shelter. 2. Security; protection from various outside forces. 3. Social; involvement in different community activities, organizations and clubs. 4. Ego; complements, strokes, verbal rewards and recognition for good works. 5. Development; improvement of self and those goals important to him.

On a slightly different scale, it has been found that certain aspects of the working place have negative and/or positive connotations for the employee. A negative connotation means that if a particular job aspect is bad the employee will be unhappy and will probably leave or do sub-standard work. However, if that same aspect is really great, it doesn't mean that much; the basic level has been met and more of the same is no longer motivating. Money is not everything, it only allows one to supply the basic needs.

There are a few guidelines. Let people set their own objectives, subject to approval by you, then use these objectives to measure their performance. Keep lines of communication open, but short. Make sure that the employees have sufficient training before giving them the responsibility. Use clear company goals and policies. Use the management team approach; if you're left out you're not as motivated as if you helped set the objectives. Set job performance based on production, not job descriptions. Be liberal with bonuses and rewards if they truly earn them; you can afford to pay a person a million dollars if he makes you two million. Remember, the buck stops with you. It is a rare event when a business goes broke because of the cowboys or other factors. You, as top dog, will determine if the ranch succeeds or fails; if you're not up to it, then get entirely out of it and let those who are better qualified to run it.[10]

MIDDLE MANAGEMENT CONTROL:

An unique situation exist on some moderate to large ranches that, have established a local historical tradition. The employees of the ranch have been there for many years and often their fathers before them. They have what they, and often the community, feel is a vested interest in the ranch. Often this middle level of management actually runs the ranch; managers come and go, but the foremen remain. The more frequent the changes at the top the deeper ingrained becomes the attitude that because they have "survived" past storms, what they do is right and what the manager does, is not. This can be an excellent way to insure continuity of ranch policy; it can also be a major stumbling block to implementing necessary changes.

The new manager, especially if he has been brought in to get things moving, is in a ticklish situation. The new manager has a "honeymoon" period, with his bosses that may last for 6 months to a year, with the employees perhaps a month or two. Assuming that he has been given authority to exercise control, he has several options: fire everyone and bring in his own team; re-organize the entire ranch structure and either bring in a few key people or attempt to fill the new positions from within; demote everybody and let

[10] If you have a bottleneck in your operation, take a beer bottle and place it on the table in front of you and take a good look at it. Where's the bottleneck? Right there on top.

the "cream" rise to the surface; attempt to re-educate and/or gain the confidence of those in key positions; create new positions and transfer your "problems"; or simply "hope" that they will see the light and "come around".

All have certain inherent problems. How do you fire someone that has been on the place all his life and maybe his father before him? Demote him and you have a very sour individual out there who, if not actually sabotaging your efforts, sure isn't talking them up. Transfer is fine, if you have someplace to put him so that he'll be reasonably content and also out of your way. Argument and re-education is fine, if there is sufficient time. Bringing in new people creates tension and firing everyone gets the community down on you. Re-organization causes turmoil. Leave them in place with the hope that they'll "come around" seldom works. The more managers that employees have survived, the less chance that they will "come around".

So what do you do? There is no "pat" answer. It depends on the relationship you have with the owners and if they perceive the problem. If they do and have given you the responsibility and authority to get the ranch productive, then re-organization and education coupled with all, or most of the above is the best approach. Nobody likes to use the "ax", but there are times when there really is no other alternative.

MOTIVATION:

In the "good old days" if you wanted to motivate someone you follow the examples of the Spanish Inquisition or the Roman legions and hung them up for a while until the right attitude re-adjustment was attained.[11] It is questionable that such motivational techniques really worked other than keeping everybody "looking busy". Fear has always been a prime mover, as has hunger and security. But what was not understood by management was that these conditions only motivated people to do a "fair days work". The problem of a "fair days work" is that it is considerably below the average potential of the human work force.

Up until fairly recently, when Japan was suddenly perceived to have the ultimate management approach, American management (and unions) were concerned primarily with working conditions. This, according to Frederick Herzberg in his book "The Managerial Choice", was addressed by: 1. Reducing the time spent on the job - "A marvelous way to motivate people, get them off the job." The simple fact is that motivated people seek more hours of work, not fewer. 2. Increase wages. Is this motivation? Yes, to seek the next wage increase. 3. Increase fringe benefits. The cost of fringe benefits has reached approximately 25% of the wage dollar, plus the cost of workman's compensation and other state and Federal programs. People spend less time working for more money and more security than ever before. Now these are no longer rewards, they have become rights.

Industrial engineers have reduced job content down to the very basics. Behavioral scientist have suggested a whole host of employee programs, ranging from human relation training, sensitivity training, one-way then two-way communications, job participation and employee counselling. Yet job productivity is low and management still cries for motivation.

According to a theory of human motivation set forth by Abraham Maslow, human motivation is dependent on a set of needs. When the first set is satisfied, they no longer motivate, the next set then becomes important; when those are satisfied, then the next set, and so on. Maslow listed these needs in a hierarchical order: physiological needs, safety

[11] "Where there's a whip, there's a way."

needs, the need for belonging and love, the need for esteem, and finally the need for self-actualization. Simply put, if the worker is not hungry, thirsty, cold or in physical distress, he then seeks to protect and safeguard his social and psychological condition. When these are satisfied he will seek rewards from interpersonal relationships, then ego satisfaction of reflected personal worth and finally intrinsic personal worth from his own sense of accomplishment.

Frederick Herzberg has proposed that motivation is composed of two factors acting in parallel, not as a single series as proposed by others. Humans may be thought of as being composed of two parts; an animal part and a human part. The animal portion has basic needs that are physiological, when these are satisfied they cease to motivate. He terms these factors "Hygiene". Hygiene factors are responsible for job dissatisfaction. If they are absent the worker is dissatisfied with the job; however, when the hygiene factors are adequate the worker is not unhappy, but he is not satisfied (or motivated) either. These hygiene factors represent security, status, salary, working conditions, interpersonal relations, supervision and company policy and administration. The human factors, which Herzberg terms "motivators" are responsible for job satisfaction. They are, growth, advancement, responsibility, work itself, recognition for achievement and achievement itself.

Hygiene factors can be thought of as negative and the human factors as positive in motivating employees. If all the hygiene factors are in place, a "fair days" work is the result. If one of the factors deteriorate, such as the introduction of a poor supervisor, or a wishy-washy company policy, production declines. However replace a poor supervisor or have a strong company policy, then work will pick up for a short period of time, then settle back into a "fair days work" effort. On the other hand make the work interesting, give the employee responsibility, acknowledge his achievements and productivity increases.

So how to implement human factors? The first step is to make sure that the hygiene factors are adequate.[12] Next start re-structuring jobs. A "good" job should contain: 1. Direct feedback; comments should be given directly to the employee by the boss in a timely and non-evaluative manner (if the performance was good he should be told, if it wasn't so good the reasons should be gone over with him as soon as possible). 2. The employee should have a customer or client to serve, either internal or external to the organization. 3. All jobs should provide an opportunity to learn something meaningful. 4. The employee should be allowed to do his own scheduling. 5. The employee should be able to "do it his way" with all or certain aspects of the job. 6. The employee should have control over resources; the employee should be provided with a minibudget to run his own operation. 7. Direct communication; if a foreman needs to talk to the computer operator, he talks to him, he doesn't have to go through his supervisor, the manager and then down the other side. 8. Personal accountability, identify the performance of the work with the individual. Because of the stress on personal fairness, society has become so afraid of holding anyone accountable for their job, that many jobs have been stripped of everything that one would wish to be held accountable for.

Jobs should be structured to be challenging, this means vertical job enrichment, not horizontal job structuring or job loading. Job loading is challenging the employee to increase his production; if he sets 50 fence post a day, see if he can set 100. Add another meaningless task to his existing one; another "Mickey Mouse" job doesn't make the work any more interesting. Rotating job assignments; string wire for a while then pound

[12] There is a saying attributed to a communist country: "As long as they pretend to pay us, we will pretend to work."

staples. Remove the most difficult part of the assignment in order that he may complete more of the "Micky Mouse" parts (this is the industrial engineering approach).

Modern motivation theory can be a powerful tool to the successful operation of the modern agri-business; regardless of what sector you're in. In order to increase production at lower per unit cost, you have to have your employees working at, or near their potential, not doing a "fair days work". This means another radical shift from traditional thinking. For those who are interested in implementing this sort of approach it is strongly recommended that you read up on modern management and motivation; a good place to start is with the "The Managerial Choice" (suggested reading section, this chapter).

OTHER CONSIDERATIONS:

There are a number of other considerations that should be taken into account during the initial planning process. These include, but are not limited to, the terms and types of lease or rent agreements, diversification into other non-ranching efforts that may be possible because of a favorable environment and contracting jobs out rather than hiring employees or doing it yourself.

THOUGHTS ON MANAGEMENT OF THE LARGER RANCH: by Earl Spence.

The fact that intensive grazing management works cannot be disputed. What can be disputed is why does intensive grazing management work for some people and not for others. A closer look at intensive grazing may shed some light on the dispute.

If an intensive grazing unit failed, what failed? Was it the electric fence? Was it the grass? Was it the cattle? If you said yes to any of these questions, take a closer look. Somewhere man played a very important role in each of these areas. If an intensive grazing unit failed, then the MAN failed for he controls or manages the unit. For an intensive grazing unit to be successful the MAN must be successful. It is your job as the owner and/or manager of a large operation to make your MAN successful, so your intensive grazing unit becomes successful and the ranch becomes successful. It is your ability to MANage that dictates the success or failure of your intensive grazing unit, not the electric fence, the grass, or the cattle.

Wellllllllllll that's easier said than done. Then let's do it.

The first step is for you to convince yourself that intensive grazing is necessary to survive in the beef industry. Secondly, make yourself aware of some of the problems that it will present; attend meetings, contact ranchers that are doing it, check with your local extension and Soil Conservation people. Then call a meeting of your management staff and discuss it with them. This meeting should be very open and all should have some say in the matter. However, you, the owner/manager must exercise firm direction of the meeting, keep it on the subject, intensive grazing management. Any negative criticisms should be considered and their validity addressed. Don't hesitate to hold more than one meeting, should it be required. At the end of this introduction phase, which may take anywhere from a week to a month or more, all in authoritative management positions should be in agreement as to the course of action to take. Any person or persons not in complete agreement should be re-located or dismissed at this time. If they are not they will be obstacles in the future and will be in a position to obstruct the success of the program.

With the firm backing of your management staff, the next step is to take it to your employees. As with the management staff meeting, this meeting should take the same format. Again stress YOUR commitment to the program as well as that of the management staff. This meeting need not be as detailed as your previous one, but again should allow for discussion and criticism. The objective of this meeting is to inform; not ask permission of your employees as to the course of action that the ranch will be taking. Total agreement of your employees is not necessary. However, remember that you are changing a lifestyle and climate that your employees may have become accustomed to; change is always hard to accept. There is no need to dismiss an employee at this point, simply because he disagrees as to the course of action, unless he is to play a key role in your program development. His attitude will likely change as time goes on, if it doesn't there still may be a place for him in the organization. It is important at this point to establish authority. If your authority is not established, then postpone intensive grazing management and re-evaluate your management position. Intensive grazing management calls for strong management, bordering on authoritarianism.

Spend the next six months to a year planning; the overall grazing unit design, organizational changes if necessary, and addressing any possible problems that may arise through continued discussions with your management staff. This is the time to work out cash flows, stock flows and other changes in the operation that will insure success. This period will also establish firmly your commitment to intensive grazing management and will convey this message to your employees that you are serious about the matter. Never let a week pass by without re-establishing your commitment through conversations, handing out or discussing magazine articles (positive or negative) or any other form of communication.

After this period you are ready to initiate intensive grazing management. Call a meeting of your top five candidates whom you feel will do well as an intensive unit manager. Again, reinforce your commitment to intensive grazing management. You may want to discuss with them the financial status of the ranch. It probably isn't anything to brag about, but its time you started to take them into your confidence, after all, they will be your future management staff. It will also reinforce the need to go into intensive grazing management. Outline your future plans for intensive grazing management on the ranch so that they can see what part they will be playing in it. You'll be surprised at the enthusiasm you can generate at this time. If done well, you will have established a core with which you can build your intensive grazing management units. Further reinforcement and training is the next order of business.

You are now ready to build your first grazing unit. Choose one person, out of the five, to be in charge of this unit. Be careful not to alienate anyone in doing this. Stress the importance of one person for accountability, but the remaining four will be involved in the grazing unit in all of the aspects, except the day to day operation of the unit.

During this time, your unit manager, yes MANAGER is involved in design and construction of the unit. This is the time that he makes the transition from cowboy to businessman.[13] Structure a program for him that will insure his complete knowledge of the workings of the ranch. Remember, up till now he was on the outside, as far as management of the ranch was concerned. He needs to be brought INTO the management of the ranch. Assume that he knows nothing of the inner workings of the ranch, educate him in all aspects, everything. This is your responsibility, you must see to it that he is brought up to speed. The unit manager will need this knowledge, or at least the ability to

[13] The masculine form is use for convenience only; women only have to be twice as good as men, but normally this is not a problem.

gather this knowledge, when he is running the grazing unit. When he feels confident in this new role, but before the grazing unit is built, take him aside and discuss your expectations of him and the grazing unit he is to manage. LEAVE NOTHING UNCLEAR.

First tell him that his salary is now 50% higher than what it used to be (bring him up to maintenance).. Then tell him that after he has run the unit for one year he will no longer be an employee of the ranch. After he regains the color in his face, explain your plan to him. The first year is to gain experience, to make sure that the performance and production targets are reasonable, to give sufficient time to work the bugs out of the contractual arrangement before putting the final version down on paper. At the end of the year (this could be extended if need be), he is to become an independent businessman. He is to provide his own vehicle, his own medical, his own retirement plan, plus all those other things that you have been providing (perks). If there are maintenance and/or service requirements to consider, how these will be worked out; either he contracts them out, or the ranch provides at some cost. Explain to him, in black and white, how you will benefit financially and most importantly, how he will also benefit financially. It should be at least double what he made before becoming a unit manager, with the possibility of going considerably higher. Tell him what you expect in the way of production; these goals should actually be arrived at between the two of you. Anything over this will be a financial bonus for him at the end of the year, or depending on the setup of the ranch, a portion to go to the service departments, a small percent to the ranch and the balance his. Go through the numbers with him. Make sure he understands exactly how those numbers are computed. It is important for him to know what the financial rewards are for excellence and that the penalty for failure is cancellation of the contract. Give him sufficient time to discuss it with his wife, girlfriend or accountant. He needs to feel confident that he is not being used, that by doing this he is helping the ranch survive, and that he stands to gain a great deal. The important thing is to be completely honest, don't get greedy, you stand to make a lot out of this also.

Begin the program of weekly meetings. Go over the concepts and ideas you discussed previously. Elicit questions from him to make sure he understands and to get a feeling of his insecurities, for there will naturally be many. Sometime during the year, begin setting production targets in greater detail. Let him know of your expectations from him and the unit he manages; and find out what his expectations are of you and the ranch. Discuss things like a 85% calf crop and 450 lb. weaning weight at six months of age. See if he agrees or disagrees. Encourage him to disagree if he feels strongly about something, remember a contract works two ways. Discuss the independent businessman aspect and reinforce your position on it and how it affects both the ranch and him.

As the big day draws closer, go over the day to day details you expect of him and he of you. As manager of this unit he has only to answer to the guidelines (policy) you have set down and which he has agreed to. As an independent businessman his time is his own. He sets up his own work schedule. He prepares a budget for the unit. You require weekly meetings with him to monitor his progress or help him solve problems that have arisen. Instill upon him the fact that he is not alone. You are there to help him. Have some of these meetings in the grazing unit itself. Let him show you what he is doing rather than have him tell what he is doing in the office. Go over his budget with him monthly to insure that he is on line. Any deviations should be discussed and the budget adjusted at this time if necessary.

At the end of the production season go over the numbers with him. Do it immediately after the calves are weaned and weighed. He is very anxious to measure himself. Perhaps more so than you are to measure him. When everything is calculated and if it's positive, congratulate him. Reinforcement at this point will outlast the low points which are bound to come. Make sure others in the organization know the numbers. If he agrees, tell them what his bonus was. This will do a lot to create enthusiasm and maybe convert some skeptics. If the results were not as good as expected it is still important to make this experience as positive as possible. Investigate problem areas and come up with plans for next year to solve them. Remember, anybody who has never made a mistake, never tried to do much. Again, let other people know about them, keep the operation open. Don't let people come to their own conclusions based on hearsay, or because they did not know the facts.

Repeat this process with other people as more grazing units are developed. Remember, although this person is running your cattle he is not your cowboy. He is a semi-independent manager and his success is your success, his failure is your failure. If there are more failures than successes look in the mirror and you'll see the reason.

This is not a recipe to be followed blindly. It is but one of many possible scenarios that can be used. Change it to fit your situation so that it will guarantee success of your unit manager. His success depends upon you and your success depends upon him.

THE LEASEHOLDER POINT OF VIEW: by J. Gordon Cran.

From the leaseholder point of view, tenure is the most important factor to consider. When a person involved in grazing takes on management of land, the time element is very important. On fee simple land tenure is not considered a problem, as the property is owned and will be available for as long as the owner desires to graze the property. It is unlikely that land in Hawaii would be purchased today to graze as the sole purpose for investment. Tenure would be influenced by taxes, zoning, and for the land owner's desire to sell for a profit.

On leased land the operator needs to be very cost conscious and business oriented. Many aspects of business are different than on fee simple land. Some points to be aware of are that leasehold equity is highest at the onset of the lease and continuously lowers as the lease runs out. Another point is that when interest rates were low (3-4%), it was easier to amortize a loan than when interest is high, 13-14%. When inflation is causing all goods and services to rise, a borrower will be paying back a loan with cheaper dollars. However if he plans on inflation and inflation slows down, a lessee may find he doesn't have time to amortize his loan should he be financed. So it can be seen that the lease term is very important.

Terms of leases vary from month-to-month to 50 or more years. The shorter the term, the poorer the stewardship. On very short leases, an operator should be aware that he may not have time to generate a cash flow if a herd is not ready to productively graze the area. Either a herd should be on the property or available to go on as quickly as possible to generate income. With cattle it is extremely slow to realize a return from production.

On short-to-medium length leases the lessee should carefully inspect the property to determine the life of the improvements. Should the improvements deteriorate faster than anticipated, costly inputs may be necessary with insufficient time to amortize the investment. Quite simply a lessee with a short-to-medium term lease will never build equity in the improvements to the land. This will result in a gradual decrease in productivity.

Obviously if land is to be leased, both the lessee and lessor would benefit with long-term leases. On government land 99-year leases with improvements owned by the lessee, as in Britain and Australia, will assure maximum incentive to a lessee.

CONTRACT AGREEMENTS:

Workman's compensation for most ranch employees is high, for example around 21% in Hawaii. This added cost, in addition to medical, dental, retirement funds and any other fringe benefits that may be offered, can very quickly make labor the most expensive item in the operation. One alternative that should be considered is to contract out certain routine jobs, such as equipment repair, maintenance, janitorial, and other service functions; as well as those once a year jobs, like making hay or silage, building sheds or fence.

FIGURE 2.2. DIVERSIFICATION - wind farm at Kahua Ranch. This 20,000 acre cattle and sheep ranch, in addition to taking advantage of an average yearly wind speed of 21 mph, also grows carnations under a half acre greenhouse, raises trout, has a group tour service, a livestock hauling service, as well as the traditional trail ride enterprise.

If the ranch is diversifying, the possibility of contracting out the entire alternate enterprise should be considered. Thus, if a trail ride enterprise is being considered, it might make a lot more sense to contract the entire operation to an existing trail ride company, or to help an individual set up a private trail riding company. This approach is being used by several ranches in Hawaii. One ranch has several recreational enterprises, each one handled by a private concession. Each concession contracts with the ranch for use of certain portions of its lands and facilities. Each concession is responsible for its own needs, in term of employees and special equipment. The ranch provides overall control in that it determines the mix of enterprises, thus providing the tourist or clients with a diverse choice of activities. The client can select one activity, or a mix, whatever turns him on. The legal and insurance requirements must be tailored to each situation and can be a real hassle, but the economic advantages can be considerable.

An extension of this concept that can be put to use by large ranches, is to contract out portions of the ranch (grazing units) to individuals, but maintain control of overall ranch policy.

DIVERSIFICATION: by John Morgan.

Most ranchers understand that it is their responsibility to generate as much revenue as possible from their cattle operation. Taken on a larger perspective though, ranchers are managers of assets, the biggest of which is land. Diversification is a tool to maximize income from assets held by ranchers.

Some forms of diversification (such as agricultural crops) require exclusive use of the land; while others (recreation) can run in harmony with ranching. The owners or managers must first set goals as to what they want and expect from their assets. If profit is a goal, then an analysis of possible diversification operations is a must. Many agricultural products are possible on ranch land other than livestock, such as forage, food, tree, foliage and flower crops.

An important consideration is how these different enterprises might fit in with the ranch. Most ranchers would not put a greenhouse right where their corral is located, even though it might be a good place for one. Limitations to consider are: water, utilities, labor and management, marketing and environmental conditions and capital. There is usually a direct relationship between profitability of any of these agricultural diversification projects and the amount of capital and management required.

Recreation is a type of diversification particularly attractive to some ranchers because (more often than not) it is compatible with raising cattle. The range of options in the field is immense. A ranch can provide one hour trail rides, week long wilderness trips, lifetime use of ranch facilities, or a multitude of other kinds of recreational "products".

Once the owners decide to get into some form of recreation, the next question is "what type"? The higher the degree of user satisfaction of the product, which is usually price related, the higher the degree of integrated planning and management required. The following are the three critical areas of concern:

1. What "product" are we going to produce? 2. Can we produce that "product" satisfactorily? 3. Can we market that "product" economically?

A rancher may decide to go into the trail ride business because he has extra horses; however if his ranch is 150 miles outside of town in a flat desert, he might have trouble marketing it, failing the second and possibly the first criterion. On the other hand, another ranch may be on the most beautiful spot on earth, have $500,000 worth of facilities and a multitude of fun, safe things to do; and still not be able to make ends meet because of poor attendance, failing the third criterion.

Lastly, assuming he has all the criteria in order, a rancher might brazenly go ahead, only to find out that his plan differs from that of the government. BEWARE of the permit process. You may move all your cattle on Japanese quarter horses (motorbikes), but you may need a permit to let someone else come with you on a ride. Actually though, a conscientious, thoughtful planner should not have too many problems. As long as there is no danger to the public, or the environment, recreational programs are likely to succeed.

However, as with marketing,the rancher must never underestimate the powers of communication. Oftentimes there may be a vocal minority that may stir controversy where there shouldn't be any; they need to be anticipated. Communication is also essential with the powers that be; those who control permits, as well as those within a ranch organization. For diversification to be successful, all employees must understand, and accept the new form of business. The owner or manager must make everyone aware that the ranch is a business, like any other business, and all projects, new or conventional contribute to the strength of the company as a whole. A stronger company means more security and greater chances of financial advancement for its employees.

SUGGESTED READING:

How to Start a Successful Business. 1984. Bevan Smith. Dunmore Press.

Lifemanship. 1950. S. Potter. Henry Holt & Co.

Management. 1974. Peter F. Drucker. Harper & Row.

Managing in Turbulent Times. 1980. Peter F. Drucker. Pan Books Ltd.

Motivation to Work. 1967. Frederick Herzberg. John Wiley & Sons, Inc.

Parkinson's Law. 1957. C.N. Parkinson. Houghton Mifflin Co.

Paths to Power: A Woman's Guide from First Job to Top Executive. 1980. Natasha Josefowitz. Addison-Wesley

The Managerial Choice: To be Efficient and to be Human, 2nd ed.. 1982. Frederick Herzberg. Olympus Co.

Up the Organization. 1970. Robert Townsend. Fawcett Crest Books.

You Can Negotiate Anything. 1980. Herb Cohen. Bantam Books,

Used with permission of NEA.

What you know or think you know, may not have any meaning in a new situation.

3 FINANCIAL ANALYSIS

UNIT OF PRODUCTION:

When land is cheap and plentiful, or when ownership is in question or impossible due to cultural or governmental policies, the animal evolves into the basic unit of production. The reason is simply that the intrinsic value of the herd or flock is perceived as being greater than the land upon which they graze. The early policies of the U.S. Federal Government in the disposition of over one billion acres through the various Homestead Acts, sales, and grants to railroads and states, as well as the administration of the public domain, did little, with a few notable exceptions, to encourage the viewing of land as anything but something to be used.

The tradition of the old west is closely tied to that of the open range. Up until 1934 grazing on the public domain consisted of intense competition among the various users. Since one individual had as much right on the area as another, there was a continuous race to get the best, first. The Taylor Grazing Act of 1934 was the first major effort to bring some control to the public domain. Presently two agencies, the Bureau of Land Management (BLM) and the U.S. Forest Service, administer over 87% of the 755 million acres of remaining federal lands. Grazing land on the public domain is now under a permit system. While there are differences of opinion as to how well this is working, there is a strong awareness that some management is better than none.

In the late 1950's other user groups, primarily recreational and/or environmental in nature, began to focus attention on the public domain. Since then the multiple use concept has dominated thinking on the public domain. Rangelands are now valuable. In the private sector farm land has always had a value, but lately that value has escalated to where it is worth much more than mere agricultural productivity would dictate. The overall effect, as far as the public is concerned, and thus the government, is that all land has value and that value is more than the animals grazing on it.

In the ranching segment of the livestock industry there's still confusion as to what is the basic unit of production. It has not been that many years ago since land was cheap, and tradition dies hard. However, tradition is not all to blame, tax laws that allow ranch loses to be written off against other profitable business ventures are as much to blame as the "cowboy myth".[1] As long as ranchers figure the cow, sheep, or other animal as the unit of production, the land will suffer, and the livestock industry will have to live with some very strange economics.

ENTERPRISE ANALYSIS:

An enterprise is a segment of the total business that could be run independently. A ranch might run cattle and also have a hunting operation. Either one could be run as a separate business; each is an enterprise.

DEFINITIONS:

GROSS INCOME: Total amount of dollars taken in by the business.

NET INCOME: The dollars left after the cost of doing business has been deducted.

[1] A myth is an effeminate moth.

COSTS: Costs are the dollars paid for materials, equipment and labor that are used in the operation of the business. All costs vary if the time period is long enough. Similarly, all costs can be considered fixed if the time period is very short. However, for the purposes of analysis, the various costs of doing business are broken down into three general categories; overhead costs, direct costs and variable costs.

1. Overhead: Those fixed, long-term, costs that are necessary to carry on the basic operation of ranching. If you want to ranch, those items that are required to ranch; salaries, FICA, telephone, postage, office expense, lease, insurance, vehicles, fuel, capital expenditures (land, buildings, perimeter fence, water, livestock), etc. In general overhead costs remain more or less constant over a period greater than one year.

2. Direct: Those fixed (longer than one year) costs that are associated with a particular enterprise and are necessary to carry on that enterprise. For a sheep operation, shearing equipment, dipping, special handling equipment, etc. In general, direct cost will remain fixed over a one year duration (direct costs are sometimes referred to as fixed cost).

3. Variable: Those costs associated with the unit of production; in a sheep operation the vet bill for each sheep, supplement, mineral requirement, etc. If it costs $10.00 per year to keep one sheep, then it will cost $100.00 a year to keep 10 sheep. Figure 3.1 shows the relationship between fixed and variable costs.

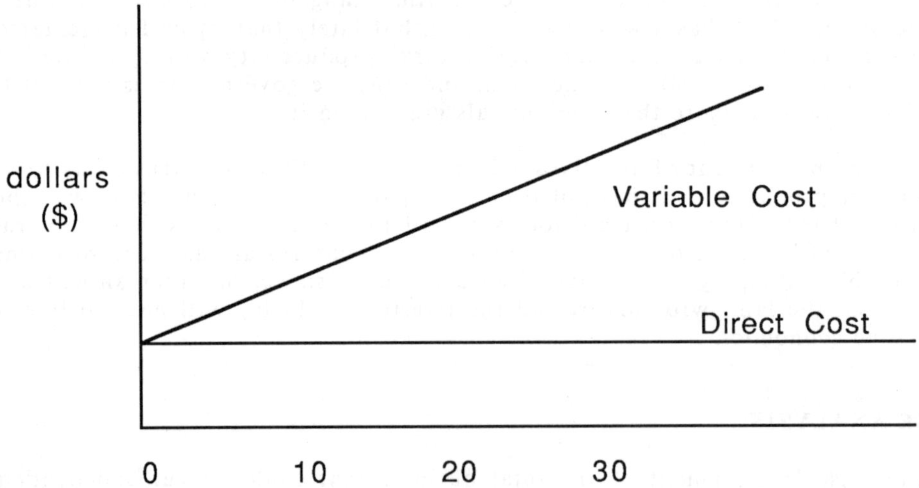

FIGURE 3.1. ADDITIVE EFFECT OF DIRECT AND VARIABLE COST.

The direct costs and the variable costs are figured for each enterprise that the ranch has. The income from that particular enterprise is then added in: (Income)-(direct cost)-(variable cost) = net enterprise return. This is the enterprise gross margin.

This is done for each enterprise that the operation has. The net enterprise returns are then added for all enterprises and the overhead cost subtracted from that. The sum of all net enterprise returns minus overhead = net operating return. If it is positive, you're still in business, if negative, you're in trouble.

Enterprise analysis allows the manager to pinpoint just where losses are occurring, which enterprise is doing well and which is not, which one is returning more per dollar invested, and which one is losing money. If an enterprise is losing money, or is not doing well, efforts can then be directed towards correcting the problem, or if not correctable, a decision has to be made as to keep it or get rid of it. It may well be that an unprofitable enterprise must be kept in order for the other enterprises to function; ie, growing hay for drought use, as the alternative cost of purchasing hay is worse. However, the manager at least knows where the losses are occurring.

Usually, variable costs per animal are already low and often should not be decreased further, except where buying in bulk allows savings to be made. Yet, because the ranching industry has traditionally viewed the animal as the basis of production, when times get tough the variable cost (cost per animal) gets cut. You doubt this? What's the first thing you cut back on in hard times? - It's minerals, supplements and calls to the Vet, isn't it? This was normal as long as the animal was worth more than the land upon which it grazed, also the animal moves and attracts the eye. The variable cost per cow is only around 25 to 50 dollars per year, however the fixed costs (both direct and overhead) may be ten times more than this, 300 to 400 dollars on a per animal basis. If the variable costs could be reduced 50% only 12 to 25 dollars could be saved, yet if the fixed costs could be reduced 50% the saving would be considerably larger.

There are two ways to lower fixed costs per unit of animal production, reduce the overall cost, or run more animals. Generally direct or fixed costs usually can not be lowered significantly except over the long haul, unless one has a lot of "Heavy Metal". However, because they are "fixed" they can be reduced on a production unit basis simply by producing more; more bushels per acre, or more animals per acre. Any increase in production normally has an associated cost connected with the increase. In the case of crops, often an increase in the amount of fertilizer used is all that is required. With livestock it may be as simple as better genetics, or as is usually the case, better management of what is already on hand. The greatest return for the dollar spent is increasing the managerial ability of the operators (since the operator is there already); the next greatest return is through sub-division of existing pastures and adoption of a sound grazing program.

YIELDS AND ASSOCIATED COSTS:

In intensive grazing the cost of sub-division maybe about 40% of the total fixed cost. Changing to IGM will often (but not always) allow increases of 100% or more of the present stocking rate, which will return twice as much gross income. Variable costs will double but fixed costs will not, the end result being a considerable increase in net income.

Figure 3.2 shows the expected increase in yield from a single input item; for example, the expected increase in production of forage due to nitrogen application alone. The same principal holds for animals; fly control is one example. An effort to control the majority of flies will yield good returns, but attempting to kill every single fly will take a great deal of effort and the additional production as a result of eliminating all the flies will be small. In the above figure, the first unit of input will yield 4 lbs/Ac, but the second unit will only yield an additional 1 lb/Ac, however the cost of the second input will be the same as the first.

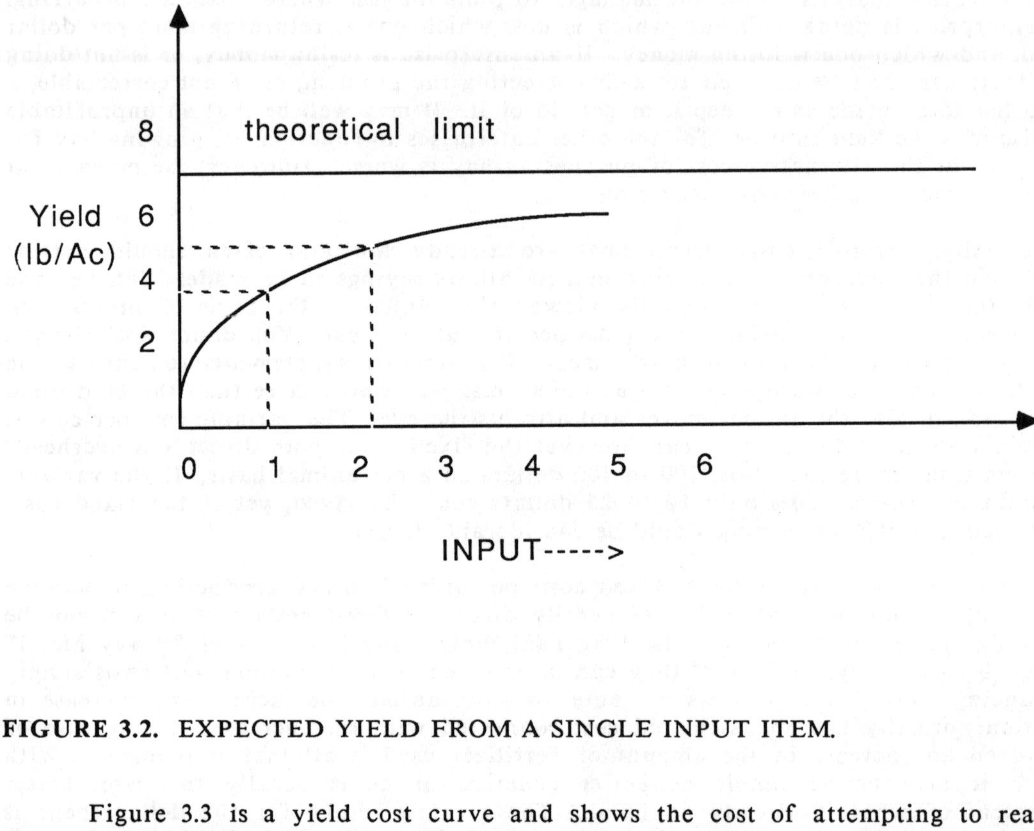

FIGURE 3.2. EXPECTED YIELD FROM A SINGLE INPUT ITEM.

Figure 3.3 is a yield cost curve and shows the cost of attempting to reach the theoretical limit of yield. Generally it does not cost much to get a big increase in yield when there is initially little or no input, however if the goal is to kill every fly on the place, the costs go straight up.

Future increases in production will come from interactions between several input items. You can only get so much improvement from genes, so much from nutrition, so much from gentle handling, from good health care, etc. However, put it all together in the right way, and you can get much more than the sum total of each, "The whole is greater than the parts". The ability to make this happen is management. To make it happen at a reduced per unit cost is effective management. Figure 3.4 shows the relationship between yield, cost and the level of management.

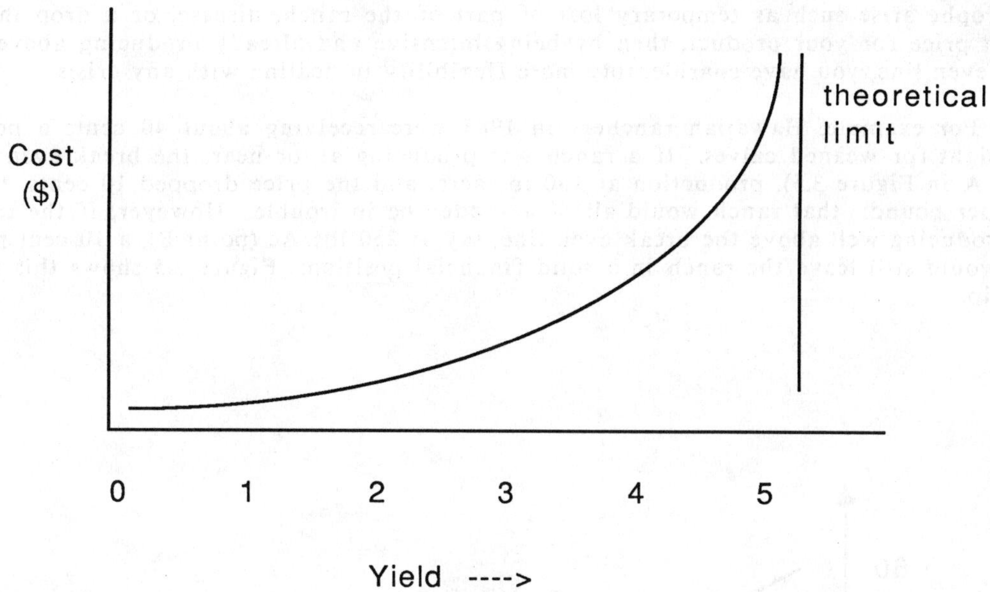

FIGURE 3.3. COST OF OBTAINING SUCCESSIVE INCREMENTS OF YIELD.

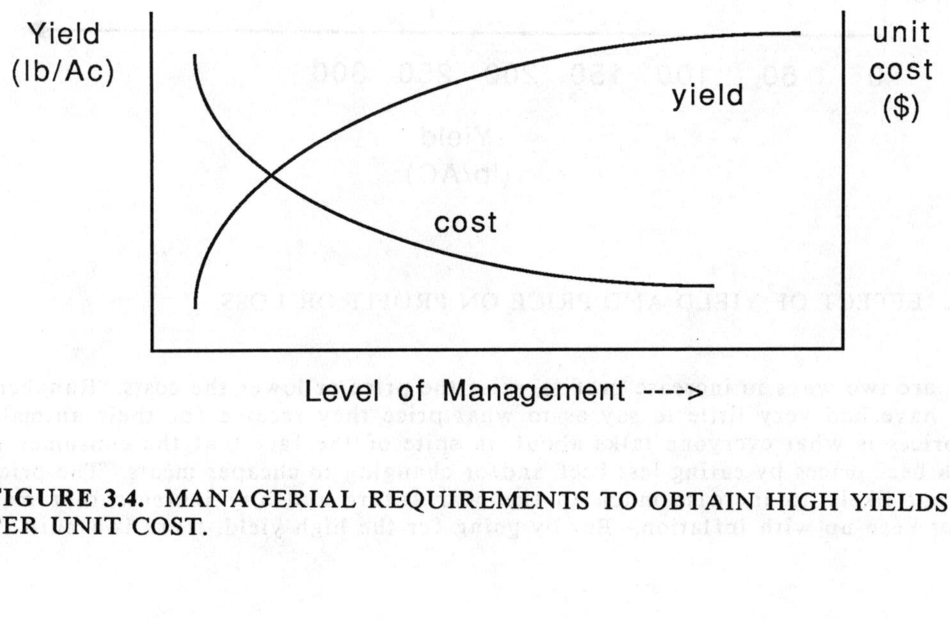

FIGURE 3.4. MANAGERIAL REQUIREMENTS TO OBTAIN HIGH YIELDS AND LOW PER UNIT COST.

The question occasionally comes up, "Why should I bust my butt going intensive? I'm doing OK, not making a lot of money, but making a living". The reason, aside from making more money, is that going intensive is like insurance. Should some unforeseen catastrophe arise such as temporary loss of part of the ranch, disease, or a drop in the market price for your product, then by being intensive and already producing above the break even line, you have considerably more flexibility in dealing with any crisis.

For example, Hawaiian ranchers in 1985 were receiving about 40 cents a pound liveweight for weaned calves. If a ranch was producing at, or near, the break even line (point A in Figure 3.5), production at 150 lbs/acre, and the price dropped 10 cents, to 30 cents per pound; that ranch would all of a sudden be in trouble. However, if the ranch was producing well above the break even line, say at 250 lbs/Ac (point B), a 10 cent price drop would still leave the ranch in a solid financial position: Figure 3.5 shows this relationship.

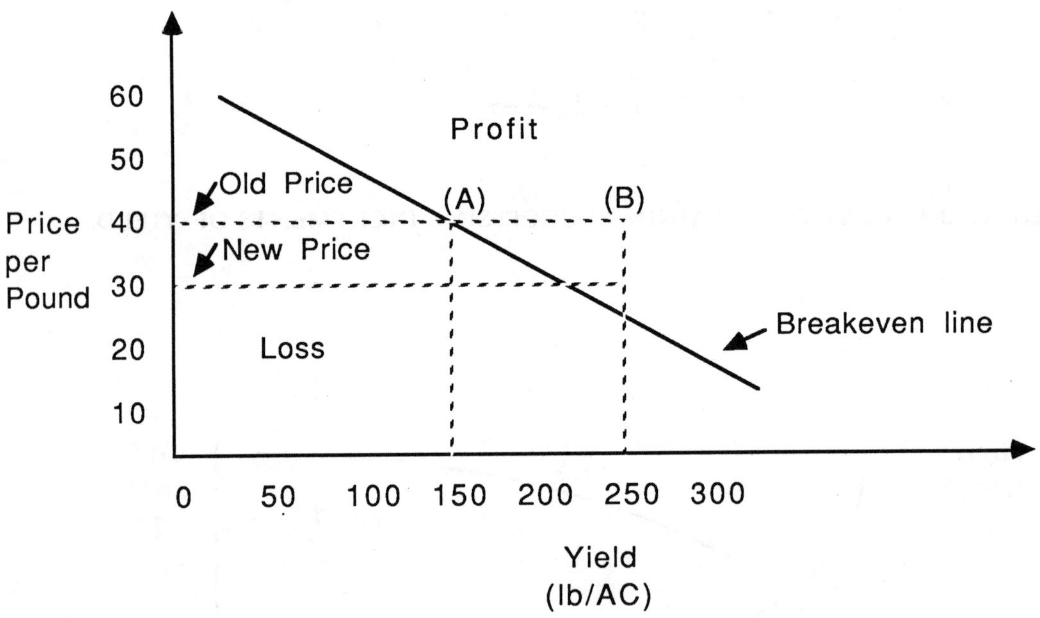

FIGURE 3.5. EFFECT OF YIELD AND PRICE ON PROFIT OR LOSS.

There are two ways to increase profits, raise the price or lower the costs. Ranchers traditionally have had very little to say as to what price they receive for their animals. Yet higher prices is what everyone talks about, in spite of the fact that the consumer is resisting high beef prices by eating less beef and/or changing to cheaper meats. The price of cattle is not likely to increase much in the near future and any increase that may occur will not keep up with inflation. But by going for the high yield, there is a built in hedge against unforeseen price drops. Under the present economic conditions, the only way to stay in the profit column is to decrease the unit cost of production. It may not be fair, and you may not like it, but that's the way it is.

TURNOVER RATE:

A rancher sells a steer to a packer for say $400, the packer sells it to a retailer for say $700 and the retailer sells it to the consumer for say $1200. The packer and retailer claim to only make around 1% on their money. So the packer makes $4 and the retailer $7 on the money they have invested. One percent, that doesn't sound too bad, that is until one thinks about turnover rate. Turnover rate is how fast a product, material, thing or dollars in this case are cycling through the system; how long it takes for the money to come back to the one who invested it. The rancher kept that steer maybe 2 years, the packer maybe 2 weeks and the retailer maybe one week. The retailer, because he is turning over that money every week is in effect making a profit of 52% a year, or $364 on the initial $700 invested. This is not to single out retail meat stores as villains, as all retail stores count on turnover to increase profits. Only livestock producers seem to have trouble with the concept.

Traditionally, turnover rate in livestock operations simply meant how fast the animal could be gotten off the ranch; i.e. increase the average daily gain (ADG). If the ADG is doubled then gross income is doubled. However, one had to watch that the cost of obtaining the higher ADG was not doubled as well, as was the case more often than not unless adequate controls (management) was in place. Again this is a case of placing the emphasis on the animal and not on the true unit of production in today's economy; the land. Turnover rate can be doubled by simply doubling the amount of liveweight produced per acre.[2]

STOCK FLOWS:

A stock flow is a diagram of all the changes in the herd structure that are expected to occur during the coming year. All expected stock sales, purchases, transfers from one category to another and deaths are recorded. Several things will be generated from this analysis that will begin to paint a "holistic" picture of the entire ranching operation. However, before a stock flow chart can be generated there are a few other concepts that must be introduced.

PERCENT CALF CROP:

<u>The number of calves weaned divided by the number of females exposed to the bulls equals the percent calf crop.</u> If you weaned 90 calves and 100 cows were exposed to the bull then [90/100]x100 = 90% weaned calf crop. Other than from a herd health standpoint it's only of academic interest to know how many cows conceived, or are pregnant, how many gave birth or how many calves were branded. However, it is of VITAL importance to know how many animals will be available for sale. Most cow/calf operations don't sell baby calves, they sell at weaning or some time thereafter. Consequently, if progress is to be made, the main measuring tool had better measure what's of economic importance.

It is often argued that by pregnancy testing the open animals can be sold and a resultant savings on pasture cost realized. Like many things, this is an over simplification

[2] Like the fly said after landing on the mirror: "It's a different way of looking at it."

and true only under certain conditions. To be economical, the testing must be nearly 100% accurate; few if any abortions or re-absorptions can result from the testing or handling; and pasture costs must be relatively high. This is seldom the case. What must be decided are the ultimate costs and benefits to be gained by pregnancy testing compared to another culling procedure. Where all heifers are bred to calve as two year old, pregnancy testing of the heifers is probably justified. However, if all cows that fail to wean a calf are routinely culled, pregnancy testing is likely not worth the cost.

WHAT IS THE PERCENT CALF CROP AND WEANING WEIGHT WORTH?

It's hard to pin down the cost to keep one cow for one year. Some charge land rent and labor and some don't. Realistically, the cows should carry the entire load, unless you have more than one enterprise. Once the cost per cow has been determined, the price one must have for the calves in order to cover this cost, is the next item of consideration. Obviously, the more and heavier calves one has to sell the more income will be generated. In general, a 10% increase in weaned calf crop will lower the break even selling price one must have to stay in business by 10 to 15%; a 10% increase in weaned weight will lower the break even price by 7 to 10%. Table 3.1 gives some idea of the percent calf crop and the weaning weights required to break even at different costs per mature cow. It should be noted that this table is a little phony, in that it assumes that all calves will be sold, and does not allow for holding any back; nor does it figure the value of cull cows that also make up part of the yearly sales. However, it does illustrate what percent calf crop is worth as well as increased weaning weights.

Since environment (nutrition) is responsible for better than 90% of the fertility of the cow herd and 70% of the weaning weights, a cheaper and more lasting solution is to improve the forage quality first, then embark on a rigid culling program. Don't blame the cow for poor performance if you're the problem.[3]

HERD STRUCTURE:

Most operations consider the livestock on the ranch as cows, bulls, yearlings and calves, which is fine if the operation is the sort that only sees it's animals once or twice a year. As discussed in greater detail in Chapter 6, the needs of an animal change as that animal grows, but also as the physiological condition of the animal changes. A lactating cow requires almost twice as much forage as does a dry cow. For a 600 lb heifer to gain one pound per day, it requires as much forage quantity as a 1000 lb dry cow and twice as much digestible protein. You can not treat them all alike and expect great things to happen. The following is suggested as a method to organize the herd so that optimum ranch performance can be realized by setting realistic performance and production targets.

A calf is a calf until it is weaned, after that it becomes something else. It is the mother's responsibility to care for that calf. How well she does that depends on how well you provide her with the proper nutrition and environment as well as her ability as a mother. At weaning the calf crop is divided into possibly three but at least two categories; HO's, SO's and possibly BO's.

BO's: Bulls less than one year old.
SO's: Steers less than one year old.
HO's: Heifers that have not been exposed to bulls.

[3] "We have met the enemy, and he is us."

As the animal ages the category changes, thus BO's would become B1's when the bull is one year of age, B2's at two years of age, and B's (bulls) when growth ceases at maturity. Steers would be the same except that there should not be too many mature steers (S's) around. The reason for separating out bulls and steers at an early age is simply that each group has different requirements and should be managed differently. Steers don't grow as fast as bulls, do not require as high a quality diet, may have hormone implants, and are not as aggressive as bulls. Steers can be added to, or taken away from the herd without causing a great deal of unrest or fighting amongst members. There will always be some social re-adjustment problems, but this is less with young animals and steers than for other classes. Bull management is discussed in greater detail in Chapter 13.

TABLE 3.1. REQUIRED PRICE PER POUND TO BREAK EVEN AT DIFFERENT PERCENT CALF CROP, WEANING WEIGHTS, AND COSTS PER MOTHER COW.

$ Cost Per Cow per year	Weaning Weight (lbs)	% Calf-Crop Weaned 60	70	80	90
		Cents Per Pound			
$100 Cost per year	300	56	48	42	37
	400	42	36	31	28
	500	33	29	25	22
$150 Cost per year	300	83	71	63	56
	400	63	54	47	42
	500	50	43	38	33
$200 Cost per year	300	111	95	83	75
	400	83	71	62	55
	500	66	56	51	45
$250 Cost per year	300	139	119	104	93
	400	104	89	78	69
	500	83	71	63	56
$300 Cost per year	300	167	143	125	111
	400	125	107	94	83
	500	100	86	75	67

NOTE: The above costs include both fixed and variable.

Females are the main reproductive units, hence age is not as important a criterion as is the sexual status. An HO has not been exposed to bulls, an H1 has been exposed for one breeding season, an H2 for two breeding seasons; after that the heifer becomes a

Cow. The use of both H1 and H2 is for the situation where calving as two year old is practiced. Females calving as two year old require a high plane of nutrition from birth until they have been successfully bred back for their second calf. They need greater attention after their first calf than do three year old. In particular, extra care during calving, early weaning of the calf and flushing prior to breeding again. A high quality diet, especially after calving (the two year old is still growing), is the single most important criterion. Consequently, H1's should be separated from H2's and C's. If females are first calved as three year old, there is no reason to use the H2 designation; it would be HO --> H1 --> C (Cow).

For females the time to make transfers from one category to another should be after weaning, for other classes whenever they occur.

HERD REPLACEMENTS:

There are many rules of thumb used to determine the number of heifer calves to keep for herd replacements. Many operators cull 20% of the cow herd, so keep an equal number of replacement heifers, plus some extra, to cover death loss. If everything worked right there would be a complete turn-over of the herd every 5 years or so. However, not all of today's heifers become tomorrow's cows. A certain percentage will die before reaching reproductive age, another percentage will die calving, another percentage will either have something wrong with them, or simply look wrong, and will be culled on that basis. Then there are those that simply fail to get pregnant. An average ranch should retain 40 to 50% of heifer calves as possible replacements.

An above average ranch should proceed a little differently. The problem boils down to the fact that what "you think" makes a good replacement and what actually does make a good replacement, may be two quite different animals. One method is to cull only those heifers that have something physically wrong with them; breed the rest and then cull on the basis of performance.[4] Chapter 6 covers selection in more detail.

The problem of determining how many replacements to keep can be resolved by dividing the estimated number of replacements needed by the percent calf crop weaned of the next youngest class. If 10 cows (C's) are to be culled and replaced with second calf heifers (H2's), you don't just replace the 10 cows with 10 H2's. The reason is that not all of those H2's will wean or even have another calf. Depending on the age and prior treatment the probable percent weaned calf crop will vary from 50 to 90%. Say the replacements are 3 years old and have had decent care, an 85% calf crop might be reasonable to expect. As to how many replacements, divide the number needed by the % calf crop (in decimal form) of that class, multiply by 100, then round off to the highest whole number; [(10/0.85)x100 = 11.7] or 12 head of H2's would be required to replace the 10 cows culled.

THE STOCK FLOW CHART:

For example, consider an operation that is running 100 mature cows (C's), has a 60 day calving season (1st quarter), sells calves at 6 months of age (end of 3rd quarter), sells any female stock that does not bring in a calf, has a weaned percent calf crop of 90% for C's, 85% for H2's, 70% for H1's and makes class transfers after weaning. All steer calves are sold as are all calves from the H1's. All heifer calves, except from H1's, are

[4] Unless you have X-ray vision and can see genes, there is no substitute for performance records.

TABLE 3.2. STOCK FLOW OF COW/CALF OPERATION FOR ONE YEAR.

Management activities --->		calving		bulling		Transfers sales.
class	year start	1st quarter	2nd quarter	3rd quarter	4th quarter	year end
C's	100	100	90 sell 10	90	100	100
H2's	12	12	10 sell 2	Transfer 10 0	12	12
H1'S	16	16	12 sell 4	Transfer 12 0	16	16
HO'	50	50	16 sell 34	Transfer 16 0	0	50
calves	0	112	112	Transfer 50 0 sell 62	0	0
B's	5	5	5	Transfer/buy 1 4 sell 1	5	5

MONTH BOUGHT (b) OR SOLD (s)

	April	May	May	June	Sept.	Oct.	Nov.
Class	HO	calf-H1	C's	H1+H2	calf	B(s)	B(b)
Number	34	12	10	6	50	1	1
Unit price	$200	$125	$300	$250	$175	$400	$1200
Total price	$6800	$1500	$3000	$1500	$8750	$400	$1200

NOTE: transfers into the herd are designated by "Transfer/buy". Transfer designates the removal of animals from one class to the next older class. Sell designates the removal of animals from the herd.

kept until breeding (2nd quarter), at which time selection is made. Table 3.2 shows one way to organize a stock flow for the coming year.

At the start of the year there are no calves; at the end of the 1st quarter (three months) all the calves that will be born will be on the ground. Based on 100 C's at 90%, 12 H2's at 85% and 16 H1's at 75%, there will be 112 calves. At the end of the 2nd quarter there will be 112 calves. Of this number there will be 12 that will be from the H1's (early weaned) which will be sold, of the remaining 100, half, or 50 will be steers and sold. The 50 HO's (transfer from calf to HO made at weaning) will be kept until selection is made just prior to bulling (end of 1st quarter next year). At this time 16 head will be selected to be placed with the bull and 34 head sold. After the HO's have been exposed to the bull they are transferred to the H1 class; the old H1 class having been transferred up to the H2's, the old H2's transferred to the C's. Sales of cull animals, those that did not calve, should be made prior to the breeding season, or if not then, at least separated from the main breeding herd. If there is ample forage they may be kept and sold whenever the market is high. The above chart could be set up in greater detail, monthly intervals rather than quarter. This would be appropriate for larger operations, or if the marketing of the animals is on a monthly basis.

The stock flow chart should be constructed at least one year in advance, as it is an aid in planning. The estimated income will be used in conjunction with the cash flow chart.

CASH FLOWS:

A cash flow is simply a record of the money spent and the money received, recorded over a period of time. A check book is one example of a cash flow. It doesn't mean very much however, unless one other step is taken, which is to keep a running balance. Most of us attempt to do this, with varying degrees of success, but it's amazing how many don't. If the money spent is subtracted from the money earned over a period of time, a net cash flow is the result. A net cash flow is of vital interest, because it relates directly to the credit or debit balance in the overall ranch account. It is used as a budgeting tool, planning new ventures or enterprises, and as a check and/or control on the current year's finances.

Say you are thinking about starting up a sheep operation in addition to your cattle and have decided on the basic physical plan (Chapter 2). You know that you are going to have to borrow some money to get it going. A cash flow will give you some of the things you should know and many of the things a "good banker" will insist on knowing about the planned venture. Such as: How much money will the venture produce; how much will it cost? When will the money come in; when will it go out? If the money that is expected to come in, does not cover the money expected to go out, how can I make up the difference? How much money do I expect to have in the bank in 5 years if I undertake the venture (goals)?

A cash flow will help answer some very important questions, not only about a planned venture or enterprise, but also about your present operation. However a cash flow is only as good as the inputs; like a computer, garbage in - garbage out.

CASH FLOW BUDGETS:

The first step in developing a cash flow budget is to decide on the scope of the budget. Is this for an existing enterprise or a planned venture, or is this the budget for the entire operation? Next, the format, or just what it is you are going to record. Obviously, income will be recorded, sales of stock, wool, crops, or any custom work. But what if you have an outside job in town, should that be recorded? That depends on the first step, what is it you're doing.

The unit of time that is to be used is dependent to some extent on the size and nature of the business. If you only make and spend money one day out of the year, then a unit of recording of one year would be all right. Most ranching operations have money coming in at different times of the year, although the bulk of it may come all at once (fall sale of calves). Money spent is another story, seldom does a week go by that some money is not spent. Again size and nature of the business will determine if the monies are to be recorded on a weekly or monthly basis. However, for most family operations, recording monthly will be sufficient.

When should a cash flow be worked up? For a new venture, prior to laying out any money or making any firm commitments. For an existing operation, at least two months before the fiscal year begins. One possible schedule, shown in Figure 3.6 for an organization whose fiscal year begins 1 January, might be: October, 1986, develop a budget for the following year (in this case, 1987), using best estimates available at that time. Late December (1986), update October estimates, making any appropriate changes. During the fiscal year (1986), the actual expenses and income are recorded, as well as any revised estimates for the future. Sometime in January 1987, review the 1986 budget.

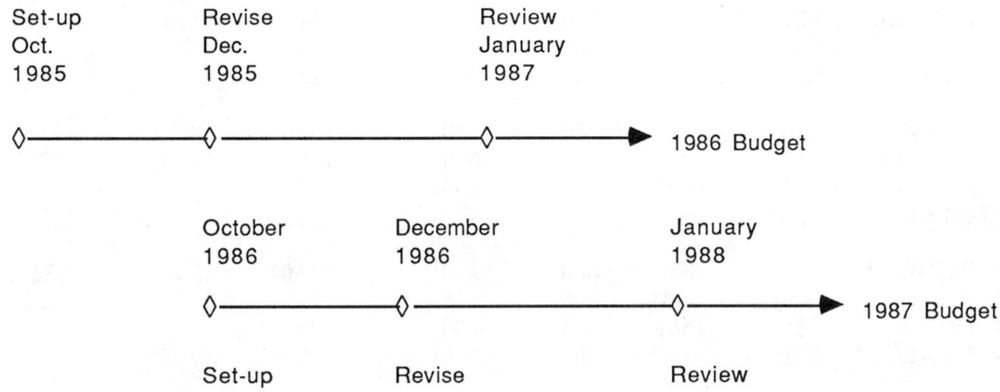

FIGURE 3.6: SCHEDULING BUDGET SESSIONS

There should be four entries under each month. These are: JANUARY BUDGET; JANUARY ACTUAL; JANUARY DIFFERENCE; SUB-TOTAL. The BUDGET is the best estimate made in October and revised in December, as to what the income and cost figures are likely to be for the coming fiscal year. The ACTUAL is the actual amounts spent or received as the fiscal year moves along. The DIFFERENCE is the difference between the budgeted and the actual amounts, spent or received each month. The SUB-TOTAL is the running total (+ or -) of the amount you are over, or under the projected budget as the year progresses. This, in effect, tells you how well your estimates are. Table 3.3 illustrates how such a cash flow budget might be set up.

The items to place along the top columns are the basic cost items that are used in the enterprise. The list of costs shown in Table 3.3 is very incomplete, actually costs could include 20 to 30 different items, depending upon the extent of the enterprise. Regardless of how many column items there are, at the far right there needs to be one column titled "TOTAL SPENT" and one "SURPLUS and DEFICIT" or (+/-) for that month.[5] There are only two entries in this column for each month, one for budget and one for actual. A third column should be labeled TOTAL. This column is similar to the monthly TOTAL, except that the entries are from the (+/-) column and represent a month to month running total for the entire enterprise. This is the place where you look to see if you can expect to make any money (budget row) or if you made or lost money (actual row).

TABLE 3.3. CASH FLOW CHART FOR FIRST QUARTER:

WIKIWIKI CATTLE COMPANY - CASH FLOW

	INCOME	-------COST--------			TOTAL	SURPLUS-	
	STOCK	LABOR	RENT	VEHICLE	SPENT	(DEFICIT)	TOTALS
JAN. BUDGET	250	500	600	250	1350	(1100)	(1100)
JAN. ACTUAL	225	500	600	225	1325	(1100)	(1100)
JAN. DIFF.	(25)	0	0	25	25		
SUB-TOTAL	(25)	0	0	25	25		
FEB. BUDGET	350	500	600	250	1350	(1000)	(2100)
FEB. ACTUAL	250	550	600	650	1400	(1150)	(2250)
FEB. DIFF.	(100)	(50)	0	(400)	(50)		
SUB-TOTAL	(125)	(50)	0	(375)	(25)		
MAR. BUDGET	2600	500	600	250	1350	1250	(850)
MAR. ACTUAL	3550	550	600	125	1275	2275	25
MAR. DIFF.	950	(50)	0	125	75		
SUB-TOTAL	825	(100)	0	(250)	50		
FIRST QUARTER SUMMARY							
BUDGET	3200	1500	1800	750	4050	(850)	(850)
ACTUAL	4025	1600	1800	1000	4000	25	25
DIFFERENCE	825	(100)	0	(250)	50		
SUB-TOTAL	825	(100)	0	(250)	50		

[5] Parenthesis () are used to mark negative amounts, how much you're in the hole.

If a little time is spent this budget and actual comparison can be used very effectively; especially where monies need to be borrowed and paid back. A cash flow is also a very good introduction to bankers. REMEMBER: HAPPINESS IS A POSITIVE CASH FLOW!

SUGGESTED READING:

Farm Business Management. 1982. S. Squire & E. Delahunty. Longman Paul.

How to Start a Successful Business. 1984. Bevan Smith. Dunmore Press Ltd.

Roadmap to Profitability. 1982. National Beef Profit Conference. National Cattlemens Association. P.O. Box 3469, Englewood, CO 80155.

The Farming Game. 1981. J.P. Makeham & L.R. Malcolm. Gill Publishing.

Where's the Beef - Profit? 1984. Beef Profit Conference. National Cattlemens Association.

Used with permission of NEA.

After romance, comes the dirt.

4

SOIL BASICS

By W. McCall and B. Smith

Plants require sunlight, moisture, support, protection and nutrients for growth; of which the soil supplies all but sunlight, and to a lessor extent, protection. The ability of the soil to supply these plant requirements depends upon the soil characteristics. These in turn are determined by the five factors of soil formation: climate, vegetation, parent material (bedrock), topography and time.

The type of parent material, kind of like genetics in animals, dictates what sort of potential the soil is capable of; once that has been set the rest is up to the environment and time. The major factor of environment is, of course, climate. The amount, distribution and rate of rainfall, in conjunction with the average annual temperature not only dictates the rate of soil formation, but also the accumulation or removal of soluble products of weathering, the type of vegetation, and has a profound effect upon the soil's mineralogy.[1]

FIGURE 4.1. SOIL EROSION: This wind swept area located in Hawaii, is the result of years continuous grazing followed by 34 years of rest. In areas under stress, grazing (energy) must be controlled. Likewise if this area is ever to be re-vegetated a controlled energy input must be in place; it can be done mechanically or with animals.

---

[1] There exists some confusion between the meanings of the words soil and dirt; soil is not found under your fingernails.

The type of vegetation growing on an area, whether forest, shrub, or grass also influences the nature of the soil beneath. Forests remove many plant nutrients from the soil profile and re-cycles them between the plants and the surface of the soil. When the forest is cut and removed the soil fertility decreases rapidly unless the plant nutrients are replaced with fertilizers or other sources. The climate suitable for forest growth also removes soluble nutrients and promotes a more highly weathered soil. Grasses, on the other hand, grow where the rainfall is generally inadequate to promote the growth of trees. The soils formed under grasses have more organic matter, more soluble nutrients, a higher base saturation, and are more fertile and productive than those formed under forest vegetation.

Topography generally dictates the amount and velocity of runoff, the movement of water through the soil, the stability of the soil material and to some extent, vegetation (north or south facing slopes). The steeper the slopes the slower soil formation takes place and the less weathered the soils will be.

CHARACTERISTICS:

Soils are the medium in which plants grow; to understand soils and soil fertility is to understand a basic prerequisite for pasture and crop growth. Plants require non-mineral nutrients (carbon, oxygen and water), which are directly involved in photosynthesis (shown in Figure 5.1, Chapter 5). In addition to non-minerals, there are 13 mineral nutrients that are essential for plant growth. These are divided into primary, secondary and micro-nutrients. Primary nutrients usually become deficient in the soil first, because plants use relatively large amounts of them. Secondary and micro-nutrients are deficient less often as smaller amounts are used. However, they are just as important as the primary nutrients; plants must have them when and where they need them. Table 4.1 lists the 13 essential nutrients for plant growth.

TABLE 4.1. ESSENTIAL MINERALS FOR PLANT GROWTH.

PRIMARY NUTRIENTS	SECONDARY NUTRIENTS	MICRONUTRIENTS
Nitrogen (N)	Calcium (Ca)	Boron (B)
Phosphorus (P)	Magnesium (Mg)	Chlorine (Cl)
Potassium (K)	Sulfur (S)	Copper (Cu)
		Iron (Fe)
		Manganese (Mn)
		Molybdenum (Mo)
		Zinc (Zn)

Soils contain organic material and minerals which are classified, based on size, into sand, silt and clay. Sands range in particle size from 0.05 to 2 mm (millimeters), silt from 0.002 to 0.05 mm and clay particles are less than 0.002 mm. When the clay is less than 0.001 mm it is called a colloid. The clay has the greatest effect upon soil properties; this is due to its small size and the large surface area each particle has in relation to it's size.

Sand also has an effect due to the coarse nature of it's particles however, silt has an indifferent effect. Soils are called sands, silt loams, clay loams, etc. depending upon the relative amounts of these three materials present. Where the amounts of the sand, silt and clay are equal, the soil is called a loam. Soils that contain a majority of organic material are termed peat, muck, or organic soils.

Soil texture and structure are important as they influence the amount of water and air that are available to plants. A soil consisting of a lot of clay (small particles) can hold much more water than an equal volume of a sandy soil. This is because the small particles will have many more pores (holes) for the water or air, than does an equal volume of a sand. When water is held in small pores, it is harder to drain out due to gravity than is water held in large pores; thus more is available to plants. Organic matter is also capable of holding more water than sands.

"Field capacity" is the amount of water remaining in a soil after all drainage, due to gravity, has stopped. The "permanent wilting percentage" is the amount of water remaining in the soil after plants become permanently wilted. The "water available" to plants is that amount between the field capacity and the permanent wilting percentage, and varies for different soils. Figure 4.2. illustrates the relationship between water holding capacity (inches of water per foot of soil) for the various soil textures.

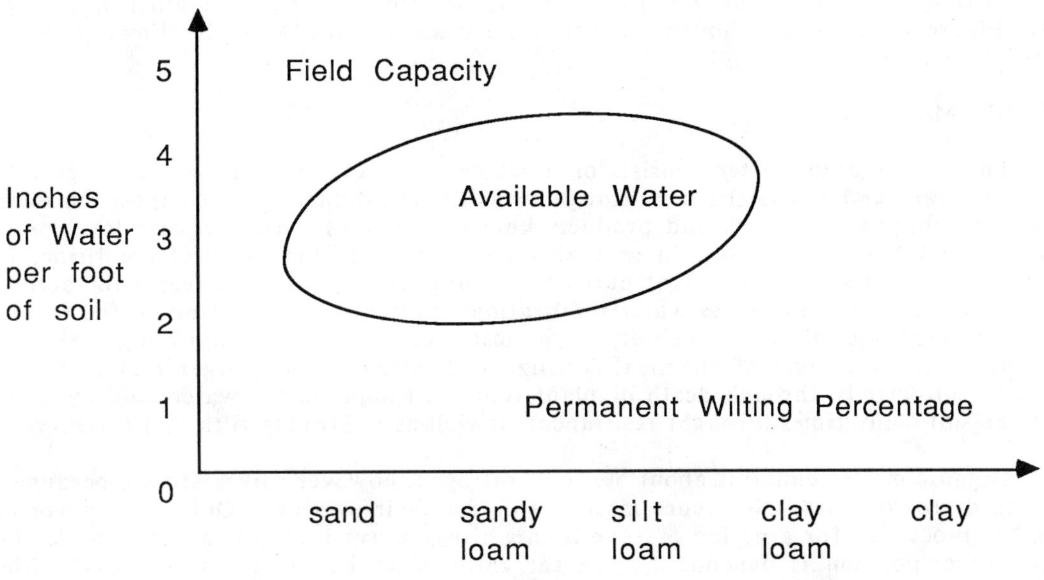

FIGURE 4.2. AVAILABLE SOIL WATER IN DIFFERENT TEXTURES OF SOIL.

The finer texture soils are more likely to form a soil crust (cap). This crust is formed much the same way that fine particles are floated up when finishing a cement floor, only in the case of soil the floating is done by impact of rain drops and wind vibration. In high growth areas this capping action is not significant as the crust is continually being broken by the action of roots, new shoots, and soil organisms. In arid environments the capping action is significant, as the crust, being composed of small particles, acts much like cement. This, in turn, cause more runoff and less infiltration of rainfall. Also, due to the small diameter of pores in the crust and the larger pores in the material underneath, the crust acts as a moisture pump, in effect pumping moisture, through capillary action, from the soil to the atmosphere. This is one of the reasons that winter wheat fields are plowed even if no crop is to be planted that year; to conserve moisture. All in all, the crust further reduces the effective precipitation from what actually fell as rain; not to mention making a very in-hospitable place for seeds to germinate.

CATIONS, ANIONS AND OTHER CREATURES:

As soils are formed during the weathering process, some minerals and organic matter are broken down into extremely small particles that can be seen only with an electron microscope. These particles are called "colloids". Colloids are either from clays or organic material. Each of these colloids has a net negative (-) charge, obtained during the formation process. The negatively charged colloids attract positive (+) charges, but repel other negative charge particles (similar to two north poles of magnets).

An element with an electrical charge is called an "ion"; if the charge is positive it is a "cation" and if negative, an "anion". Potassium, sodium, hydrogen, calcium and magnesium are all cations, while chlorides, nitrates, sulfates and phosphates are all anions. Consequently, the more colloids a soil has the better it can hang on to cations (unlike charges attract) and the easier it is for anions to be leached out of the soil (like charges repel). Figure 4.3. illustrates how a soil particle attracts cations (+) while allowing the anions (-) to pass by.

ORGANIC MATTER:

The soil organic matter consists of a whole series of products which range from undecayed plant and animal tissues through transitional products of decomposition to the fairly stable brown and black end products known as humus. An adequate level of organic material benefits the soil in several ways: improves the physical condition; decreases erosion losses; supplies plant nutrients; improves soil tilth; increases the activity of soil organisms; and increases water infiltration or retention, depending on the texture. In fact organic material offers the only hope most ranchers have of improving their soils, other than costly infusions of chemical fertilizers. Incorporation of organic material into soils, especially sands, through death of plant roots will improve the water holding capacity of the soil many times (drought resistance), as well as improving tilth and fertility.

Organic matter contains about 5% total nitrogen, however soil microbes, because of their rapid growth rates, will secure most of this for their own use. Only after decomposition has proceeded for a period of time is this nitrogen available for higher plants. The rate of decomposition is dependent upon the ratio of carbon to nitrogen (C/N). Most agricultural soils have a C/N ratio of about 10. Plant materials low in nitrogen, such as wood chips or sawdust, that are incorporated into the soil will raise the C/N ratio; this will decrease the relative amounts of nitrogen available for the soil organisms. As a result of lower nitrogen supplies in the added material the soil organisms will tie up other soil nitrogen, or use nitrogen fertilizer that was meant for the crop. This effect is responsible for the poor showing of many crops when they are planted immediately after plowing in crop residues (usually low in N) unless adequate nitrogen fertilizer has been applied.

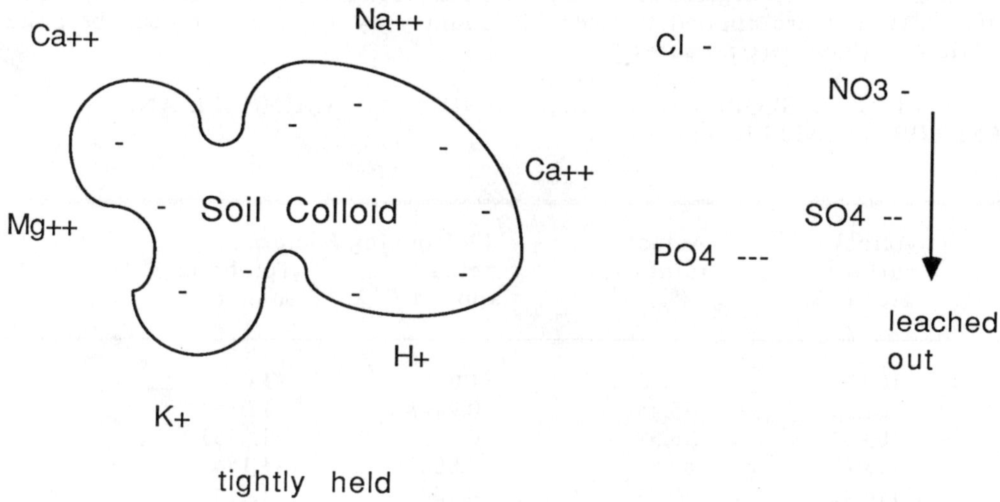

FIGURE 4.3. SOIL COLLOIDS ATTRACTING CATIONS BUT NOT ANIONS.

OTHER SOIL FACTORS:

The depth of the soil is a major factor in determining productivity. Roots of crop plants will often extend to 6 feet, kikuyu grass roots have been found to a depth of 16 feet and alfalfa roots will extend to 30 feet. Soil depth determines how large a nutrient reservoir is available for the plant to exploit. Topography is another factor, the steeper the slope the less productive a soil is. If a slope of 0 to 1 percent is given a value of 100, a slope of 5 to 8% will only be 60% productive, if it is not easily eroded. If it is easily eroded it will only be 30% as productive.

Another aspect of topography that is often overlooked is that there is more surface area in hilly country than what a map usually shows. If you took a square foot and placed it horizontally over a slope of 45 degrees and projected that square foot down on the slope, you would actually cover an area of 1.41 square feet on the slope. So what? Well, Table 4.2 shows how slope affects the amount of actual surface area, the amount of actual rainfall on that area, and how much fertilizer or other material that must be added to obtain the same rate as on level ground, if applied by air.

If one inch of rain fell (remember rain is measured by how much falls in a level container, only 0.71 inches would have fallen on an area with a slope of 45 degrees. This is one of the reasons that different species of plants are found on slopes (usually more arid kinds). Another reason is, of course, the slope's position with respect to the sun. When fertilizing hill country by air the rate for the slopes should be greater than that used for flat land. To find out how much more, take the flat land rate and multiply it by the reciprocal of the cosine of the slope in degrees. For example, if the average slope was

30 degrees and the rate of fertilizer recommendation was 200 lbs per acre, you would need to apply: (200 lbs per acre)x(1/cos 30 degrees) = (200)x(1.1547) = 231 lbs per acre on 30 degree slope land. For most farming applications the above is somewhat academic as the errors in slopes under 15 degrees are less than 3.5%. However, when fertilizers, herbicides or other materials are applied to steep hill country by air, the error can be quite large, especially for those vertical acres.

TABLE 4.2. AFFECT OF SLOPE ON ACTUAL LAND AREA, RAINFALL AND AERIAL FERTILIZER APPLICATION.

Slope in degrees	Actual surface area (%)*	Actual rainfall (%)	Multiplying Factors Actually applied #	To obtain same rate~
0 (flat)	100.0	100.0	1.0	1.0
10	101.54	98.48	0.9848	1.0154
15	103.53	96.59	0.9659	1.0353
30	115.47	86.60	0.8660	1.1547
45	114.42	70.71	0.7071	1.4142
60	200.00	50.00	0.5000	2.0000
75	386.40	25.88	0.2588	3.8640

* This is the actual surface area that most maps do not account for.

This is actually the cosine of the slope.

~ This is the reciprocal of the cosine (1/cosine slope).

NOTE: To the determine the affect of slopes other than shown, use a calculator that has a cosine function and take the cosine of the slope (in degrees); this will give the multiplying factor to find out how much rain fell, or material was actually applied, based or aerial applications. To determine how much material to apply by air to obtain the level surface rate, take the reciprocal (1 divided by the cosine of the slope).

SOIL ORGANISMS:

The various groups of soil organisms do not live independently of each other. They form an interlocked system more or less in equilibrium with their environment, which includes the above ground portion as well as the below. The best known soil organisms are the earthworms; their importance was first stressed by Charles Darwin in 1881. The principle food of earthworms is dead or decaying plant material, including both leaf litter and dead roots. There is a difference in acceptability of material; litter with a nitrogen content greater than 1% is readily consumed whereas lower nitrogen content is not. Litter high in polyphenols, such as pine needles or oak leaves, are little eaten. Dung is a preferred food for many species of earthworms, while others seem to prefer protozoa and bacteria found on soil particles. Worms are intolerant of drought and frost and favor well drained soils rather than heavy clays and sands. Earthworms are absent on soils low in calcium or very acidic, and are seldom found in soils with a pH less than 4.5. Researchers in New Zealand found that their most productive pastures had 2.8 million worms per acre with a weight of 2,100 lbs. They noted a close correlation between the productivity of the pasture and the weight of the worms, there being about 375 lbs of worms for every 2,200 lbs of annual dry-matter production during the course of the year.

The other soil invertebrate fauna, like the earthworms, need a well aerated soil for active growth; they can tolerate drought far better than waterlogged conditions. The food supply determines numbers and consists primarily of dead plant tissue; however living plants, the soil microflora and other invertebrates may also be considered as food by some. The most frequently recognized invertebrates are: nematodes, mites, collembolas, pauropoda, symphyula, thysanura, protura, enchytraeid worms, millepeds, centipedes, flies, beetles, ants, termites, crustacean and gastropods and various other insect orders. The weight of other invertebrates besides earthworms, has been measured in England to be between 36 to 169 lbs/acre, depending on soil.

The soil micro-organisms consist of bacteria, actinomycetes, bacteriophages, fungi, algae and protozoa. In a soil that contained 3% organic matter the estimated dry weight of just the bacteria is between 270 to 620 lbs/acre. The fungi are an important group as they are the most efficient users of cellulose and lignin. Most fungi require some source of nitrogen, however there are some which can fix atmospheric nitrogen. The protozoa are mostly rhizopods, flagellates and a few ciliates and their major food source are the bacteria. Up until recently protozoa were thought to be mostly harmful to the bacterial population; however, it is now known that beneficial soil bacteria are more efficient when there are protozoa around. Another example of the benefits of having a predator regulating a consumer group.[2]

Algae affect plant growth in four ways: they add organic matter, help bind soil particles on the surface, improve the aeration of swamp soils and fix atmospheric nitrogen. The nitrogen fixing blue-green algae are of great importance in swamp rice or paddy soils. This power of nitrogen fixation is not dependent upon light as these algae can fix nitrogen in the dark if supplied with sugar and a small amount of molybdenum. Most soils contain algae that can fix nitrogen.

pH OR SOIL REACTION:

The pH of a soil is a measure of the acidity or alkalinity of the soil solution; to the soil chemist it is the amount of active hydrogen-ions present in the soil. The pH scale runs from 0 to 14, with 7 being the neutral point (pure water). A soil is said to be more acid as the pH drops below 7, and more alkaline as it raises above 7. The pH scale is not a linear scale, but is logarithmic. What that means is that if the pH of one soil is 6 and another is 5, the pH 5 soil is 10 TIMES more acid, than the pH 6 soil.

In a very general way, the amount of rainfall an area receives pretty well determines if a soil will be acid or alkaline. The reason is that the bases (calcium, magnesium, potassium and sodium) are soluble in water and more so in slightly acid soil water.[3] So if there is sufficient rainfall to cause water to infiltrate to any depth in the earth, say to the water table, that water as it passes through the upper layers of soil will dissolve out the base elements and carry them down to the water table; where, for all effective purposes, they are lost to plant production. As the base elements are leached out, their place is taken by hydrogen and aluminum ions. Hydrogen ions are present in water, as well as being released by plant roots and the process of decay. Aluminum is released by the weathering process of various soil minerals.

In arid regions, the same leaching action occurs, however the water only infiltrates to a fairly shallow depth. Consequently, the base elements are not lost to the system and

[2] It's doubtful that the consumer being eaten feels good about it, regardless of how good it may be for the entire group.

[3] Rainwater is slightly acid, even without pollution.

over time tend to accumulate, making the soil alkaline. The removal of the upper base elements and deposition at a lower depth, causes a build up of these minerals and a resulting cement-like layer. This layer, called caliche, varies in depth depending on the average rainfall of the region. In arid regions the depth to the caliche may only be 6 to 12 inches or so. The caliche has the effect of reducing the amount of soil volume that plant roots can exploit, it also tends to keep water from penetrating. In an unusually wet year this layer holds the water above it and may turn an otherwise dry desert into one big bog hole; as occasionally happens in the spring in the American intermountain west after a heavy winter snow melt.

Agricultural soils tend to become more acidic with time. The reasons are the leaching losses associated with rainfall, irrigation, removal of the base elements through harvest of plant materials (one ton of alfalfa hay will remove 25 lbs of calcium and 5 lbs of magnesium from the system), increased presence of organic material in the soil (forming organic acids), as well as fertilizer application, particularly nitrogen. Most nitrogen is applied in the form of various ammonia compounds. Ammoniacal nitrogen is absorbed by the soil and after a few weeks is converted to the nitrate form releasing hydrogen ions to the soil, which in turn tend to make the soil more acid.

The pH of a soil can be changed. Acid soils can be made less acid (sweeten) by the addition of the various base elements. However, limestone is the most usual additive. If magnesium is also required, a dolomitic limestone will supply about 18% calcium in addition to 12% magnesium. In Hawaii, as in many tropical islands, ground coral, which contains 32 to 39% calcium, is used. Alkaline soils can be made less alkaline by the addition of gypsum or sulfur. The amount of lime or sulfur to add depends on the pH of the soil, the type of soil, the crops to be grown, the particle size of the additive, and if it is to be broadcast on existing pasture (topped-dressed), or plowed in.

Different species of plants are adapted to different soil pH's; in fact, different varieties of the same species may also be so adapted. Many weeds, such as Spiny Amaranth, have a wide range of pH tolerance (4.0 to 8.0) while many important agricultural crops do best in a fairly narrow pH range. In general, grasses do best at a pH range from 6.0 to 6.5, as does alfalfa. However, certain grasses do well at low pH's, such as St. Augustine and Hilo grass, partly because of adaptation to acid soils and partly because of reduced competition from other, less adapted species. As a general rule, grasses and other plants that require a lot of water for optimum growth are more acid tolerant than are those adapted to drier climates.

The type of soil is important in so far as it determines how much lime must be added to bring about a given change in pH. Sands generally require less lime than do clays to bring about the same amount of change in pH. The resistance to change in pH is termed the buffering capacity of the soil. Another important criteria in changing pH is the particle size of the material (lime) added. If the soil test calls for a ton of lime per acre, placing a one ton hunk of limestone on each acre will not get the same result as spreading a ton of finely ground lime over it. The finer the grind the faster the material will be incorporated into the soil.

The fastest method of obtaining results from the application of lime is to mix it into the soil prior to planting seed. In most grazing situations this is not feasible so top-dressing is used. Surface applied lime takes longer to react and does not do so as completely as lime that is mixed in with the soil. The length of time before effects of top-dressing will be noted will vary, depending upon the type of soil, the vigor of the plants, the number of soil organisms, the amount applied, and the initial pH of the soil.

In parts of New Zealand response has been noted in as little as a month, while in Hawaii 6 months or longer may be required.

IRRIGATION:

Irrigation is a special problem and close attention needs to be paid to just what the irrigation water contains. If a particular soil is irrigated with the same water source for sufficient time, the chemical analysis of the soil will begin to look suspiciously like that of the irrigation water. How long? That, of course, depends on the soil and what the water has in it; on coarse textured soils that do not contain lime, significant changes can occur between 10 to 20 years. Certain irrigated lands on the eastern side of the San Joaquin Valley in California have been turned into the equivalent of high rainfall soils (acid). Drip irrigation, because it wets a lesser soil volume, may increase this rate.

Irrigation water can be very acid or alkaline, or anywhere in between. That same water may contain significant amounts of different minerals. Some can be of benefit, some can be toxic. It is important that a chemical analysis of the irrigation water be made. To find out the amount of nutrients supplied by an acre foot of irrigation water: first determine the amount of the nutrient in parts per million (ppm), then multiply by 2.7 (an acre foot of water weighs 2.7 million pounds) to get the pounds of nutrient supplied per acre foot of water.

The suitability of irrigation water is essentially determined by the concentration of three materials: sodium, chloride and boron. However, the concentration of lithium, fluoride, cadmium and other heavy metals, and bicarbonate may also make the water unsuitable. Table 4.3 list the three major toxic materials and their concentrations in irrigation waters. This classification does not take into account the total salt content; in general if irrigation water is high in total salt it will also be high in sodium and/or chloride. Just because the water condition is poor, does not always mean that it is unsatisfactory. There are four conditions where poor quality water might be suitable: 1. Very low evapo-transpiration rate (cool, overcast climates); 2. Where drip irrigation is used; 3. A well drained soil kept continually supplied with water; 4. The cultivation of tolerant crops.

SOIL FERTILITY:

Soil fertility is one of those things that if you have it, keep it at all cost; if you don't have it, you can go broke trying to get it. Building up the fertility of sub-fertile grazing land through fertilization can be very expensive. Generally, you can buy fertile land a whole lot cheaper than you can fertilize to make it fertile. It is very important that you have the grazing management in place prior to launching a major fertilization program. Proper grazing management will increase the amount of organic material in the soil. However, if one or more necessary minerals are in short supply or missing, these will have to be added either directly to the land or indirectly by feeding mineral supplements to the animals.

Soil nutrients get into the soil in two ways: 1. The weathering of bedrock or parent material. 2. Transportation of material from other outside systems. The weathering process requires considerable time. This can be speeded up, but only at a price in terms of energy inputs. In Australia proponents of the "keyline" system of irrigation claim to turn poor, infertile subsoils into highly productive soils in just a few years through proper irrigation and cropping practices. In Hawaii good pastureland has been made out of lava flows (see Figure 4.4.)

TABLE 4.3. CLASSIFICATION OF IRRIGATION WATER BY TOXIC CONSTITUENTS.

Classification	Sodium	Chloride	Boron
		(ppm)	
Excellent	less than 4	less than 3.5	less than 0.3
Good to very good	4 - 6	3.5 - 4.5	0.3 - 0.5
Fair to good	6 - 7	4.5 - 5.5	0.5 - 1.0
Poor to fair	7 - 9	5.5 - 7.5	1.0 - 2.0
Very poor and possibly injurious	9 - 12	7.5 - 10.0	2.0 - 3.0
Very poor and generally injurious	12 - 16	10 - 14	3.0 - 4.0
Usually injurious if used over a prolonged period of time	over 16	over 14	over 4.0

Adapted from Ideas in Soil and Plant Nutrition, by J. Traynor.

The major way that nutrients get into an agricultural system is by transporting of material. When movement of soil nutrients is considered the general tendency is to think only of commercial fertilizers. However, soil nutrients can arrive on the pasture from a wide variety of non-commercial sources; rain water and dust can bring in surprising large amounts of most nutrients. Areas that have frequent thunder showers may deposit as much as 17 lbs of nitrogen per acre per year. Boron, one of the trace elements that most commonly limits yields, may be present in rain water near coastal areas at levels of 5 grams per acre, which is more than sufficient to maintain the boron requirement of most crops. Even phosphorus may arrive on the wind, ranging from a low of 0.4 ounces per acre per year in Hawaii to 3.5 lb/acre in India, to as high as 87 lbs/acre in parts of Puerto Rico.[4]

Plants tend to concentrate nutrients with trees and other deep tap rooted plants acting like nutrient re-cycling pumps. Deep roots pick up nutrients from the sub-soil, ground water, or even parent material and incorporate it into tissue. Whereupon death or leaf fall these nutrients are returned to the surface, decomposed and made available to other plants. Organic matter in the soil gives the soil better water holding capacity (defense against drought) as well as providing a ready source of nutrients for both plants and animals. Any supplement brought into a pasture for livestock is, in effect, a fertilizer. A mature cow on free choice mineral will consume about 50 lbs per year. Of this amount, 90% or more will pass through the animal in the form of dung and urine. If phosphorus is present in the mineral at a 10% level and the stocking rate is one cow to the acre, 4.5 lbs/acre of P will be returned each year. At an 85% weaned calf crop only 3.3

[4] These high values are usually downwind from some dust source.

lbs/acre would be lost by marketing a 450 lb calf; thus leaving a net of 1.2 lbs phosphorus returned to the acre each year. Not a great deal but at least in the right direction.

FIGURE 4.4. SOIL FORMATION IN ACTION: A 14 acre pasture that was carved out of the 1955 lava flow near Hilo, Hawaii. A few years after the flow the lava was crushed, fertilized, planted to pangola grass and utilized as a bull and horse pasture. Fertilization continued and today (1986) little gravel can be found and for all intent and purposes the site is, if not soil, at least a very good growing medium.

Animals also cause a re-distribution of nutrients within a grazing system. Investigations in Hawaii found that in continuously grazed paddocks dung tended to accumulate around water, shade areas, and bed-grounds. In radial designed intensive grazing units (wagon wheel) the nutrient flow was concentrated about one third the distance of the radial, away from the water; there being around 27% more dung found here than in other areas. In a strip grazing program where animals grazed the first day on half the paddock near water and the second day grazed the back half, but had access to the same water supply, there was 127% more dung in the first half than in the second. In dairy operations where night paddocks are used, unless the night paddocks are frequently changed, the level of nutrients that are transferred in, via dung and urine, can quickly build up to toxic levels. In both New Zealand and Hawaii it has been noted that in small intensive (one day grazed) block designs, there did not appear to be any re-distribution of nutrients; dung was spread fairly evenly over the entire area.

Anything entering a pasture, in effect brings some mineral with it, ie. birds using a tree to nest in at night. Irrigation water, as discussed above, often contains high amounts of various minerals and may change the nutrient status, given sufficient time. A similar effect occurs with livestock drinking water; the sulfate content of water may be as high as 23 ppm (Pecos River). Brackish water contains all sorts of minerals, some at toxic levels; some water contains high levels of nitrates which, under certain conditions, may contribute to nitrate poisoning. Coastal regions that are subjected to spray are continuously covered with minerals contained in sea water. In volcanic regions sulfur emissions may be sufficiently high to result in loss of plant leaves or even death. Then there is the man-made pollution, high concentrations of lead along major highways, acid rain and all the rest.

FERTILIZATION:

Any time a crop is harvested and hauled away, nutrients leave the system. For example, a ton of alfalfa hay may remove 12 lbs of phosphorus (P), 55 lbs of potassium (K), 25 lb of calcium (Ca) and 5 lbs each of magnesium (Mg) and sulfur (S). A ton of bermuda grass hay may remove 14 lbs of P, 42 lbs of K, 6.5 lbs of Ca, and 4.5 lbs each of Mg and S. Unless these nutrients are replaced, production will eventually decline to levels where the outflow of nutrients just balances the in-flow from natural sources. Table 4.4 lists some of the more common fertilizers and the percent content of each element.

TABLE 4.4. COMPONENTS OF VARIOUS FERTILIZERS.

MATERIAL	N	P_2O_5	K_2O	OTHER MATERIALS
		(%)		
ammonium sulfate	21	-	-	24% S
urea	46	-	-	-
nitrate of potash	13.5	-	44	-
di-ammonium phosphate	17.5	50	-	-
rock phosphate	-	34	-	33% Ca
normal superphosphate	-	20	-	20% Ca; 12% S
treble superphosphate	-	46	-	14% Ca; 1.4% S
muriate of potash	-	-	60	47% Cl
sulfate of potash	-	-	50	18% S

The major fertilizers come in many different chemical compounds and combinations. However, commercial products are usually listed according to the percentage (by weight) of the individual elements listed on the label as N-P-K, or more likely as N-P_2O_5-K_2O. Thus, 16-10-12 would mean; 16% of the material is nitrogen, 10% is phosphate (P_2O_5) and 12% is potash (K_2O). To convert from phosphate to P multiply by 0.43; from P to phosphate multiply by 2.29. To convert from potash to K, multiply by 0.83, and from K to potash multiply by 1.2.

Some fertilizers move through the soil quickly, others do not. Generally, nitrogen, potassium and sulfur move through quickly, while phosphorus, calcium and magnesium move slowly. Banding of the slow moving fertilizers is one method of getting the fertilizer to the root zone fast. However, for pastures, economics generally dictate that fertilizers will either be applied and incorporated into the soil at time of planting or will be top-dressed.

A word about nitrogen. Nitrogen, perhaps more than any other nutrient, is considered THE fertilizer. This is because a single application will often yield such significant returns, even on otherwise poor soils. However, it should be pointed out that unless that poor soil is fertilized with other nutrients the second application of N will usually yield very little. A shot of N given just prior to the dry season, or cool spell, will often continue plant growth as a much higher rate than if not fertilized; in effect smoothing out the low end of the production curve often seen at these times. In Hawaii 250 lbs of N per acre doubled the production of kikuyu grass (other nutrients were in ample supply), another 250 lb increment (given at the same time) doubled it again. In fact, a doubling occurred for every 250 lb increment added, up to 1000 lbs of N per acre. That sort of grass production is not attained by the similar addition of any other single element.

Nitrogen is available in the soil from organic material. The process of release from organic material to forms available for plant growth is termed mineralization; the reverse process is immobilization. Mineralization and immobilization occur simultaneously in soils. The major direction the process is going depends on the carbon/nitrogen (C/N) ratio in the soil. If a lot of dead material has just been plowed into the soil the C/N ratio may be over 30, in which case immobilization will be favored. A ratio below 20 favors mineralization; between 20 to 30 the processes are about equal. At high ratios, above 30, there may be no nitrogen available for plant growth. This is why, after plowing a field of dead sod or other organic trash, it is best to wait a month or so before planting, either that or broadcast fertilize with nitrogen, or by banding near the roots of the crop (30 lb/acre of N for each ton of material turned under).

Two other processes that are important are nitrification and denitrification. The soil contains bacteria that convert ammonium-N to nitrate-N (nitrification). Nitrate-N is used by all crops as well as microorganisms. However, nitrates move freely with soil water and can be leached from the system. Nitrates can also be lost by denitrification; the conversion of nitrates to gas, usually elemental nitrogen. Denitrification usually occurs in soils high in organic matter, under extended periods of waterlogged conditions and at warm soil temperatures.

Nitrogen is also made available through the process of fixation. Biological fixation, the association between Rhizobium bacteria and legumes, is perhaps the best known method. Bacteria form nodules on the roots of the legumes. These nodules fix N from the atmosphere and make it available to the legume. The legume, in turn, makes carbohydrates (energy) available to the bacteria. The amount of N fixed varies from a few pounds per acre to over 500. Little of this N is immediately available to other plants growing near the legume, however. Nitrogen is made available to other plants through the process of decomposition. Roots and nodules of the legume die in the normal process of growth; these decompose and N then becomes available to other plants. Leaf and litter fall is also another pathway to transfer N from the legume to other plants. The primary value of growing legumes in pastures is for the nutritional value to the animal and only of secondary importance (though of considerable value over time) to the health of the associated grasses.

There are two other kinds of fixation; natural oxidation and industrial. The heat generated by lightning causes oxygen (O) to react with N in the air forming nitrate-N. In the U.S. rain and snow will add an average between 5 to 10 lbs of N per acre per year. The industrial process is the way we make nitrogen fertilizer, the most common of which is the synthesis of ammonia from atmospheric N and hydrogen (H).

Fertilizer N can be lost by a process called volatilization; the loss of N as ammonium gas. This happens when applying an ammonium form to alkaline soils or on recently limed soils. To avoid such losses on alkaline soils use a nitrate form or incorporate into the soil soon after application. Volatilization of urea can also cause considerable loss of N. In Hawaii as much as 70% of the N contained in urea has been recorded as being lost through volatilization. In one trial as much yield was obtained using 25 lbs of N per acre from ammonium sulfate as by using 50 lbs of N per acre from urea. Urea, to be effective, must get into the soil profile fast. If it lays around on the surface for even a day or so, much will be wasted.

MANURE:

In the past manure was considered "good", recently (the feed yard era) it was considered "bad" and now it is almost, but not quite, "good" again. Manure hasn't changed all that much, it's just that in our "throw away" society, manure doesn't throw away very gracefully.

In feed yards, the term manure is the combination of dung and urine. However, in grazing situations manure is usually considered as dung only. The quality (mineral content) and quantity of manure and urine varies, depending on the quality and quantity of the feed, the water consumed by the animal and the animals age and class (milking or dry). Typically the manure (both dung and urine) of grazing animals will contain higher concentrations of many nutrients (especially nitrogen, phosphorus, calcium, magnesium and sulfur) than is found in the diet they are eating. Thus a young steer on a diet containing 1.51% nitrogen and 0.18% phosphorus will have a fecal analysis of 2.67% of N and 0.81% P. Table 4.5 shows the results of one study on the mineral content of different animal manures (dung and urine combined).

TABLE 4.5. CHEMICAL COMPOSITION OF DIFFERENT ANIMAL MANURES*

ANIMAL	P	K	Ca	Mg	S	Cu	ppm
			(percent)				
Chicken (on litter)	4.3	1.6	1.6	3.6	0.3	0.4	26
Dairy cow	2.7	0.5	2.4	1.6	0.6	0.3	28
Steer (feed yard)	3.5	1.0	2.3	0.6	0.5	0.4	23
Hog	2.0	0.6	1.5	2.0	0.3	0.5	18
Horse	1.7	0.3	1.5	2.9	0.5	0.3	19
Sheep	4.0	0.6	2.9	1.9	0.6	0.3	16

* Contains both feces and urine

In swine, the urine contains no calcium, all calcium being voided in dung, however in horses the urine will contain 44.4% of all calcium voided by the animal. The distribution (the percent of the total voided) of the different minerals continued in dung and urine vary for each species; Table 4.6 shows this distribution between dung and urine of several minerals for some common livestock.

The benefits of manure to crop soils has been well documented. Once in the soil the manure (similar to other organic materials) acts to resist soil compaction, aids in the prevention of soil crusting (capping), increases the physical characteristics of the soil as to aggregation, friability and tilth and also increases water infiltration and retention (manure actually holds more water than a similar amount of organic material). There are also biological benefits to manure, reductions of root-knot nematodes and other plant root pest have been reported. The reason that manure is not used more in crops is economics; it cost money to load, haul and spread manure, it's simpler and often cheaper to buy it in a bag.

Animals will generally avoid grazing around dung patches made by their species, but will graze close to dung of another species. This avoidance is apparently due to odor. Under normal conditions it requires two to three weeks for this odor to leave; this time can be hasten by breaking the dung into smaller pieces, burying and/or rain. In continuous grazed pastures these dung pats are characterized by taller, ranker grass growth which may persist long after the odor has gone. Since it takes a couple of weeks for the odor to disappear, the plants are not grazed during this period. By the time the odor has gone, the plants are ranker and less palatable than forages from other areas; consequently they tend to remain ungrazed unless the animals are forced to eat them.

TABLE 4.6. DISTRIBUTION OF VARIOUS MINERALS CONTAINED IN DUNG AND URINE: Percent of the total voided.

	SWINE		SHEEP		HORSE		CATTLE	
	urine	feces	urine	feces	urine	feces	urine	feces
Ca	0	100	16.7	83.3	44.4	55.6	1.1	98.8
K	55.4	44.6	84.9	15.1	58.0	42.0	70.2	29.8
P	16.9	83.1	6.2	93.8	0.1	99.9	2.6	97.3
N	27.7	72.3	61.6	38.4	35.6	64.4	52.4	47.6

The above values are averages and will vary depending on diet and class of animal.
Adapted from Farm Animal Manures.
Ca = calcium; K = potassium; P = phosphorus; N = nitrogen

The forage around dung pats are greener than surrounding forage, this is particularly noticeable in cool weather. The obvious reason for this is the fertilizer effect (primarily nitrogen). The less obvious is that the texture and color of dung makes it warmer during the day and into the night than the surrounding grass, similar to asphalt on a sunny day; the heat released through decomposition may also contribute. Soil temperatures one inch under a dung pat have been recorded 12 to 15 Fahrenheit degrees higher than soil temperatures one foot away. This warming of the soil and grass roots, under and around the cow pie would, at cooler times of the year, allow a higher growth rate than grass growing some distance away from the dung pat.

Beef cattle will defecate about 12 times and urinate 9 times daily; this of course varies depending on diet. The average size of a dung pat is approximately one square foot and a urine patch, 3 square feet. During a 24 hour period a hundred cows will cover an area of approximately 3,800 square feet with some 2000 lbs of urine and 5600 lbs of dung. The mineral composition of this will be around 43 lbs of N, 10 lbs of P_2O_5 and 35 lbs of K_2O in addition to other minerals. In grazing situations all this is deposited on the surface of the ground; yet, in IGM units, little of this is visible a few days after the herd has been shifted. There are two reasons for this rapid disappearance. The first is the physical impact of the animals themselves on the cow pies through treading and trampling actions; the higher the animal density the greater this effect. Then there is a whole host of soil organisms (dung beetles, earthworms, etc.) all working to incorporate manure into the soil. In a healthy soil this is accomplished remarkably fast; in fact, the length of time manure remains visible is a fair indication as to the health of the entire system.[5]

SUGGESTED READING:

Darwin on Humus and the Earthworm. 1945. C. Darwin, Faber and Faber.

Environment and Plant Ecology. 1975. J. R. Etherington, John Wiley & Sons.

Farm Animal Manures. 1974. J. Azevedo and P.R. Stout. Manual #44, Univ. of Calif. College of Agriculture publication.

Fundamentals of Soil Science. 5th ed. 1984. H. D. Foth, John Wiley & Sons.

Ideas in Soil and Plant Nutrition. 1980. Joe Traynor. Kovak Books.

Soil: The 1957 Yearbook of Agriculture USDA. Superintendent of Documents, Washington D.C. 20250.

Soil Conditions and Plant Growth. 10 th Ed., 1973. E. W. Russell, Longman.

*Soil Fertility Manua*l. 1978. Potash & Phosphate Institute, Suite 401, 2801 Buford Hwy., NE. Atlanta, GA 30329.

Water for Every Farm. 1974. P. A. Yeomans. Murray, Sydney, Australia.

[5] In some intermountains region of central Nevada some of those cow pies must be at least 10 years old. Of course that place has 9 months of winter and 3 months of damn cold weather.

Used with permission of NEA.

Plants and animals may co-exist, but they don't always like one another.

5 PLANT BASICS

Plants are an extremely diverse kingdom, full of opportunists and other rule breakers. So bear in mind that when limits or ranges for a certain group of plants are given, there will almost always be one or more species that didn't get the word.

FIGURE 5.1. Hawaiian Silversword. Some plants, like people, will try to make a living anywhere. This one is found growing on volcanic cinders at an elevation 10,500 feet.

METABOLIC PATHWAYS:

All plants are capable of manufacturing their own food. Most plants do this through a chemical process called Photosynthesis (Figure 5.2). Photosynthesis is the trapping of the sun's energy, utilizing one or more of the different chlorophylls, and with carbon dioxide and water as the base material, building complex carbon compounds, with oxygen released as a by-product. For every gram atomic weight of carbon fixed (12 g), 112 Kcals of energy is stored. Every year plants fix, in organic compounds, about 6-7% of the carbon dioxide contained in the atmosphere or dissolved in bodies of water. About one sixth is returned through respiration of the plants themselves and about one thousandth is temporarily withdrawn from circulation by being buried in sediment, locked in underground water, or deposited as peat. The rest serves as fuel (energy) for animals, fire or decomposition.

$$\underset{\text{Carbon Dioxide}}{12CO_2} + \underset{\text{Water}}{24H_2O} \xrightarrow{\text{light}} \underset{\text{Oxygen}}{18O_2} + \underset{\text{Carbohydrates (112 Kcal)}}{12(CH_2O)}$$

FIGURE 5.2. EQUATION FOR PHOTOSYNTHESIS SHOWING ENERGY STORED.

Like animals, plants also respire, which is basically the reverse process of photosynthesis. Oxygen is taken in through openings in the leaves and young stems (stomata) and carbon dioxide is released. This process goes on, night and day, growing or non-growing, and ceases upon death of the plant. When a plant is actively engaged in photosynthesis there is an additional respiration taking place, but unlike normal respiration it only occurs during daylight hours. The way that a plant puts all these processes together is called it's metabolic pathway. There are three distinct pathways found in the plant kingdom; the Calvin-Benson Cycle (C3), the Hatch-Slack Cycle (C4), and Crassulacean Acid Metabolism (CAM).

THE C3 CYCLE:

The Calvin-Benson Cycle is often referred to as the C3 pathway. The C3 refers to the number of carbon atoms that are contained in the first carbon compound built by the plant. This first 3 carbon compound is the first step in the building of all the various carbohydrates, cellulose, proteins and other essential components found in plants. The C3 pathway is generally considered to be older, from an evolutionary point of view, than either of the other two types of metabolism. Plants with the C3 pathway are very good in high carbon dioxide, low oxygen environments and in fact the higher the carbon dioxide and the lower the oxygen the better they do. All legumes have the C3 pathway as do the majority of the grasses; in fact the C3 cycle is found in the vast majority of the world's plants.

With respect to grasses, the C3's are found in cooler and/or temperate environments and seem to do best at average temperatures between 68 to 77 degrees F (20 to 25 degrees C). Their lower limit ranges between freezing and 39 degrees F (0 to 9 degrees C) and their upper limit lies around 86 to 95 degrees F (30 to 35 degrees C). Each species has it's own temperature preference and it's unlikely to find one that does equally well at both extremes. Most C3 plants are operating at full steam when light intensities are only one fourth that of full sunlight; this is called light saturation. Typically C3's require about twice as much water as C4's to make the same amount of dry matter. However, animals generally find the C3's more digestible than comparable C4's, especially so when older, more mature plants are compared.

THE C4 CYCLE:

The C4 pathway is found in many of the grasses and some members of Euphorbiaceae (spurges), Chenopodiaceae (goosefoots), and in the genus Amaranthus. This pathway does contribute some very special advantages, but generally only in hot and/or dry environments. The C4 plants are much better than C3's at using low concentrations of carbon dioxide. For example, the low levels of carbon dioxide released by the respiration due to photosynthesis is rapidly re-used by the C4 plant. Thus, under drought (shortage of water) conditions the plant does not have to open its stomata as wide to obtain carbon dioxide to continue to function as does a C3 plant. This ability to close the stomata, yet still function, allows the plant to use water to greater effect than C3's by decreasing the amount of water lost through transpiration. In fact, the C4 plants generally hit their peak photosynthetic efficiency at a concentration of around 190 parts per million (ppm) of carbon dioxide while the C3's require close to 600 ppm. The average carbon dioxide content in the atmosphere is about 326 ppm, although this varies by 8 ppm on either side depending on season and latitude.[1]

The C4 plants also function better than C3's as the concentration of oxygen increases; both are equal at zero oxygen, but the C4's maintain a higher level of photosynthetic efficiency at 20% oxygen (present atmospheric concentration) and even up to 50% oxygen content. Under full sunlight the C4 plants are not light saturated as are C3's, which is one of the reasons that they typically out-yield C3's by two to one. The C4's also operate at a higher temperature level than the C3's; some can produce at temperatures as high as 113 degrees F (45 degrees C). The optimum range is around 95 degrees F (35 degrees C) when most C3's are under temperature stress. However, at low temperatures the C3's have an advantage as most C4 plants turn off when the average daily temperature drops to 59 degrees F (15 degrees C) and are usually injured at 40 degrees F (10 degrees C); many C3's are at peak performance at this temperature. Soil temperatures also play a significant role. Many C4 grasses slow down production when the soil temperature drops into the high 60's Fahrenheit degrees or high teens Celsius degrees. Consequently if the soil temperature is high they can withstand a cool snap better than if the soil temperature is also low. Figure 5.3 shows the ability to produce (photosysntetic rate) between C3, C4 and arctic lichens at different average daily temperature ranges.

THE CAM CYCLE:

The remaining metabolic pathway is the Crassulacean acid metabolism or CAM cycle. Essentially this is a combination of the two other pathways. The plants take in carbon dioxide at night through widely open stomata and fix it to intermediate compounds of various organic acids (mostly malate). These acids decompose only in the presence of light. During the night these acids accumulate and are passed into the cell sap where they are stored in large vacuoles found in the cell. The following morning the plant closes it's stomata and with light, these acids decompose to release carbon dioxide, the process then continues as either the C3 or C4 pathway, apparently depending on the concentration of malate. The advantage to the plant of the CAM cycle is that by only opening it's stomata at night it can greatly reduce water loss through transpiration. Many succulents, especially those of the lily, orchids, cactus, ironwood, milkweeds and iceplants families, use the CAM cycle. This pathway is inefficient compared to straight C3 or C4 metabolism, and is generally considered as an adaptation for life in very harsh environments; sort of allows the plant to hang in there.

[1] The reason for this variation is due to the uptake of carbon dioxide by plants during the spring and summer months in northern temperature latitudės; there is a greater land area in the northern hemisphere, hence more plants.

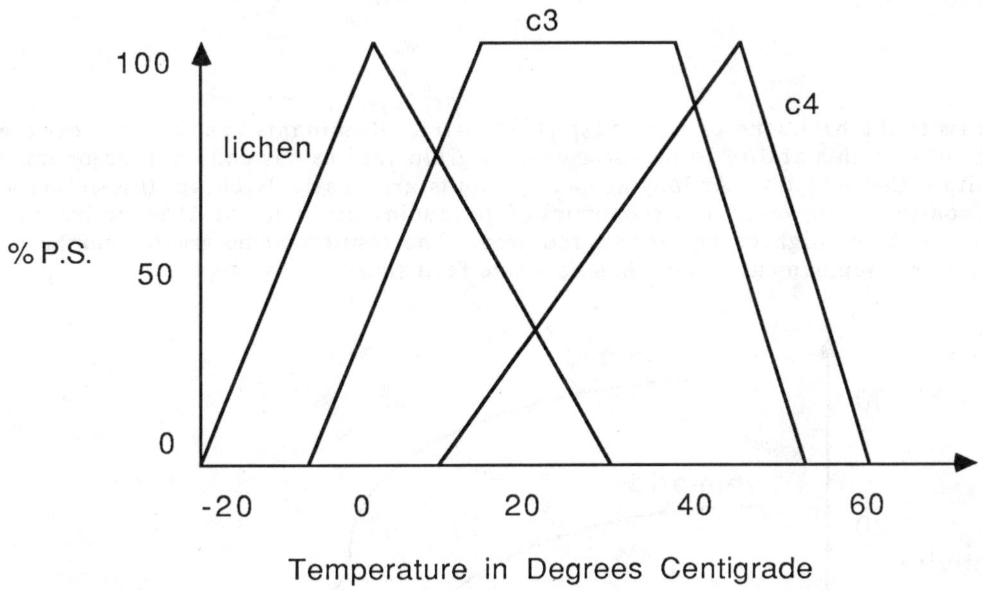

FIGURE 5.3. PERCENT PHOTOSYNTHESIS RATE (P.S.) AS A FUNCTION OF TEMPERATURE FOR THREE TYPES OF PLANTS.

LEAF AREA INDEX AND CANOPY COVER:

The amount of production that a plant can accomplish in a given time is dependent on the amount of the sun's energy that it can capture and convert to tissue. A plant does this by putting out a lot of leaf and absorbing as much of the sun's radiation as possible. When the area of leaf exposed to the suns ray equals the area of ground the plants are growing on, the leaf area index (LAI) is 1; when the area of leaf is twice the area of ground the LAI is 2. After the leaf area index reaches 1, overall production declines for a single plant growing alone. However, the case is a little different for stands where many plants co-exist. For corn, and most cultivated plants, when the leaf area index is about 4 maximum production will occur if nothing else is limiting.[2] Beyond a leaf area index of 4 production tends to decline somewhat, both for crops and natural systems. In temperate regions a canopy will rarely get above a leaf area index of 11; in tropical areas around 12 to 13 due to the higher light intensities. The amount of light striking the canopy is also a function of latitude and the local water supply. Yields are reduced at high latitudes (arctic) due to lack of light and heat. In low latitudes (tropics) it is reduced more by lack of water and frequency of distribution. Figure 5.4 illustrates this relationship.

Another measurement similar to LAI is the percent canopy cover. This measurement is normally used for trees and can be easily measured in the field. Make a frame with your hands, look straight up towards the sky and estimate the percent of cover in the frame; those wearing glasses, use the glass frame as the reference area. Take several measurements over the area (at least 10) and average. If the percent canopy is over 30% generally production of the understory will begin to decline; much over 90% (except in tropical regions) understory production is virtually eliminated.

[2] Corn, sugarcane, sorghums, bamboo and all cereals such as wheat, oats, etc. are members of the grass family.

GRASS:

Grass is the backbone of the livestock industry. Ruminants can, and do, eat other materials; in fact this ability to perform well on grain rations has caused a major mis-direction within the industry. As long as energy inputs are relatively cheap, this effort will no doubt continue. However, the economics of producing grain in the U.S. are beginning to catch up with the high energy inputs required. The result will be greater emphasis in using grains as a supplement rather than as a sole feed source.

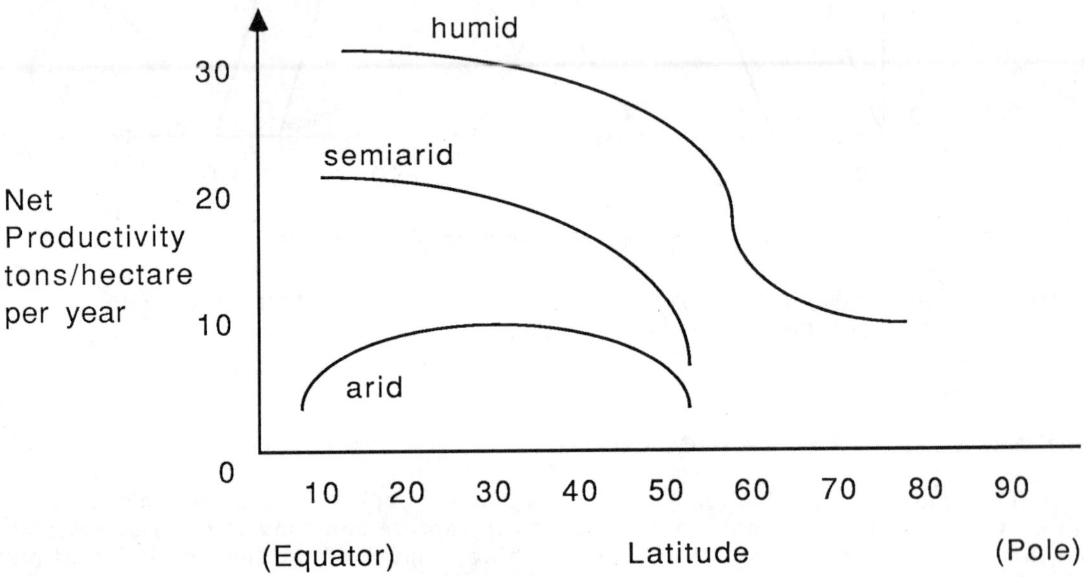

FIGURE 5.4. DEPENDENCE OF PRODUCTIVITY AS A FUNCTION OF LATITUDE.

CLASSIFICATION:

There are two broad classifications of grasses, annual and perennial. Annual means living only one year and perennial more than one. Annuals are typically found in arid environments or in disturbed areas and are generally classified as pioneer plants. Perennials are considered to be a measure of the maturity, or "climax" of the system. However, many perennials often act like annuals, this is especially true in wet/dry tropics where there is more dry than wet. In many areas that have a more or less uniform climate many annuals often live longer than one year. In fact, if an annual grass such as wheat is kept in a vegetative state (no flower stalk allowed to develop) through clipping, it may live several years. This trait often messes up attempts to identify some grasses, as many keys use annual or perennial as one of the criteria of identification.

There are several other classifications, some of which are used for plants in general as well as grasses:

TROPICAL: these plants are killed by freezing.

SUB-TROPICAL: plants that can handle light and/or occasional freezes.

TEMPERATE: plants which can tolerate freezing.

WARM SEASON: usually refers to grasses that grow best in warm or hot summer conditions, usually C4 grasses.

COOL SEASON: grassess that grow best during the cooler portions of the year; spring, winter and/or fall, usually C3 grasses.

SOD GRASSES: grasses that send out runners or stolons, whose growth form is along the ground, rather than upright.

BUNCH GRASSES: grasses whose growth form is upright, rarely if ever send out runners.

C4 GRASSES: those grasses that have the C4 metabolic pathway.

C3 GRASSES: those grasses that have the C3 metabolic pathway.

TROPICAL AND TEMPERATE GRASSES

Unfortunately there is no sharp distinction between the various classifications and where they grow or even their growth forms. For example, C3 grasses may be found in the tropics and C4 grasses high in the temperate latitudes. Sub-tropical grasses occur in both tropical and temperate latitudes. A few bunch grasses send out runners, etc. However, for sake of this discussion, tropical, sub-tropical and warm season grasses will be assumed to have the C4 pathway and temperate and cool season grasses the C3.

Many tropical grasses (especially pasture grasses) have coarser stems with longer and broader leaf blades than temperate grasses. As a result the grazing animal, under light set stocking rates, finds it easier to keep to one species of grass. This selective grazing places a stress on the more palatable grasses and contributes to many tropical pastures ending up with only one species making up the bulk of the forage. Another contributing factor is that most of the sod grasses produce few if any fertile seeds and spread vegetatively, through stolons (runners) and/or rhizomes; the resulting mass at ground level makes a poor place for seedlings to develop. Of course selective grazing also occurs on temperate forages. However, in dense swards the ability to select one species from another is minimal. In arid regions where plants are physically separated from one another, selective grazing pressure can be intense.

The growth rate for tropical grasses is nearly twice that of temperate grasses; rates of 267 lbs/acre/day have been reported with some C4 grasses. The maximum annual yield of a tropical grass was 37.9 tons of dry matter per acre (Napier grass in Puerto Rico). The maximum dry matter yield in Europe is about 9 tons/Ac/year; over three times this amount is commonly reported for a large variety of tropical grasses, many on commercial operations. These forage yields translates into liveweight cattle gains of 1158 lbs/Ac/year for trials in N.W. Europe on rye grass (fertilized with 450 lbs of N/Ac), to 2460 lbs/Ac/yr in Queensland, Australia on irrigated pangola grass (600 lbs N/Ac). All of the above were on research stations; on temperate commercial farms, fertilizing at 450 lbs

N/Ac, liveweight gains of 890 lbs/Ac can be expected; on tropical pastures commercial gains of 1470 should be attainable with similar fertilization levels.[3]

FIGURE 5.5. FAST GROWTH OF TROPICAL GRASSES. This can sometimes cause a problem if one parks one's car a bit too long. Because of the fast growth rate of C4 grasses, it doesn't take very long before the pasture is "out of control". This lack of control has been the major stumbling block to livestock production in the tropics.

The C4 grasses (warm season) are able to utilize nitrogen fertilizers in the production of dry matter at least twice as efficiently as C3 grasses (cool season). However, the C3 grasses contain more tissue nitrogen than do the C4's. This characteristic seems to be an inherent trait in the C4 metabolism that is related to their ability to survive under conditions of less soil fertility than comparable C3's. The C4 grasses also use water better, producing more pounds of dry matter per pound of water than temperate grasses.

[3] One Hawaiian ranch has produced 955 lbs liveweight per acre per year fertilizing with only 160 lbs/Acre of 16-16-16.

There is a difference in the type of carbohydrate that each type of grass stores. Temperate grasses (C3's) contain mainly sucrose and fructosan which is stored mostly in the stem. On the other hand, tropical grasses (C4's) and all legumes produce mostly starch and sucrose. Legumes store starch in both leaves and stems while the tropical grasses store high concentration of starch in leaves and most of the sucrose in stems. Both types of grasses contain lipids (fats) and levels may reach 10% in young leaves. However, these levels drop off rapidly as the leaf ages. In general the calcium, phosphorus and cobalt content of cool season grasses are higher than in warm season ones; but the content of magnesium, copper and zinc is usually higher in warm season grasses.

The amount of sodium (salt) contained in forages is of importance if for no other reason than all that cost and bother of buying, hauling and putting out salt for livestock. High or low sodium accumulators (ability to store sodium) can be found in both warm and cool season grasses. The sodium content of most green temperate grasses is adequate for animal requirements, however about 27% of the tropical grasses (according to one report) contain sodium levels that are below that required for animal nutrition. Tropical legumes are also reported to be low in sodium content. When the plant has seeded (matured), or turned brown, the sodium content is such where it no longer meets animal requirements.

The tiller density of tropical grasses is much lower than temperate, although some tropical grasses such as pangola are similar to the coarser temperate grasses (eg. orchard grass).[4] However, the tiller density of most tropical grasses is stimulated not only by fertilization but also by grazing in which the lowest parts of the stems are trampled into the soil; new tillers grow from the nodes and so increase sod density.

The leaf to stem ratio in temperate grasses is much higher, around 75%, while the average for tropical grasses is nearer 50%. However, new varieties of dwarf elephant grass have a leaf to stem ratio of 75% and properly managed kikuyu grass will also approach temperate levels. The importance of leaf to stem is discussed in the section on forage quality.

Temperate grasses often flower (go to seed) during the long day light days (late spring, early summer), but require a period of low temperature before onset of flowering, this is called vernalization. Most Tropical grasses are able to flower practically the entire year; however, some flower more during the winter than summer (guinea), while others, such as molasses only during the fall. This ability to flower all year round creates some problems, as flowering (heading) often occurs a couple weeks or so after heavy grazing with little to no new leaf growth.

Finally, the gross energy content of both tropical and temperate grasses is about the same (4.3 Kcal per gram dry matter). Also both types of grasses, if grown under identical conditions, especially with reference to temperature, will have similar digestibilities. It has been found that high growing temperatures are a major factor in the lower digestibilities associated with tropical grasses. It has also been noted that the digestibility of tropical grasses may be greatly improved if the stem formation is suppressed. There is not a great deal that one can do about growing temperatures or climate, but there is a great deal one can do about keeping stems from forming on tropical grasses. We call it IGM.

---

[4] A tiller is the basic vegetative unit of the grass, not the flower stem, but the stem where the leaves grow from.

GROWTH AND YIELD:

Grass growth is similar to that of other living organisms in that it follows the classic growth curve, shown in Figure 5.6.

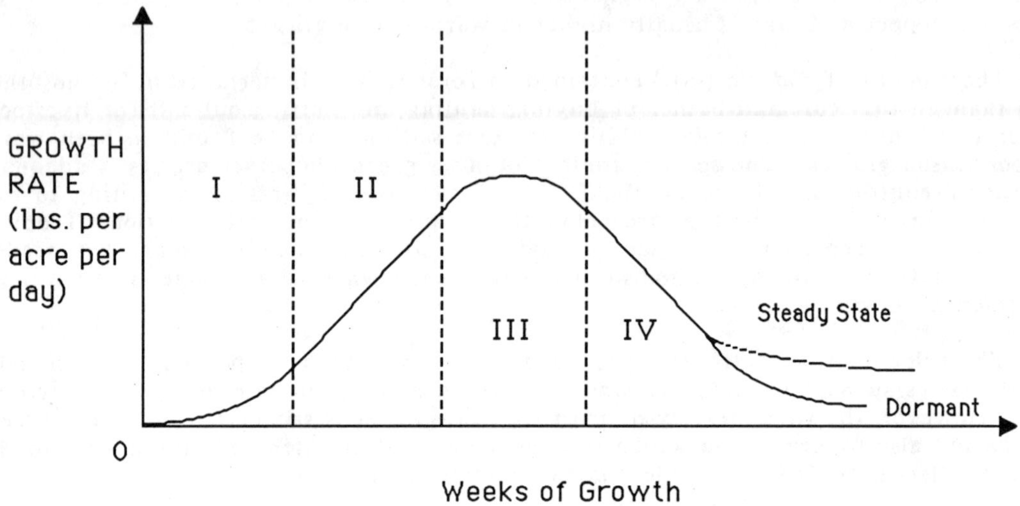

FIGURE 5.6. BASIC GROWTH CURVE. NOTE: This figure adapted from Pasture: The Export Earner.

This curve is basic to almost all biological growth and is used here to illustrate either: individual leaf growth, entire plant growth, or pasture growth; with growth continuing on (steady state), or ending in a dormant period (drought). The curve is broken into four parts, or phases of growth, starting from a pasture that has just been mob grazed, or mechanically harvested. The first phase (I) is that period when the pasture is recovering after harvest. The plants have been stressed and have lost much of their ability to capture sunlight. They have also lost a lot of carbohydrate and as a result growth has slowed down. Phase II is the period when the pasture is making it's most rapid growth. The area of green leaf exposed to the sun is high (LAI of 4 or so) and most of the pasture plants are in a vegetative state (no seed heads). Phase III is when the growth rate slows down, plants flower and go to seed. The yield of the pasture will be at its highest at this stage, but as you will see later overall nutrient quality is low. In phase IV the growth rate slows and may flatten out (steady state) for a period of time, or go to zero (dormant period). The first three phases of the growth curve, the sigmoid growth curve, is what is usually used to illustrate growth, However, in the final phase (IV), the decline, has special relevance to pasture management.

As the grass plant grows new leaves are continually being formed with the older leaves becoming darker in color, eventually turning brown and finally falling to the ground and decomposing. The young leaf tends to accumulate nitrogen (N), phosphorus

(P), potassium (K) and other minerals for later use. As the leaf ages much of the N, P and K are moved (translocated) to areas of new growth, but calcium (Ca), sulfur (S), iron (Fe), manganese (Mn) and boron (B) remain behind. In some young grasses (i.e. kikuyu) the K content may be as high as 4% and the Ca about 0.2 %, but in an old leaf the K will drop to values between 1 or 2% and the Ca (still being moved in) may increase to 0.5%. This high level of potassium is common in many grasses and is one of the reasons that grass tetany may become a problem with young rapidly growing grass.

New leaves, like many new things in life, are dependent upon their parents, roots in this case, to supply them with nutrients until such time as they can make it on their own. Depending on the species and if sun (growing in full sun) or shade leaves (which are usually a little larger than sun leaves), this can be as short as 3 days or as long as 20. One rule of thumb is when the leaf reaches about one fourth to one third the size of a mature leaf the compensation point is reached; ie. it is making as much food as it is using. As the leaf grows beyond the compensation point it sends a greater amount of its production (carbohydrates) to the roots, or other storage locations, Phase III. Soon, usually within one or two months, the leaf senescences (old age), at which point it is "over the hill".[5] The leaf is again using more carbohydrates for respiration than it is producing by photosynthesis; finally, after a life span ranging from 20 to 80 days, it dies. Also each grass tiller only supports so many leaves at any one time; the actual number depends on the species of grass and fertility of the soil. In rye grass the typical number of leaves is three; when a new leaf is produced, the old one dies.

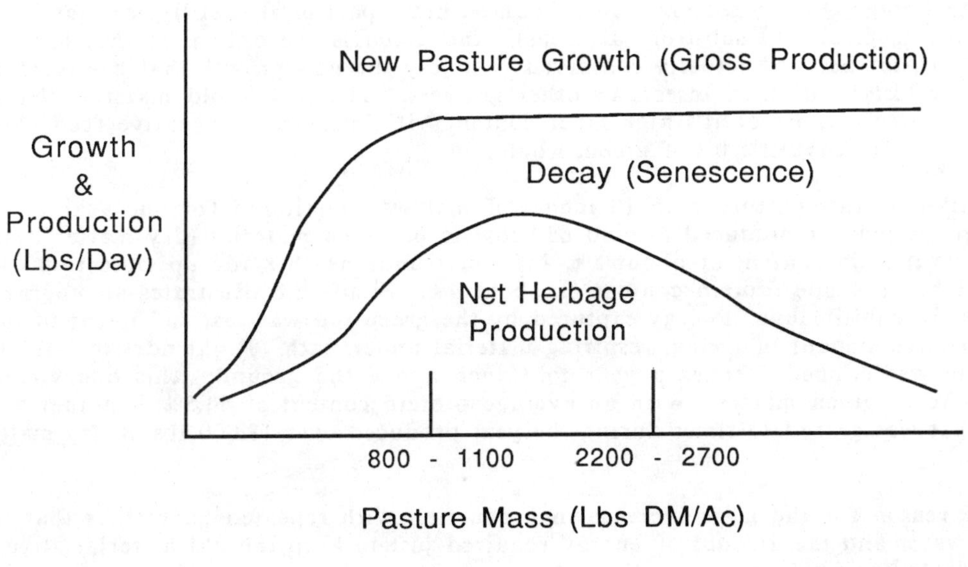

FIGURE 5.7. DECLINE OF NET PASTURE PRODUCTION WITH INCREASING PASTURE MASS.

[5] It's not so much being over the hill that's bothersome, it's just that I can't remember being on top.

In pastures and/or ranges the process of senescence occurs along lines similar to individual plants. Following a heavy grazing, mowing, or after a dormant period, gross pasture production increases then flattens out at some level of total plant (herbage) mass. Net production increases up to some point, different for different species combinations, then decreases to zero. Decay (senescence) starts out slow then increases to equal the rate of gross production. Figure 5.7 shows this relationship for a rye grass clover combination.

If everything stayed the same a steady state condition would be reached with new growth equaling the death rate and the over all pasture mass remaining the same over time. In nature this seldom happens as there are variations in weather, insect pests, diseases, fire, etc. that act to remove some of the old and/or new growth, thus stimulating the plants towards some net production. Discounting disease, insects and other grazers, the rate of decay is a function of moisture and temperature; faster in warm humid environments and slower in dry cool ones. In a grazing situation this pasture mass is removed periodically by the grazing animal. This removal of material, if timed to take place during phase II, the period when the pasture is growing at it's fastest rate and if sufficient residual is left (see Residual Dry Matter section), will yield the maximum amount of high quality forage.

EFFECTS OF GRAZING:

Herbage, unless removed by grazing or other defoliation process, will eventually die, decay and disappear from the pasture. The idea of conserving forage by letting a pasture go rank appears to be firmly embedded in the minds of many. Obviously something is there at the end of a growing season or the idea wouldn't have gotten fixed in the first place. If the pasture is saved until all of it is dead or brown, it has little to no quality, (see Forage Quality section). If it is saved but a part of it is still green (as is the case in many tropical and sub-tropical areas); the green is the only part that has any quality. What, in fact was actually conserved is simply the new growth that has replaced that eaten or lost by disease, insects or other graziers. The rest is old material that is resistant to decay because of it's high fiber content; it almost has a negative feed value and if it wasn't for the little bit of green, would.

A kikuyu grass pasture with 10 inches of mat was monitored for one year to determine production. It produced around 650 lbs/Ac of green material (dry matter basis) with a crude protein content of about 8%. The mat underneath made up close to 10,000 lbs/Ac and had a crude protein content of about 4%. Figure 5.8 illustrates an ungrazed area that is in equilibrium. Energy captured by the green top was just sufficient to support the massive amount of living, respiring material underneath. Right next to this plot another one was clipped 7 times a year to 1 inch above the ground; this one yielded 11,000 lbs/Ac of green material with an average protein content of 11.5%; another area close by, that was grazed 13 times during the year produced over 18,000 lbs of dry matter per acre.

The reason for the large increase in production with repeated harvest, is that the nutrients, water and the amount of energy required just to keep the old material alive is made available for productive growth, rather than just maintenance. This accumulation of old material also reduces the amount of effective rainfall. After a dry period, one inch of rain fell. In the area where the mat had been removed the moisture penetrated 5 inches in the soil; in the area where there was the heavy amount of old material (9 inches), it never even reached the soil.

Management has little input as to climate, but it does have a great deal of input as to frequency and the severity of grazing (degree of defoliation). The degree of defoliation is the amount of material that has been removed. However, a better way to look at it is as the plant does; how much is left to carry on.

When green grass is harvested the plant is put into a state of shock. The length of time required for the pasture to recover depends on several factors, such as soil fertility, season of the year, soil moisture content, temperature, but most importantly, the degree of defoliation, etc. Grass roots are continually growing and dying, this is normal; in fact the amount of roots that die during the course of a growing season may be three or more time the amount of living material. One reason why grasses are used to put organic material back into the soil in a crop rotation program.

FIGURE 5.8. SENESCENCE.

In bunch grasses, heavy defoliation stops new root growth. The reason for this stoppage is that the plant no longer has the necessary amount of leaf to capture sufficient energy to support the same amount of new root growth; also the plant must start to build new leaf to replace that lost, this takes energy. The degree of root stoppage depends upon the grass species and amount of green material removed. Typically for bunch grasses, if 40% of the green material is removed there is no stoppage of root development. At 50% green removal a slight amount of root slow-down occurs, however at 60% green removal the amount of new root growth is slowed by 50%; at 80% removal, all root growth stops. At these heavier rates of defoliation, it may take three to four weeks for the plant to start growing new roots; new leaf growth has priority. Grass that has had repeated heavy defoliations will have a shallow root system, perhaps less than 6 inches in depth.

This is not necessarily a "bad thing" if there is ample fertility and soil moisture. However, if soil moisture is low and/or the plant is heading into a long dormant season (fall or dry season) heavy defoliation may kill the plant.

Contrary to popular opinion, overgrazing is a one on one situation; one plant versus one animal. The animal, like most predators, will win most of the time. After a plant has been grazed it will send out new leaves, these are the best the plant has to offer. The grazing animal know this; after all it spends one third of its life looking for good things to eat. So as soon as the new leaf is high enough, about a half an inch for cows, the animal is there to eat it, if allowed to. This hurts the plant.[6] The end result is that the plant uses up it's root reserves building new leaf but the leaf is never allowed sufficient time to "pay back" the loan, eventually Mother Nature forecloses and the plant dies. The animal then moves on to the next plant. It doesn't happen quite like this, but close enough. This is why in continuous grazed, or lightly stocked pastures, you often see areas grazed right down to the ground, next to tall rank growth (spot or patchy grazing); or after a fire the animals will camp on the new regrowth and ignore the older unburned material.

In addition to selective grazing and the trampling into the ground of stolons, the effect on tillering (mentioned above) and the influence of the grazing animal on the sward density and composition (Chapter 8), the animal exerts other more subtle effects on plants. In areas of low fertility the thiamine contained in the animals saliva causes grazed grasses to recover faster than those that were mechanically harvested. The fact that some of the effects of grazing should prove of some benefit to the plant should not prove too surprising; after all they did evolve together.

RESIDUAL DRY MATTER:

Residual dry matter is the amount of forage, on a dry matter basis, remaining in the pasture when the grazing animals are removed. How much of the above-ground part is removed will determine the length of the recovery period; or the harder a plant is grazed, the longer must be the rest period for recovery. The New Zealanders have a saying "grass grows grass". What is meant is that if you leave sufficient residual, the pasture will recover considerably faster and produce more usable forage than if hard grazing is practiced. The moral being, for fast recovery, don't graze it into the dirt.

Research from both New Zealand and Florida indicates that when the total residual dry matter (DM) remaining, after the grazing animals have left, drops below 500 lbs/acre, the time required to restore vigorous growth, or the time before the paddock can be grazed again, is considerably longer than if 1000 or more lbs/acre are left. For example, if the residual is 500 lbs/acre, dwarf elephant grass will take close to 64 days to reach a yield of 2500 lbs/acre compared to only 15 days for a residual of 2000 lbs/acre. Different species of grasses will vary in the time required as will the same species grown under different environmental conditions. This concept applies to stands of either temperate or tropical grasses, bunch or sod, range or pasture situations, or where the forage is mixed with many ungrazable forbs and woody species; only the values will be different.

Now you may be thinking: "For Pete's sake, of course the more you leave the faster you'll reach a given level of yield." True, if you want a pasture level of 2500 lbs/acre, and you pull out when the residual is 2500 lbs/acre, you don't have any waiting

[6] Hurt may not be the right word here, but plants do not respond to having leaves removed and this can be detected with the proper electrical meters (see the never to be released book "I have to scream, but I have no mouth.")

time at all. What we are really talking about is the amount of new growth that occurs between grazings. The more new growth that occurs in a given period of time, the greater the annual forage yield and the higher the forage quality. This, in turn, means more animals per acre and the better their performance.

Well, that sounds pretty good. The more you leave the more you get. Unfortunately, a point is soon reached where the recovery time stays about the same, regardless of how much is left. For dense pastures this seems to be around the 2000 to 2500 lbs/acre level for all grasses, even the large tropical bunch grasses such as guinea and dwarf elephant. For example, preliminary work in Florida with dwarf elephant grass found that residuals of 1500, 2000 and 2250 lbs/acre and 42 days rest yielded 600, 900 and about 800 lbs/acre respectively.[7] Residuals much over 2000 lbs/acre are not recommended unless there is a specific reason for lax grazing. As will be pointed out in the section on quality, there are other reasons for keeping the residual below the 2000 lb/acre level.

The reasons for the slow rate of recovery (low growth rate) at low residuals, an increase in growth rate between 1000 and 2000 lbs/acre and then a tapering off of the of the growth rate at high residuals, all have to do with the following: 1. The amount of green leaf that is left after grazing to capture sunlight (leaf area index). 2. The fact that old leaves are less efficient producers than younger ones. 3. The ratio of non-photosynthetic material (brown but alive and thus respiring) to green material (photosynthetically active). 4. The amount of the green material that is shaded. 5. The leaf to stem ratio. Green stems also manufacture carbohydrates, but usually just barely enough to provide for stem respiration; mature stems, yellowish in color, use up more energy than is made.

In heavily grazed paddocks with little residual (flogging) the amount of photosyntheticly active material (green) is low; as a result there is little energy being stored. However, the roots and remaining stems are still respiring. Consequently, energy must be translocated from root and stem storages to take care of the respiration and also build new leaf. The overall effect is to stop root growth and, depending on the overall vigor of the plant, even root death. All this takes time to correct and get back on track.

In lightly grazed paddocks (high residuals) mostly leaf and stems are consumed, the highly palatable young leaves more so than the older ones. The older and ranker materials, which consist of a high stem to leaf ratio, are left behind. As a result there is a high rate of respiration and consequently a relatively low production of net photosynthate available for new growth. In addition, the old leaves tend to shade the younger ones, further reducing production.

Where forage is sparse, as in arid environments, the residual concept is only applied during the growth period. The concept is based on the height left, rather than on so many pounds per acre. The ideal height to pull out for fast re-growth will vary depending on the species, but will generally be around 50 to 60% of the before grazed height of the green plant. If more than this amount is to be removed there should be ample soil moisture, also the recovery time will increase. Of course once the above ground part of the plant has gone dormant or died, it make little difference how much is removed. Remember that old adage, "Take half and leave half"? The original concept was based on the green plant. Somehow or other the concept got twisted up and came to mean dead material as well, saving half of the dead material has little validity as far as plant health is concerned.

[7] Very little work has been done with the residual concept with the large tropical grasses; consequently, the upper limits may change as more information becomes available.

ESTIMATES FOR MINIMUM GRAZING:

There are a couple of ways to estimate the minimum point to graze a pasture and still get a reasonably fast recovery. The easiest but less accurate, is to estimate the percent green material that covers the area. This can be done simply by taking a "hike" through the paddock with a pencil and paper (after you have done it a few times these won't be needed). Plan your "hike"so that it covers all the major areas of the paddock; do not just concentrate in the area around the gate. Every so often (try to do this at least 10 times during the "hike") stop and look down, make a frame with your hands,[8] then estimate the percent green in the area made by your hand frame. After you have finished, figure out the average percent green. A variation of this is to note the number of times the toe of your foot (either one or both) hits green material; also keep track of the total number of steps. Divide the number of times you hit green, by the total number of steps, multiply by 100 and you have the percent green.

The percent green should be 50% or better. If the average is less than 50%, the pasture probably has a LAI less than 1, and was grazed too hard to expect a fast recovery. This approach is reasonably well adapted for actively growing set stocked or long term rotation pastures where the concern is when to move the animals out, before damage to the pasture occurs. It does not indicate residual dry matter, for that you need a clipping frame.

ESTIMATING FORAGE QUANTITY:

Measuring the amount of forage present in a pasture is difficult at best, but there are times when it is very important to know, such as when in a drought or in break grazing. One old standby method, but most time consuming, is to make a frame of known size. The frame is usually larger than one square foot; for sod grasses pastures, a two square foot frame is sufficient, for bunch grass pastures a four square foot frame is better. Frames can be made out of almost anything, wood (poor), reinforcing rod (better), etc. They should be open on one side so they can be slid into place in tangled areas. The frame is placed or tossed randomly around the paddock and the forage clipped to whatever height you are going to allow the animals to graze down to. But no closer than a 1/2 inch above the ground. In clipping the sample only take what the animal will be expected to eat (cow clip); do not include roots, litter and other debris. Include in the clip only that material that falls within the border of the frame. If a grass stem lies outside the frame, but the leaf falls within the frame, then include the leaf but not the stem. Collect the cuttings in a paper sack, not plastic.

The number of samples to take depends upon the kind and size of the pasture being sampled. If the pasture is fairly uniform, as to height and density of sward, then usually 5 samples are sufficient. If there is a lot of variation in the pasture, many brush areas, rocks, different kinds of plants, or of large size, then more samples are required; however, 15 samples will usually be enough if taken in a random manner over the entire paddock.

After the clips have been taken they can be grouped together and dried. The material can be spread out on screens, in the sun or in a protected place and air dried; air drying, by itself, only gets the moisture level down to around 20%, unless in arid climates then to 10 or so percent. If the material is spread around on a tray and placed overnight in an oven at the lowest temperature setting possible (try to keep it below 200 degrees Fahrenheit, 150 is best), this will usually be sufficient. REMEMBER, dry grass will burn,

[8] Like you used to do while playing hide and seek when you cheated and peeked.

so be careful. Another method of drying is to place in a microwave oven for 7 to 10 minutes; the exact time will depend upon the size of the sample.[9] Also forage dried in a microwave should not be used for chemical analysis. The forage is dry enough if it breaks when squeezed in the hand. After drying, the different clips can be lumped together and weighed.

Generally, you will not have large quantities to weigh, so use a fish or postage scale. After you have the weight divide by the number of samples that were grouped together; this gives the average weight. The next step is to divide the area of your clipping frame into the number of square feet in an acre (43,560); then multiply this number by the weight. If your weight is in pounds, you will have the average number of pounds per acre in that pasture. Multiply this value by the number of acres in the pasture and you have the total amount of forage available in that pasture.

An alternative to drying, but less accurate, is to estimate the percent dry matter (DM) of the clipped forage. Young, tender, green, leafy material is usually between 15 to 20% dry matter, older, ranker material between 20 to 35%, it is brown, but still standing about 55 to 65%, cut hay will get up to 80 to 85% DM. Weigh the clipped material, multiply by the estimated dry matter content, and use this value in determining amount of available forage.

The above method is very time consuming and a bit tiresome, there are, however a number of electronic devices on the market that are able, with varying degrees of accuracy, to estimate the amount of forage present. All require calibration under actual field conditions, however their use looks promising. They are in-accurate, but considering the way we estimate forage quantity at present, a definite improvement. Their major value appears to lie in their ability to quickly develop a "graziers eye". Pasture meters are discussed in Chapter 13.

FORAGE QUALITY:

In the "good old days" it was enough to say that your pastures were belly deep to a horse, or that you got 10 tons of hay to the acre; big was beautiful. However, in those "good old days" production was low and so were costs, but prices were often good, so it didn't matter too much. Other than for dairy people, quality didn't account for much. "Tain't" so today. Today's costs have taken much of the slack out of livestock operations; half a pound a day gains are no longer acceptable. The quality of the feed on offer can spell the difference between profit and loss, and it's just as easy to harvest quality forage as it is poor quality stuff. With poor quality forage you have to buy quality in a bag; it's a whole lot cheaper to grow it then harvest it at the right stage.

THE PLANT:

As far as the grazing animal is concerned, forage plants are composed of four parts; young green leaves and stems , old green leaves, dead or brown leaves, and mature stems. Given the opportunity (light stocking) the animal will select it's diet in that order. There are a couple of very good reason for this. To the animal the young leaf and stem tastes better and probably smells better, and because it does, the animal searches it out.

---

[9] Don't use a high setting and watch closely. A prairie fire in a microwave is exciting but messy; we know.

This is nice, because the young forage contains a lot of crude protein and also has a high digestibility; all of which means good animal performance.[10]

New young green leaves will have a digestibility between 70-75%; young stems are often higher. Old green leaves will be between 60-65%, dead leaves about 45%, mature green stems around 50% and dead stems close to 35%. Because of the higher digestibility of leaves over stems, leaves will remain in the rumen approximately 24.5 hours compared to 33.3 hours for green stems and up to 72 hours for dead stems. This faster turnover rate means that for a given size rumen, the amount of material an animal can eat (intake) during a 24 hour period increases. The greater the intake, the greater will be the gain in weight. Intact Holstein bulls grazing young leafy rye grass pastures have been recorded as gaining in excess of 3.1 lbs/day over a 14 month period in New Zealand.

The crude protein content of green leaves is somewhere around 20% but this decreases with age and for dead material will be around 4%. The thing about crude protein content is not the total content; it is how much of that protein is available to the animal (digestible protein). The higher the crude protein content the greater percentage of it is digestible protein. Forage with a crude protein content of 15% may have 80% or more of it as digestible protein. On the other hand, forage with a crude protein content of 4% only has about 40% of it actually digestible. This is because at the lower crude protein levels much of the protein is bound up in fiber (lignification) and not available for digestion.

As the leaf ages, nutrients and energy are transferred to young growing leaves, roots or stems; the old leaf becomes basically fiber and finally dies. This process may take anywhere from 20 to 80 days, depending on species. Shortly after the leaf dies, molds, fungi, bacteria begin their work and remove or degrade the nutrients further; leaching further removes protein and energy and eventually the leaf is fully decomposed. It is suspected that a number of decomposition products may actually be toxic to animals. This may be one of the reasons that animals will select almost any green material over dead. The death of leaves may be hasten by shading of mature leaves by young ones, low soil fertility, nitrogen transfer, dry weather, or seed head production. When the plant begins to go to seed, nutrients are rapidly transferred from stems and leaves, in effect sacrificing them to get the necessary energy and nutrients to grow seed.[11] As a result, when seed heads appear, dead leaves accumulate rapidly and the overall digestibility and protein content declines sharply.

All grasses lose quality (protein and energy) the older they become, tropical grasses more so than temperate ones. Figure 5.9 (see also table 5.1) shows the loss of quality, in this case crude protein, as a function of length of re-growth for a well fertilized kikuyu grass and for an unfertilized range kikuyu. What these curves show is not only the fairly rapid decline of crude protein content with increasing age, but the effect of fertilization. Location also affect quality; plants growing on the sunny side of hills in winter were found to be more digestible that those on the shady side, the reverse was true in the summer. High soil fertility will increase the quality of tropical grasses significantly, both as to crude protein and digestible energy, whereas the effect on temperate grasses is much less dramatic. Not only will grass from fertile soils have an initial higher quality, but it will keep this edge longer. However, for a growing animal, even the fertilized kikuyu is

[10] Digestibility is the percent of the material eaten that is actually digested by the animal. If a cow eats 10 lbs of dry matter and her manure contains 5 lbs of dry matter the digestibility of forage is 50%; more in chapter 6.
[11] From a plants point of view, the only reason to grow is to produce seed so as to continue the species; not, as many in the industry think, to provide food for animals.

of questionable value after 9 weeks of growth, and range kikuyu has little value after only 6 weeks. When the crude protein content drops below 10%, it has value only as a maintenance ration (bull feed) due to the low digestible protein content.

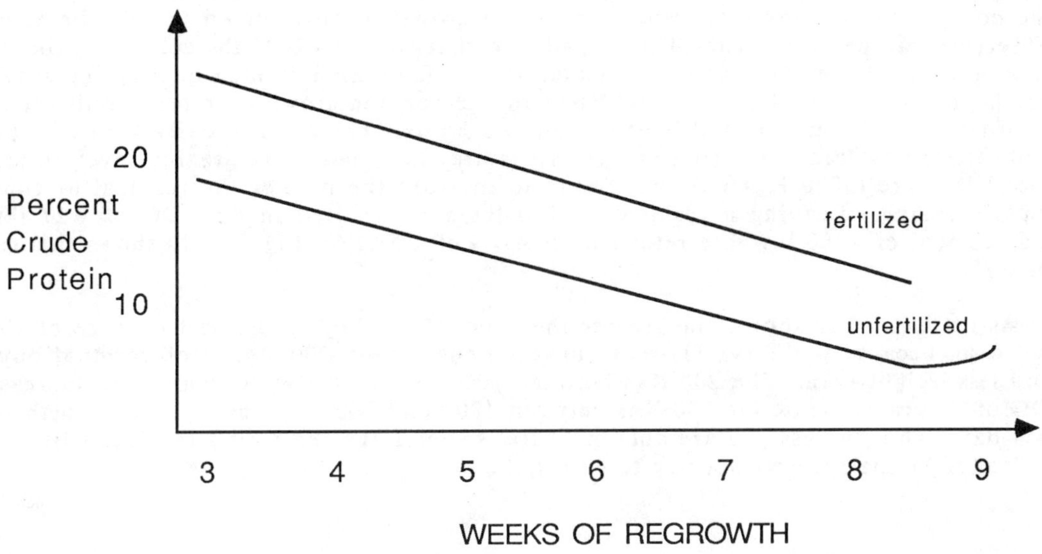

FIGURE 5.9. DECLINE OF CRUDE PROTEIN CONTENT IN KIKUYU GRASS AS A FUNCTION OF RE-GROWTH TIME.

The only exceptions to the above decline in quality is when the plant is harvested, or if leaf and stem death occurs due to some environmental cause, such as freezing or a chemical application like paraquat. If the weather remains cold, decomposition is slowed and the losses are mostly leaching or degradation by sunlight. Even during winter, every time leaf temperatures rise above freezing the decomposing organisms get to work and further lower the feed value. Also, leaching is not to be taken lightly; one inch of rain on dead leaves may remove over 50% of the carbohydrates and reduce the overall feed value by 30%. A light drizzle removes more nutrients than does a heavy rain; even dew is an effective leaching agent. Young leaves are less susceptible to leaching than are mature ones and leaves that have been injured will suffer greater leaching losses than healthy ones. About the only things that are not leached from a plant are fiber and lignin.

THE PASTURE AND/OR RANGE:

In spite of the definitions given out by agronomists and range scientists concerning pastures and ranges, the major difference between the two is that one has more water than the other. There is no magic line that separates the two environments or the plants that grow on them. The grasses that grow on each area all do the same sorts of things that grasses do everywhere. The two environments form what is called a continuum, sort of like going from slow to fast.[12]

A major factor in pasture/range quality is that the first 500 lbs of forage removed from an acre is of higher quality than the next 500 lbs, and so on down to last 500 lbs. In a rapid growth situation, on fertile soils, this difference will only amount to a few percentage points of crude protein. However, where growth is slow, or on less fertile soils, the difference can be very great. Thus a paddock that contains 2000 lbs/acre of available forage might only have 500 lbs/acre of material that is of high enough quality for growing steers (stocker feed). The rest might be suitable for non-growing animals (bull feed), but in order to utilize it you will have to run two herds. The average daily gain of New Zealand late spring lambs entering rye grass pastures that had a pre-grazing level of less than 2600 lbs/acre (3000 Kgs/ha), was found to increase the greater the residual at completion of grazing. Leaving a residual of 890 lbs/acre resulted in an ADG of 0.27 lbs, while a residual of 1780 lbs/acre returned an ADG of 0.58 lbs. Figure 5.10 shows this relationship.

As FIGURE 5.10 shows, the greater the residual the higher the performance of the lambs. Also, because the curve flattens out, each succeeding 200 Kg/ha of residual buys less and less weight gain. The 200 Kgs between 1000 and 1200 Kgs/ha bought an increase in ADG of 40 grams, while the 200 Kgs between 1800 and 2000 Kgs/ha only was worth 10 grams a day. Thus, unless you are able to utilize some of the remaining residual with another class of animal, it may not pay to go for the max.

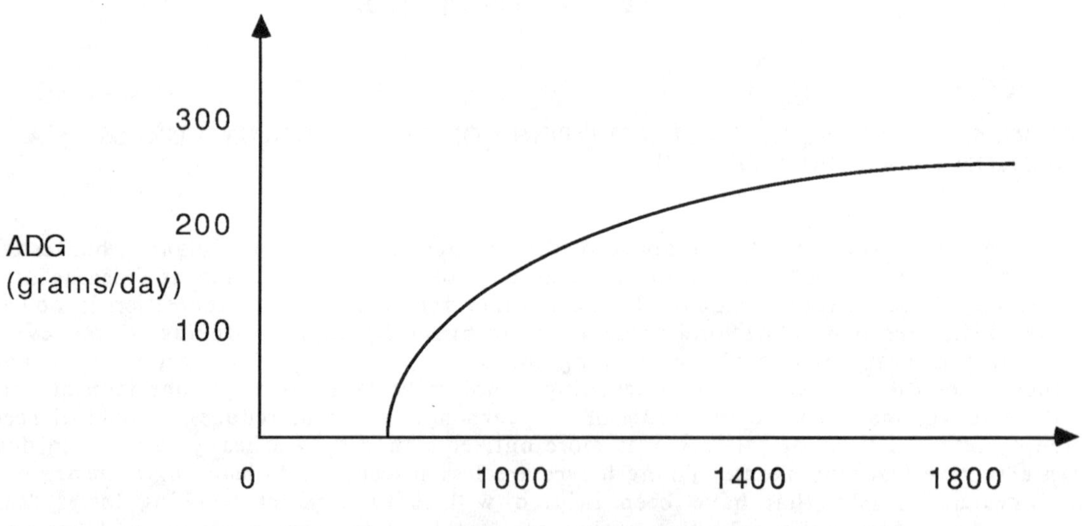

FIGURE 5.10. RELATIONSHIP OF RESIDUAL DRY MATTER AND AVERAGE DAILY GAIN FOR LATE SPRING LAMBS.

[12] Many things in nature form continuums, but humans, being what they are, like to put boundaries around things then give it a name. This is why females are either pregnant or not, never just a little bit pregnant.

Another factor affecting pasture quality is the amount of green grass compared to the amount of dead material present in a paddock. This has a direct bearing on the amount of material an animal will eat, which in turn determines the amount of gain. Grazing animals select their diet as best they can; if they are allocated sufficient area, or amount of forage so that they can select. As a result, with light stocking, even when the amount of dead forage is 80% of the feed on offer, the diet that they eat will only contain about 20% of the dead material. Only when the dead forage is 100% of the feed on offer will the diet contain 100% dead material. However, the amount of dead forage eaten will be much less than if it was green; close to 2% of the live animals body weight for dead, compared to around 3%, or better, for green (dry matter basis).

A couple of interesting effects have to do with the amount of feed on offer. Again, with New Zealand lambs, if the lamb was allowed only 1 Kg DM per day (2.2 lbs/day) no gain was recorded. However, 2 Kgs/day on offer resulted in a gain of 90 grams/day (0.2 lbs/day) but this peaked at a gain of 180 grams/day when the feed on offer reached about 5 to 7 Kg DM/day. Depending on quality, intake and thus gains reach their maximum when the animal is offered 3 to 5 times more than they are actually expected to eat. The other effect has to do with the amount of total feed in the pasture, the pasture mass. The greater this is, the higher the rate of gain at a constant stocking rate. Figure 5.11 shows the relationship for New Zealand ewes, between liveweight gains and intakes, at different pasture masses and amounts of feed on offer.

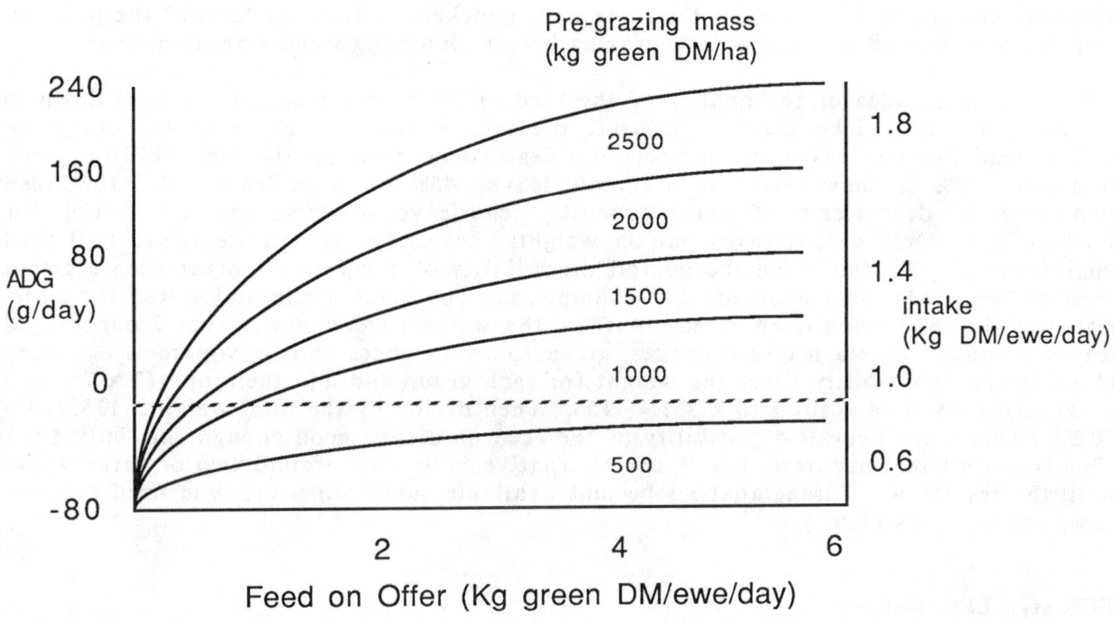

FIGURE 5.11. PREGRAZING PASTURE MASS, FEED ON OFFER, INTAKES AND DAILY GAINS OF EWES.

What also appears to be happening is that as the pasture mass increases, the amount of feed that has to be offered to the animal to obtain maximum gains, decreases. The reason seems to be the ease of which the animal can obtain feed. At the lower pasture mass (low grass height) the animal has difficulty in obtaining enough food, even when large areas are given to roam around in. Rate of gain is related to rate of feed intake which is related to pasture mass, density, length of tillers, size and number of the bites and is covered in Chapter 6.

LEAVES AND STEMS:

As mentioned before, the average leaf to stem ratio of temperate grasses, as a whole, is on the order of 75% over the growing season, but only 50% for tropical grasses. These values change during the growing season with high ratios prior to setting seed and low values afterwards. The ratio is also lower depending on the length of time animals have been in a paddock, since animals tend to eat the leaves first.[13] The major factor in the lower digestibilities and protein content between the temperate and tropical grasses is that the temperate grasses have more leaf than the tropicals. If the goal is to increase the quality of the feed on offer in tropical pastures, then the key would be a grazing program designed to keep stems from forming on tropical grasses.

The effect of the leaf to stem ratio on daily weight gains was dramatically shown in a trial conducted in Louisiana on a set stocked, sub-tropical pasture. Cattle entered a young leafy pasture and gained close to 2.0 lbs/day for the first 45 days. At this time the majority of leaves had been consumed and only stems, old leaves and a little new leaf growth was available. The gains dropped to less than 0.5 lbs/day over a similar time period. How does one maintain a high ratio of leaf to stem? There are three ways, mowing after the animals leave, use more than one herd (stockers in front to "cream" the paddock and dry cows behind to clean up), or grazing before flowering stems start to appear.

To get an idea of the quality of the feed on offer, cut a sample of feed down to the level that it will be grazed. Separate the sample into new green leaves, old green leaves, dead (brown) leaves, green stems and dead stems, then use the digestibilities given in above; 75% for new leaves, 65% for old leaves, 45% for dead leaves, 50% for green stems and 35% dead stems. If you have mostly dead leaves or stems, you're in trouble for stocker feed. Only green leaves put on weight; brown leaves and stems are bull feed (maintenance). To determine the overall digestibility of the feed on offer each segment must be "weighed" and combined. For example, say you made a clip and sorted the material into five piles then oven dried. Assume the weights were; new leaves 2 ounces, old leaves 5 ounces, brown leaves 6 ounces, green stems 4 ounces and brown stems 6 ounces. Multiply the digestibility times the weight for each group and add them up: (2 x 75) + (5 x 65) + (6 x 45) + (4 x 50) + (6 x 35) = 1155. Then divide by the total weight: 1155/23 = 50.2%. This is the overall digestibility of the feed on offer; good enough for "bull" feed. This is not all that accurate, but if the alternative is to wait around two or three weeks until the results of a forage analysis become available, much superior; you need to know now, not when it's history.

FORAGE LEGUMES:

There are some 11,000 different species of legumes in the world and all have the C3 metabolic pathway. They have adapted very well to both temperate and tropical environments. Without getting into the arcane academic world of botanical classification, the

[13] In a pasture that has mostly all leaf and is heavily stocked, the animals will remove the forage in layers, starting at the top. Each succeeding day the pasture will decrease in height, in some cases looking almost as if someone came in with a mower set at ever decreasing heights.

forage legumes can conveniently be divided into three general groupings: climatic (tropical, sub-tropical, temperate); growth form (upright or trailing); quality (good, poor, poisonous). There are a large number of good forage legumes and a great deal of material in the form of books, circulars, manuals and pamphlets on different species, varieties, cultivars, methods of cultivation, harvesting, etc. What follows is some information that is pertinent to IGM.

Legumes generally require higher amounts of phosphorus, molybdenum, sulfur, cobalt, copper, zinc, potassium and boron than other plants. For the most part these increased requirements are due to the unique relationship legumes have with certain kinds of bacteria and their resultant ability to "fix" nitrogen. This "mutual benefit" society with certain bacteria (rhizobia) is called symbiosis. The rhizobia get a nice place to live and the legume gets about 75% of its' total nitrogen requirements for growth. Rhizobia invade the roots of the legume and cause nodules to form. Within these nodules the rhizobia "fix" atmospheric nitrogen into forms that are available for use by the bacteria as well as the legume. About 70% of the nitrogen produced by this association is eventually returned to the soil.

It is a common misconception that grasses growing in and around legumes get a direct benefit from the "fixed" nitrogen. They do not. Legumes or rhizobia do not breathe "fixed" nitrogen into the soil. Secretions from the growing regions of the legumes roots, called root exudates, are the closest thing to a direct transfer of nitrogen to the soil, this may account for between 2 to 11% of the total nitrogen "fixed"; generally bacteria get to these before higher plants can. It should also be mentioned that not all legumes have rhizobia, not all legumes that have rhizobia fix nitrogen, and not all plants that fix nitrogen are legumes.[14]

The benefit to the grass is indirect and results mostly from the decomposition of the nitrogen containing parts of the legume; roots, nodules, stems and leaves. In legumes with taproots, about one third or more of the total nitrogen supply contained in the plant is returned to the soil by decay of the root system; whereas legumes with fibrous root systems return only about one fifth. For example, alfalfa will have between 34 to 50% of its' total nitrogen supply below ground, in roots and nodules, while white clover will have a much smaller portion of its' total nitrogen below ground, only 16 to 19%. Leaf litter is the second major pathway by which nitrogen gets into the soil. This is accomplished by leaching, decomposition and various soil insects or others that bring the leaf material below ground, such as ants, earthworms, sow bugs, etc.

Generally speaking, legumes do not lose quality with age nearly as fast as grasses do. Also their feed quality is generally higher except when compared to very young grasses (Table 5.1). This is good, but also bad. Good, in that as long as it is kept green and growing loss of quality is not a major consideration; as a result, set stocking, or lax grazing can be used without the penalty that such practices extract with grasses. Bad, in that it causes some management problems when grass/legumes combination are grown.

The quantity, or yield, of forage produced by temperate legumes, temperate grasses and tropical legumes is about the same if grown on soils of similar fertility. However, the yield of tropical grasses is about twice that of temperate forages or tropical legumes (Figure 5.12). In other words, on the average, an acre of grass in the tropics can produce twice as much forage as an acre of tropical legumes or a similar acre in temperate regions

---

[14] Certain grasses [mostly tropical] form a weak association with a bacteria to fix nitrogen. Some trees and shrubs such as the ironwood tree has a similar ability; also several kinds of algae fix nitrogen.

of either grass or legume. The reasons have to do with the lack of cold weather and also that tropical grasses have the C4 pathway which gives them a leg up in hot weather. It is this rapid growth of C4 grasses, often in excess of 270 lbs(DM)/Ac/day, that causes problems when tropical grasses and legumes are grown together.

TABLE 5.1. CRUDE PROTEIN CONTENT OF THREE COMMON HAWAIIAN SUBTROPICAL LEGUMES AND THREE GRASSES AT DIFFERENT STAGES OF RE-GROWTH.

	WEEKS OF RE-GROWTH					
SPECIES*	4	5	6	7	10	18
	(crude protein content in percent)					
LEGUMES						
Greenleaf	18.8	19.4	15.0	16.9	15.6	11.9
Silverleaf	21.3	20.0	18.8	19.4	17.5	13.8
Glycine	23.1	23.1	18.8	21.9	16.9	18.8
GRASSES						
Kikuyu grass	22.4	18.8	17.5	15.0	11.3	6.3
Green panic	20.0	18.1	15.0	11.3	8.1	6.9
Buffel	12.5	11.9	10.0	9.4	9.4	5.0

* All heavily fertilized; grasses 794 lb/acre of 10-10-10 and legumes 397 lbs/acre of 0-20-20.

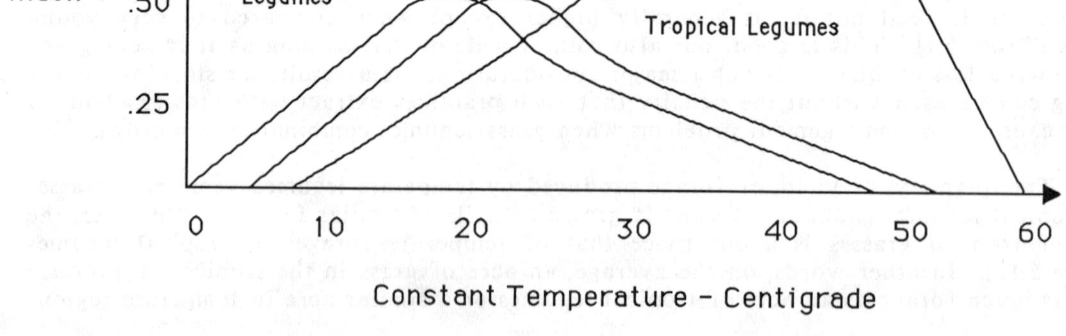

FIGURE 5.12. COMPARISON OF RELATIVE YIELDS OF TEMPERATE AND TROPICAL GRASSES AND LEGUMES.

Most legumes have the upright growth form although this can vary from a one or two inch high forb, to shrub, bushes and trees. One of the more promising forage varieties of tropical leucaena has the ability to grow 20 feet a year.[15] The other growth form are the trailing legumes; these are almost all found in the tropics or sub-tropics. The closest thing to a trailing legume in the temperate zone are the common white clovers which spread by means of stolons.

MANAGEMENT CONSIDERATIONS:

In theory, the legume should make an ideal forage companion in a pasture, especially in wet-dry environments. The legume, with its deep tap roots, will utilize moisture found below the root zone of the grass and will stay greener longer into the dry season. Because of it's ability to maintain a relatively high quality over a long period of time it can extend the life of the grazing season, or increase the overall quality of the forage on offer. This theory appears to be fairly reliable when dealing with upright legumes, however trailing legumes are something else.

White clovers, because of their stolon forming habit and the resultant ability to withstand heavy grazing, appear to be the best temperate legumes for intensive grazing. The preferred mix of clover to grass, according to New Zealand work, is on the order of 30% clover. In a grass-clover mix, lax grazing will tend to favor the legume, while hard grazing will favor the grass. Continual grazing of a unit with cattle will also favor the legume, whereas grazing with sheep will favor the grass.

In the tropics Hetero <u>Desmodium heterophyllum</u> is similar to white clover in that it is a creeping perennial and resistant to heavy grazing. Hetero has attracted notice because of its ability to co-exist with pangola and other sod grasses. Under intensive grazing, where the mat is controlled, it appears to be the ideal companion legume. The only problem with Hetero is the difficulty of obtaining commercial seed, however the plant does establish from cuttings.

Trailing legumes are prone to overgrazing. Unlike upright legumes, trailing legumes do not have a true crown, from which re-growth after grazing begins. The length of the remaining stem determines the rate of re-growth; the less stem remaining the longer the rest period required. Little work has been done on rotational systems to determine when to remove the animals, but by knowing something of the character of the trailing legume it is suggested that trailing legumes be grazed no shorter than 9 to 12 inches. There are two warning signs used to determine if the legume is being overgrazed: 1. The leaves of trailing legumes are usually very large and of good color. If the leaves look small and feel leathery, grazing pressure should be reduced. 2. The amount of legume in the pasture is an indicator of stocking rate, there should appear to be as much legume as grass; more or less, indicates too low or too high a stocking rate, respectively.

Many trailing legumes are very bitter when young which makes them relatively unpalatable early in the growing season. Thus, if lightly set stocked the grass will be grazed first and then the legume, resulting in a lengthening of the grazing season. However, if the "proper" stocking rate is not closely maintained, what usually happens is that the grass is ultimately overgrazed and the legume takes over, or more likely, the whole pasture is replaced by less desirable forages. Trailing legumes are very sensitive to stocking rate, at high rates the grass takes over and at low rates the legume becomes king.

---

[15] Needless to say, you don't plant this kind of stuff unless you have good control or run giraffes.

Management is easier if the companion grass is slightly out of it's element, so that its growth rate is slowed down to where it's more in line with the legume.

The tropical legumes are not as prone to causing bloat as are the temperate legumes. Because of bloat problems with most temperate legumes not many graziers use an all legume pasture. This is a waste of a valuable resource. If it pays to feed alfalfa or sweet clover hay to stock, then the same forage can be grazed a lot cheaper. Of course, if animal control is lax major problems will develop.

Generally speaking, if the animals are introduced to the legume at the end of a grazing cycle, a half hour the first day, an hour the second and so on, the danger of bloat is minimized. In high bloat risk pastures the animals should not be introduce early in the morning or while there is dew on the legume. The reason for this is under these conditions the stomata of the legume is open thus allowing digestive juices to enter and increase the rate of digestion, which in turn causes more gas and foaming agent to be released. Selecting animals for bloat resistance is discussed in Chapter 6 The quality of the feed on offer should, after introduction, remain as near the same as possible. If a central watering source is used one of the several bloat additives may be added to the water to reduce the danger. Another technique is to handle the legume as living hay, discussed in Chapter 12.

A word about alfalfa. The growth habit of alfalfa is such that all sprouting is from the crown. New shoots develop at about the time the older ones start to flower. If the plant is cut or grazed at this time, little damage results. However, if the harvest is delayed until after the new tillers have obtain some height, then both the old and new tillers will be harvested together; plant vitality decreases. Harvesting the new tillers too soon will, like overgrazing in grasses, eventually kill the plant.

The alfalfa of the tropical world is probably leucaena. Leucaena, like many upright legumes, is quite palatable when young. As a result it does not work well in continuous grazed situations. In Australia leucaena had a lot of bad press due to it's high mimosine content. It was recently found that their ruminant animals did not have a certain rumen microbe that was capable of degrading mimosine. The appropriate microbe has been introduced and mimosine poisoning is no longer a problem.

Hawaiian ranchers have always suspected leucaena of causing reproductive problems, however not in all years or seasons; nor does there appear to be any problem with market animals. Because pastures containing leucaena are some of the most productive pastures in the state (726 lbs per acre produced in one year with steers on a 22 inch rainfall, 16 paddock, leucaena/guinea grass grazing unit), ranchers are not eager to give them up. The reason for the problem is still undefined but it is suspected that it involves the phyto-estrogens that most legumes produce during periods of rapid growth. Until more is known it is suggested that leucaena, especially during periods of rapid growth, should make up less than 25% of the ration for breeding animals.

TOXIC FORAGES:

The words "toxic forages" conjures up all sorts of gruesome pictures of dead and dying animals. However, the word toxic only means capable of causing problems. The grazing animal has a limited tolerance to almost any chemical substance - it is a matter of degree. Even essential nutrients can be harmful if consumed in excessive amounts; for example, vitamin A, sodium chloride, copper, etc.[16]

The list of toxic substances is very extensive and will certainly be increased as the exact nature of the toxin(s) is unknown for many species. The most important groups related to grazing animals are given below.

1. ALKALOIDS: These are a varied group of nitrogen containing substances often having pharmacological effects that have been exploited in human medicine (eg. morphine, quinine, atropine, curare, strychnine). In rangeland plants these alkaloids are prominent, usually producing chronic intoxication associated with impaired liver functions. Some plants that contain these are the lupins, rye grass, phalaris, crotalarias, members of the Solanaceae family (apple of sodium, apple of Peru, tobacco, tomato, jimson weed, poha, datura, angel's trumpet, etc.); also a fungus produces these substances that may flourish in highly productive pastures.

2. CYANOGENIC GLYCOSIDES: These are not themselves toxic but after digestion yield hydrocyanic acid, which is. Some plants which contain these are; lablab, cassava leaves, star grass, sorghum species, white clovers, passion fruit leaves, seeds of cherry, chokecherry (Prunus species) peach, plum, elderberry and arrowgrass (Triglochin maritima). The majority of these plants must undergo some stress, such as drought or freezing before harmful effects are likely to be noted. A sulfur mineral supplement will usually counteract the effects of these substances unless high concentrations are encountered.

3. HORMONES: While all plants produce various hormones the legumes seem to be the major plant family that produce estrogens similar to those secreted during the oestrus cycle of animals. The two types of estrogen compounds are isoflavones and coumestans; coumestans being the more potent of the two. Where spring breeding is practiced these hormones may reinforce the natural hormone. However, large intakes of this plant hormone interfere with the animals reproductive cycle and/or may cause abortions in pregnant animals. Sheep appear to be affected more than cattle. Many legumes are suspected of this tendency: alfalfa, leucaena, sweet clovers, red clover and white clover (if infected with the fungus "Pseudopeziza tirfolii") are some examples. However, adequate soil phosphorus levels are thought to reduce the potency. Dilution of the legume with another forage, such as mixed grass/legume swards, will also help.

4. PHOTOSENSITIZERS: Some substances found in plants cause animals to become highly sensitive to sunlight. This sensitization includes both an increased tendency to sunburn and dazzling of eyes. The hypersensitivity is due to substances which are capable of absorbing light and transferring the energy to surrounding tissues. There are two types of photosensitizers of plant origin; primary and hepatogenic. When St. Johnswort (Hypericum species) is eaten substances are absorbed and carried through the blood stream to the skin and eyes. Lantana is the major hepatogenic photosensitizer. Hepatogenic plants have no substances which are absorbed by the body, as such. However, they do disrupt liver functions, which in turn cause some products, that are photodynamic, to enter the blood stream rather than be secreted as bile. These substances then move to the skin and eyes. Because the hepatogenic

[16] Consider the furor being raised over feed additives and other substances used by the livestock industry by over-zealous health advocates. Yet, they conveniently ignore some of their preferred foods that are known to contain harmful substances, for instance high level [1.5%] of canavanine in alfalfa sprouts.

plants attack the liver they are much more dangerous. One Hawaiian ranch, in one month, in 1985, lost a total of 12 cows due to lantana poisoning. Other known photosensitizers are: broomcorn millet, Japanese millet, rape (Brassica species), various clovers (Trifolium species), horsebush (Tetradymia species) and lupins.

5. GOITROGENIC: These substances contain both nitrogen and sulfur and are mostly found in plants of the mustard family. The major exception is Leucaena which contains mimosine. However, ruminants in Hawaii and many other places have a runen bacteria that is capable of degrading this substance.

6. AMINO ACIDS: In addition to the 22 amino acids found in most proteins many others have been recorded in the non-protein nitrogen components of plant tissue. Most are non-toxic. Two exceptions are indigospicine, a hepatotoxin found in creeping indigo and also mimosine, mentioned above.

7. OXALATES: These are found in a wide range of plants and occur as soluble or insoluble compounds. In the rumen the soluble oxalates are decomposed; toxic effects occur only when the concentration is very high. Two effects occur: 1. A reduction in the concentration of calcium in the blood, causing a deficiency of calcium. 2. Deposition of calcium oxalate in the kidney tubes, leading to kidney failure. Oxalates are commonly found in the families Chenopodiaceae, Oxalidaceae and Polygonaceae, they also occur in many other plants, such as; buffel grass, setaria grasses, lambs quarter, beets, spinach, greasewood (Sarcobatus vermiculatus), halogeton (Halogeton glomeratus) and all members of the genus Oxalis (yellow wood sorrel).

8. NITRATE: This is a normal plant metabolite. Some species tend to accumulate nitrate and a few accumulate nitrites. When the nitrate level in forages rise above 0.5% on a dry matter basis, there is a potential source of trouble. Many common forage plants (oats, corn, alfalfa, kikuyu grass, rye grass) can accumulate dangerous amounts of nitrate under certain conditions of weather and soil fertility. Nitrate is itself a rather innocuous substance, but in the rumen the bacteria will reduce nitrate to nitrite which may be fatal. Under normal ruminal conditions, nitrite is further reduced to ammonia. Toxicity occurs when the rate of production of nitrite exceeds the rate of conversion to ammonia and the concentration of nitrite in the blood rises to high levels. High levels of nitrite in the blood converts haemoglobin to methaemoglobin and this results in oxygen starvation; staggering, vomiting, labored breathing and quite often death, unless treated. On autopsy the blood is brown. Treatment should be by a veterinarian and consist of a 4% solution of methylene blue in either a 5% glucose or a 1.8% sodium sulfate solution, administered IV at the rate of 100 cc/1000 lbs body weight. Sheep are less bothered by nitrite poisoning than are cattle. Pigs are not bothered by nitrate because they do not convert nitrate to nitrites. Researchers in New Zealand could find no evidence that sublethal levels of nitrates contributed to abortions, vitamin A nutrition, decreased milk production or lowered birth weights.

9. OTHER ELEMENTS: Fluorine in trace amounts is an essential nutrient for mammals; excessive intake interfers with the metabolism and impairs energy production. Members of Acacia tend to concentrate fluorine as an acid. Copper is another essential element for both plants and animals. Excessive copper intakes can occur when the soil is very rich in copper or deficient in molybdenum. Clovers tend to accumulate copper. Molybdenum is also essential in both plants and animals and acts as an antagonist to copper (ties it up). Excessive amounts of "Moly" can occur in almost any plant, but more so in legumes than grasses.

10. OTHER THINGS: Many plants have all sorts of mechanical defenses against grazers in addition to chemical defenses. Thorns, spines, hairs, tough leathery leaves and growth habits are a few of the methods that have evolved to protect the plant from the animal. Under normal circumstances these do not cause problems as the animal simply avoids mechanically protected plants and concentrates on those that don't have such defenses. Under drought conditions or where animals are left in a pasture in which there is little forage available, or where the forage is closely intertwined with such plants, problems can occur. During drought, cattle will often eat prickly pear cactus. The thorns will not only penetrate the mouth but may puncture the stomach. Sheep and goats will also eat cactus but because of smaller mouths can avoid most of the spines. Spiny Amaranth will cause ulcers in the mouth and on the tongue, especially in cattle and horses. This makes eating painful and results in lowered performance. Signs are slobbering, kind of a hang-dog look, loss of weight and light nibbling of feed. Check the mouth, if ulcers are present, you'll see them as they can get quite large. Treatment consist of removing to a "clean" pasture or stall feeding; call your Vet.

SUGGESTED READING:

A Guide to Better Pastures for the Tropics and Sub-tropics. 1980. L.R. Humphreys. Wright Stephenson & Co. Pty. Ltd.

Environmental Adaptation of Tropical Pasture Plants. 1981. L.R. Humphreys. MacMillian Ltd.

Forage Legumes for Energy-Efficient Animal Production. 1985. *Proceedings of Trilateral Workshop held in New Zealand, 1984.* USDA, Agricultural Research Service Washington D.C. 22005.

Forages: The Science of Grassland Agriculture. 1973. M.E. Heath, D.S. Metcalfe & R.F. Barnes. Iowa State Univ. Press.

Grazing Animals. 1981. Ed. F.H.W. Morley. Elsevier Scientific Pub. Co.

Grasses of Hawaii. 1968. Peter P. Rotar. Univ. of Hawaii Press.

Hawaii's Poisonous Plants. 1979. Roger E. Baldwin. Petroglyph Press.

Intensive Grassland Management in the Humid Tropics of Puerto Rico. 1974. Bulletin 233. Univ. of Puerto Rico, Mayaguez Campus, College of Agricultural Sciences, Rio Piedras, Puerto Rico.

New Zealand Farmers Guide to Grazing Management. 1981. Keith Milligan. New Zealand Farmer. Box 56-064, Auckland 3, N.Z.

Nutritional Limits to Animal Production from Pastures. 1981. J.B. Hacker (Ed.) *CSIRO Proceedings.* Commonwealth Agricultural Bureaux, Farnham Royal, UK (Pub.)

Pastoral Industries of Australia: Practice and Technology of Sheep and Cattle Production. 1979. Ed. G. Alexander and O.B. Williams. Sydney University Press.

Pasture: The Export Earner. 1984. New Zealand Institute of Agricultural Science.

Physiological Basis of Crop Growth and Development. 1984. Ed. M.B. Tesar. American Society of Agronomy & Crop Science Society of America, Madison Wisconsin.

Physiological Plant Ecology. 1975. W. Larcher. Springer-Verlag.

Rangeland Plant Physiology. 1977. R.E. Sosebee. Pub. Society for Range Management.

Roberts, C.R. Several papers on tropical forages. Distributed by J.H. Williams & Sons Pty. Ltd. P.O. Box 102, Murwillumbah, 2484, Australia.

The Whatawhata Way. 1982. Ed. Margaret Shimmins. Pub. Agricultural Promotion Associates, Ltd.

Tropical Crop: Vol. 1 (Monocotyledons) and vol 2 (Dicotyledons). 1972. J. W. Purseglove. John Wiley & Sons.

Tropical Forages in Livestock Production Systems. 1975. Ed. E.C. Doll & G.O. Mott. American Society of Agronomy, special pub. #24.

Tropical Legumes: Resources for the Future. 1979. National Research Council. Pub. National Academy of Sciences.

Used with permission of NEA.

If you want super genetics, first get super nutrition.

6 ANIMAL BASICS

Historically the major emphasis in the livestock industry has always been on the animal and realistically, probably always will be; this book not-with-standing. The only concern with this is that the majority of the emphasis seems to be placed on high energy input schemes, rather than attempting to learn how to manipulate what we have. The genetics of our animals have been developed to the point where about the only way they can be expressed, is in a feed yard. This may be fine if the animal of the future is going to be little more than a standing converter of feed stuffs to meat. However as long as animals graze, some of those genes are going to have to do other things; it's mostly those other things that this chapter is all about.

FIGURE 6.1. DRUTHERS: If they had their druthers they would rather be on level land and not perched on a side hill. These animals are grazing this hill side, not because they want to be here, but because the dairyman wants them there and knows how to graze animals where he wants them, when he wants them and for how long he wants them. It's animal control and it's called IGM.

BEHAVIOR:

One can pick up a great deal of information on animals and their behavior from books and talking with others, and this can save a great deal of time. However, to turn that knowledge into understanding there must be a reference frame on which to hang all those pieces of information. Cow, sheep or horse sense are things that have to be learned by close association with animals; one has to spend a great deal of time looking at, trailing, searching for, caring and working with animals before an understanding is gained as to how the animal views the world.

The majority of people in the livestock industry today do not have the same degree of animal sense that was present just 30 years ago. The reasons are twofold: There are more things that can make the animal do what we want it to do without us having to be smarter than the animal, such as hot shots, squeeze chutes, etc. Secondly, there are just too many things in this modern world clamoring for our attention; TV's, fast transportation, community activities, etc. Modern technology does not make animal sense out dated, it just makes it harder to acquire.

The fast pace of life in the city is rubbing off, we tend to jam animals rather than easing them around. Because they do jam, or at least the majority of them, we continue to do it, not realizing what it is costing us in lost production. Anybody that has ever milked a cow, knows, or should know, that if you treat her mean, she'll hold up her milk. If you were using that milk check to buy groceries, you figured out real quick how to make her do what you wanted her to do without her "getting on the prod".

Cows walk about 3 miles per hour and horses around 5. Now if your chasing wild cattle that difference is an advantage. But if you're driving or trailing a bunch, you end up pulling back on the reins a lot, either doing a lot of zig-zag riding, playing stop and go, or jamming the rear enders; course you could get off and walk. What generally happens is that there is a lot of whooping and hollering, a lot of herd quitters, the horses get all lathered up, the riders get stove-up, and the cows (the ones you still have) come in on a high trot with manure all over their rear ends. So who won?

It takes time, caring, plus a real interest, to learn about animals. You have to sit up a couple of nights with sick or dying animals, worry through a few difficult births, spend a few days easing a herd along, look for strays in a snow or dust storm, and spend some time just sitting on a hill watching; you have to be close to the animal. Is it worth it? Yes, both economically and for your own piece of mind and body.

GENERAL:

A British breed cow will walk a minimum of 1.5-2.0 miles per day if the the paddock size is greater than one half acre.[1] Dairy cattle will walk about half of this amount, and Brahman cows about twice. In mixed herds the average minimum traveled will be more towards the minimum breed distance than the maximum. The distance traveled in paddocks less than one half acre, drops off fast as the area decreases, but appears to level off at around one half mile for paddocks of 0.1 acre. The maximum distance traveled depends on many things, distance to water, amount of forage, weather conditions, topography, placement of salt and mineral, location of preferred forages, and breed, and most importantly size of the paddock. Typically, Brahman or Zebu type cattle are walkers, in 1500 acre or larger paddocks they have been reported as walking over 7 miles in a day, British breeds near 5 miles and dairy less than 4 miles.

Cattle see, hear and smell differently than humans do. There have been reports of cows finding their calf as far away as 10 miles (the calf was up wind). Cows hear high frequency noises much better than humans, that's why the cracking of whips can drive

[1] British type cattle does not include the European breeds.

them up a wall. A one way gate that clangs will scare the next animal in line and he will not want to go through. With their heads down cattle can see all around them, however they don't see as we do, their vision is more like looking through a "fish eye" lens (no straight lines). They have trouble judging depths and distances, which is why they are often balky and skittish. Bright spots and shadows confuse cattle. If cattle have to cross a shadow, it's best to have solid walls on the chute, it makes them feel more secure. All chutes and crowd pens should be circular and with solid walls.

SOCIAL

All herd animals have a social dominance and leadership order; this is similar to the "peck" order in chickens. Calves reared in automated feed systems with a limited number of feeding positions, develop an order based mainly on size and age. Social orders established in groups of dairy calves will generally last until the animals enter the herd. In some breeds, especially those of Zebu type animals, social facilitation (doing what others are doing) at feeding is strong, so that sufficient feeding points need to be supplied so that the entire group can feed together. Otherwise the fed calves will move away from the feed and these will be followed by others that have not yet suckled.

In semi-wild cattle, cows and calves tend to stay together for about 2 years. The males will begin to dominate their dams at about 18 months of age and shortly thereafter leave and join the all male group. From then on they have to fight their way up the social ladder until they reach "king bull" status and gain mating rights. Bulls tend to become more territorial with age. Bulls under 2 1/2 years of age are fairly well herd adjusted, by about 4 years of age a clear dominance order is established and around 6 years of age they become very independent, territorial and dangerous to move from their regular paddock positions. Bulls form "friendships" as calves and tend to keep these relationships into maturity. Cows also form close associations as calves but these tend to weaken as they have calves of their own and form their own kinship lines.

Cattle tend to form social groups of between 40 to 80 head, this size appears to result from the ability of the cow to recognize, identify and deal with (sort of a first and second name basis) approximately 120 other individuals.[2] In dairy herds it has been reported that increases in milk production on the order of 10% resulted when large herds, over 100 head, were split in two; 70-80 head appear to be about right. Under range or wild conditions the social group will split up during the day into several grazing groups which will consist of 1 to 20 or more individuals, depending on availability of forage. In good years grazing groups will be larger than in poor years. These grazing groups will usually form up into the social group around water, shade or some preferred area for a portion of the day. In high density grazing where animals are being shifted frequently the grazing and social group are typically one and the same. The behavior of the social group is different from that of a grazing group in that there is a greater concerted activity within the social group. This concentrated activity results in a different pattern of use on both the vegetation and the land. This is the so called "herd effect"; which is nothing more than a concentrated animal energy input to a small area for a short period of time.

In herds there are three very general groupings, leaders, dominants and followers. In a moving herd the leaders are generally not the top dominant animals, the dominants take more of a hi-middle position. The lead animal will seldom be the same one at all times, the lead appears to shift around within the leader group. However, the ones at the

[2] Chickens can identify with about 180 other chickens; man can typically handle about 110 other humans. And we call chickens "dumb clucks".

tail end of the herd always seem to be the same subordinate individuals. This natural grouping of cattle on the move is what causes problems for city cowboys.[3] By positioning themselves at the rear of the herd and "flogging" away at the followers they are attempting to push the followers right up into the midst of dominants. They don't want to go, the dominants don't want them there, it is socially un-acceptable. Those followers know one thing, as mean as you are, you won't be there tomorrow, but those dominants will be. So what do you do? Split the herd just behind the dominants and if you feel you have to flog something, flog the dominants, they'll push the leaders and the followers will follow, just like they've been doing all their life.

Heifers are low on the social pole. One study showed that 2 year olds were driven away from a range supplement by older cows. When the 2 year olds were removed from the herd, the 3 year olds were driven away. Dominance is a factor of age, size, aggressiveness, agility, and whether horns are present or not. Horned animals are generally dominant over polled; however once the social order has been established, the removal of the horns does not always result in a loss of social position. Generally, Angus dominate Shorthorns who in turn dominate Herfords. A strange animal introduced into a herd must fight to define its place, this often causes considerable unrest and stress in the herd; however any animal may be removed from the herd without causing social problems.

All herd animals have a "flight distance". This is the closest distanced a dominate animal of the same species can approach a less dominant one, before the less dominant feels threaten. This social distance is approximately 2 to 3 body lengths; the more dominant the animal the greater its flight space. There is an exception to this and that is when a dominant animal approaches a less dominant with the head low and outstretched; this is the signal for grooming. Grooming often occurs between animals of wide differences in social position; groups formed for fly protection and shade will also have animals of different social positions standing close together. However, small grazing groups tend to be composed of animals of similar social status. There is another distance that small groups of animals tend to maintain (gregarious distance) to each other within the herd; 25 to 30 yards for sheep and 140 yards or so for horses. Then there is the distance that the animal can recognize members of the same species; 220 yards for sheep and 550 yards for horses. Sheep and goats can recognize humans over a half a mile away and wild horses can spot a horse and rider about a mile away. The flight distance for approaching humans depends to a large degree with what the animal has experienced from human hands in the past. For example, mustangs in central Nevada are "run" from horseback, they will flee from a mounted rider at distances of over half a mile, however a rider leading his horse can approach within a couple hundred yards before the herd gets uneasy.

Cattle and sheep will generally graze apart, but cattle and horses will often be together. Sheep will often use the same area but at different times. Horses will often allow sheep to take supplementary feed from between their legs, but cattle will not be allowed this close. Both calves, cows and horses will kick at lambs that are in their way, adult sheep will generally move aside. Horses will often run cattle for the pure joy of doing it. The horse is clearly the dominant, then cattle, then sheep. There have been occasions where, with no assistance from man, an animal of one species will adopt, or attempt to adopt, a young of another species by keeping the natural mother away. In Australia a Holstein cow adopted and successfully raised a Merino lamb.

---

[3] The difference between a cowboy and a stockman is one of maturity.

SEXUAL:

Nothing so stimulates a bull as the sight of another bull at work. Yet, a power far greater than mere sexual excitement makes young bulls stand away from cows in heat, kicking dirt with their hooves. That power is the dominance of an older bull. When a young bull joins the herd the older bulls quickly put him in his "place" by lowering their heads threateningly, head butting, or out and out fights. Once the young bull has found his rung on the social ladder there are seldom any more fights. However, like young everywhere, their obedience last only as long as the "old man" keeps an eye on them. While the "old man" is off at work, or even when his back is turned, the young will swoop in and serve other cows with amazing efficiency. The young will mount once and serve instantly, while the "old man" may fumble around and only serve once in 20 mounts, while trying to keep the young at bay.

The effect of age in the bull battery was graphically illustrated in observations of a herd with four bulls in South Africa. The four bulls were of different ages; bull A was 10, B was 4, C was 3 and D was 2 years of age when first grouped together. That first year A sired over 70% of the calves, B about 16%, C about 7% and D about 5%. The second year the results were about the same, however the third year things started to change. The old man (A) went over the hill and only sired about 12% of the calves as did C and D. Bull B was now dominant and sired more than 63% of all calves. Bull A was removed and in the fourth year bull B sired 72%, C about 12% and D, 15%. In the final year bull B (age 8) only sired 25%, C (age 7) became dominant and sired 62% and D (age 6) sired 12%.

If there are relatively few cows in heat at any one time, the dominance of one bull over the others will affect the overall fertility of the herd in so far as grouping the calves are concerned. Also if that bull is sterile or sub-fertile a lower calf crop is the result, even though the other bulls are fertile. For best results bulls of similar age should be used in each herd.

The dominance of a bull within a particular breed is primarily determined by age or seniority, followed by size, strength, agility and of course horns. There is a difference between breeds, Brahman blooded bulls are timid when working with Herford or Shorthorn bulls; other breed factors have been mentioned in the previous section. Sometimes a cow will seek out a bull, but usually this is the dominant bull. Also bulls generally do not play favorites when serving heifers. However with cows there is a tendency to spend some time "courting"; if there are just a few cows in heat, courtship may begin 48 hours before onset of estrus. Courtship consist of the bull laying his chin on the rump and back of the cow, sniffing and tasting the cow's urine and giving the flehmen response by curling the upper lip with the head lifted and neck stretched out. (The flehmen response is sometimes given when meeting other bulls.) This courtship has been found to be of some practical value. Cows that were run with "Gomers" (vasectomized bulls or bulls made sterile or non-functional through various methods) prior to the breeding season were found to have a significantly larger percentage of calves dropped at the beginning of the breeding season than in a control group; the overall percent calf crop was also slightly higher.

The heifer will come in heat as early as 6 months for some dairy breeds, but most beef breeds will reach puberty between 8-12 months, or longer depending on nutrition. The cows estrus cycle is between 18-23 days and varies from 6-23 hours. On the day of estrus the cow will be much more aggressive with many head-to-head butting encounters, will be walking more than usual and eating and resting less. During estrus the cow will typically join up with one or more other cows in estrus plus other non-pregnant females to make a sexually active group. These groups will seldom be larger than six head and spend their time riding, being ridden or trying not to be ridden. These groups serve as a visual signal to attract bulls to cows in estrus. When a bull joins the group, female mounting usually stops, however the estrus females does not usually stand for the bull until further courtship has taken place, brief as it is.

Animals that are exposed to high temperatures and/or solar radiation will pick up more heat than they can get rid of during the day, so their body temperatures rise. Generally they will lose this heat during the night. However, if the rectal temperature of cattle rises above 102.2 degrees Fahrenheit, conception rates will go to near zero. Those thinking about breeding during very hot weather, should keep this in mind.

REPRODUCTIVE AND MATERNAL:

Sheep and horses are "follower" species whereas deer, goats and cattle are "lying out" species. What this means is that new born lambs will follow anything that moves, but new born calves generally won't. Cattle and goats, as birth approaches, tend to isolate themselves and "hide" the calf or kid. Goats more than cows, will return to the spot where they gave birth and expect the offspring to be there. Sheep tend to locate a birthing area and give birth there, the flock generally moves on and leaves them there. The ewe identifies with that area, and if she has a difficult labor and wanders off, or is disturbed and gives birth elsewhere, she will return to the original area to look for her lamb. The lamb may or may not follow. It takes about three days for the follower instinct to become fixed on it's mother (lamb to ewe bond). However, mother to offspring bonding, requires relatively little time to complete, generally a minimum of four hours for sheep. Cows will eat the afterbirth and this assists in bonding with the calf, they also lick the calf clean, bonding with cattle often requires but a few minutes. The time it takes a calf to first suck varies from a few minutes to several hours, the average for both temperate and tropical areas is around 110 minutes. Calves, whose dams have large pendulous teats, will require a longer time before completing the first suck. Beef calves will take the first suck faster than dairy calves. The average time for foals to take the first suck is about the same as for beef calves; lambs average about one hour.

Feeding pregnant heifers in the evening, rather than in the morning as is commonly done, has been found to result in 17% more daytime births. The removal of calves from mothers 24 or 48 hours prior to turning the bulls in is of questionable value. Several studies have failed to notice any significant improvement in percent conception on cows in normal flesh and only minor improvement in cows in thin flesh. Often overlooked is the effect on the calf, a study in Louisiana found that calves that had been removed were over 6% lighter at weaning than those that had not been removed from their mothers. For a 450 lb calf that's a 28.5 lb difference at weaning. Apparently the calves were never able to compensate for the distress of the temporary early weaning. Flushing cows, especially thinner than normal cows has been found to increase conception rates.

There is always a concern about death's associated with birthing and there are all sorts of numbers that have been reported; most of them bad. One study conducted in Colorado, on a commercial ranch involving 7500 females, recorded the percent of calves born dead and also those that died during the first 90 days, for heifers calving as two's and three's and four year old cows. As might be expected the heifers calving as two's had the highest loss, 5.4% born dead and 8.9% dead during the first 90 days; those calving as three's, 2.5% born dead and 5.8% later, those calving as four's, 1.8% born dead and 4.2% during the next 90 days. The largest cause of death at birth in all groups was dystocia (difficult birth); the largest cause of death after birth in all groups was scours followed by injury.

HERDING:

Animals have what is termed a balance point. For cattle this point is just behind the front shoulder, each side, and at the center of the head. In a corral you will generally have to penetrate the animals "flight space", this distance depends on how much handling the animal has had and how much trust she has in the handler; in a feed yard animal this distance is small, in a "sookie" right off the range, pretty big. If you slowly approach an animal head on, she will tend to back up or whirl, however if you approach head-on, but slightly to one side she will move to the other side. If the animal is broadside to you and you slowly approach (right angles to her body), but slightly in front of her front shoulder, she will go backwards, if you approach behind her front shoulder she will go forward, directly at her balance point and she will either whirl and go directly away from you, or go forward or back. The maximum impact is when approaching at an angle about 45 degrees off of the balance point. Animals that have eyes on the sides of their head can see all around them when grazing, if the forage is short. However, with the head raised there is a blind spot directly behind them and down the muzzle. Consequently you can not drive animals by positioning yourself directly behind them. They can't see you back there, so they keep turning their heads around, and as they do, they tend to move in that direction.[4] To drive animals directly forward you must position yourself about 20 degrees off to one side and behind them. This positioning with reference to the animal has its greatest application under range or large pasture conditions. Along fence lines, roadways, small paddocks or corrals, less knowledge of balance points is required as the choices the animal has is limited.

These same principal applies when moving a herd. However the balance point is not as well defined for a herd as for an individual; as it depends on how strung out the herd is. Also important is how close you are when they first spot you, much more than a quarter mile and they will do whatever; closer and you can start to bend them in the desired direction. Also, if you're coming at them "hell bent for election" they'll scatter like sheep that just got jumped by a mountain lion. Just ease in on them and position yourself so a straight line is made between you, the group and where you're trying to take them. Once they are going in the direction you want, stay behind them, but similar to an individual animal, off to one side by 20 or so degrees. British cattle are easier to herd (push) than are Brahma's. With Brahma's it is best to wait a bit and allow leader's to get into position before crowding; Brahma's, because of their strong herd instinct (social facilitation), will pull better than push. In this respect they are similar to sheep, when one goes, all go.

When moving from one side to another, do so quickly but not on a high trot or gallop. Individual animals will vary some and of course the terrain will play a big role as to which way they want to go. Animals generally prefer to run on level ground and will usually go down hill rather than up; they also prefer known trails to making new ones. However, all bets are off with those wise old wild ones. They generally have some special place where they got away from riders once before and generally that's where they go. All of the above are based on a calm, quiet slow moving herd. If they're on a high trot, they spend more time looking where they're going, consequently you have to be slightly in front and closer in. If they are in a flat out run, best to let them run it out, then move in when they're winded.

---

[4] Much like automobile drivers often do when they look out the side window a little too long .

Under typical range conditions one rider can handle 25 head, two riders 200 or more and three riders should be able to handle any reasonable size herd. These numbers can go much higher if the animals have a trail to follow, or know where they're going. In the not too distant past it was not uncommon to see one or two riders trailing herds of well over 400 head to and from winter and summer pastures. Too many riders, especially the rodeo cowboy types, are counter-productive; they cause confusion, break equipment, waste time and run off a lot of pounds. Dogs, unless they are in absolute control, do more damage than good; this is especially so where cows and calves are being moved. In heavy bush country a lot of riders and dogs may be justified, in open country they are an economic liability.

ENVIRONMENTAL:

Most animals experience little to no discomfort at air temperatures between 59 to 77 degrees F. At low temperatures (freezing) Holsteins require about 14% less energy than do Brahma's; at high temperatures Brahma's are under less stress. Low temperatures will cause intake requirements to increase up to 30%; however if the animal is wet, intake will be depressed by up to a third. High temperatures and humidity will depress intakes from 20-40%; but if the humidity is low the loss will be less. High humidity and high temperatures cause animals to rest more; sheep, with no shade, will spend more time standing in tight groups. On cold days animals will orient their bodies so that they are at right angles to the sun, but with moderate wind, horses will turn downwind and cattle upwind. Wind chill affects the animals energy requirements and may require considerable energy just to maintain body temperature. The condition of the animal and hair coat also affect the energy requirements due to high or low temperatures. Thin animals do not have the insulation that fatter ones have, consequently require more energy. Animals with sleek hair coats can withstand high temperatures better than rough coated ones; but sleek hair coats in the winter will result in a greater heat loss.

Animal color plays a significant role in how animals are able to adapt to their environments. The heat flow on a hot sunny day will be 30% greater into a black steer than into a white one. In pastoral areas of Kenya it was found that as the heat stress decreased (from the arid low lands to the cooler, wetter highlands) there was a change in the predominant coat color; more cattle had lighter coat colors in the hot arid low lands and there were more darker cattle in the higher wetter country. A study conducted during the 1974 drought in Kenya found that at an elevation of 4500 feet dark cattle absorbed more solar radiation, drank more water, lost less weight during the drought and gained it back faster than light colored cattle. During the drought more light colored cattle died at this elevation than dark colored ones. However, the reverse was true at lower elevations in a more arid environment. It would appear that light colored cattle are better able to exist in hot arid environments than dark colored ones. However, at altitude, where the nights can get cold, the dark colored cattle apparently regained the heat lost during the night faster than the light ones, this lessens the stress on them and resulted in better performance.

Seeking shelter from the weather is a natural behavioral trait in animals. Sheep in Australia had greater wool weights and heavier lambs when provided with 1 meter high windbreaks than without. Cattle in gulf states were found to gain less weight when provided with artificial shelter than those with no shelter (they didn't go out and graze in bad weather). However, dairy and feed lot animals performed better with shelter than without. Also getting out of a cold wet, wind is one thing, getting out of the sun, another. Animals prefer natural (vegetational) shelter to artificial shelters. A study on effect of shade and type floor in feed yards in Florida, using crossbred heifers, found that the average daily gains for those in a dirt pen with natural shade was highest (1.74

lbs/day), compared to dirt pen with artificial shade (1.41 lbs/day) and concrete pen with artificial shade (1.21 lbs/day).

Pastures should have a combination of brush and trees, but the canopy cover should be no more than 30% of the total pasture area. One or two trees per acre, or an artificial shelter will result in accumulations of dung and urine under the shelter (nutrient transfer) and result in a lower performance under stormy conditions. It would appear that for shelter to be effective it must resemble natural shelter in that it should be spread out and a lot of it. If something resembling a savanna appearance can not be maintained, it is best (from an economic and production point of view) to have no shelter, except under extreme conditions of heat, cold and wet.

STRESS:

Stress is what is called a continuum, like going from slow to fast, or in the case of stress, from boredom to distress. Too much, or too little stress is counter productive; a certain level of stress is required for normal development. Too little stress (boredom) can often be found in horses that are kept in stalls, swaying or wood chewing; shipping fever is one example of distress. Behavioral signs of stress are often displacement activities, such as when animals in a middle of a fight, stop to graze.

Virtually anything can become stressful if continued too long or if the level is high enough. Stress interferes with the normal secretion of hormones, this in turn results in a lowering of the bodies defense system to other diseases. The point at which a stimulation or situation becomes stressful is, to a large extent, determined by how the individual perceives the situation. A calf running across a pasture in play may expend 100 Kcals of energy; however, if a cowboy is making that calf run the situation is now stressful, even though the energy expended by the calf may be the same.

Anytime an animal experiences anything new there is stress. The first time an animal enters a corral or chute, it is stressed. If no harm comes to the animal, the next time it enters the same corral, there is less stress and so on. However, if every time the animal enters the corral it gets hurt, very soon the mere sight of the corral will cause the animal to show signs of distress. Don't get the mistaken idea that because you penned them up to "help" them (like doctoring sick ones) that the animals know and appreciate this. As far as they're concerned, you hurt them, why you did it has no meaning for them.[5] Animals handled often will be less stressed by handling than those not handled as frequently, but only if the handling is gentle.

Animals can usually handle one stress factor for a limited time without a large drop in production; however, serious problems can arise when these factors start piling up on top of one another. Everything about being corralled and worked is a stress factor; the act of corralling, sorting, being driven into and out of chutes, being caught, being treated, the noise, the dust, the crowding, the lack of water and feed, the disruption of the daily routine, plus any hot shot or clubbing the animal might receive from over zealous cowboys. It's no wonder that gains may take two to three weeks to get back to normal after a day in the sweat shop.

To keep the stress levels within tolerable bounds the best solution is to train the animals so that the surroundings or certain non-painful experiences become "routine"; much the same way dairy cattle are trained to enter the milking parlor. This can be done

[5] Let's face it, from their point of view, the only time you have ever been nice to them is when you fed them; the rest of the time you've been down right mean.

by letting the animals explore the area on their own, or using the corral for water and/or minerals, or to "ease" them through the facilities once or twice a month. The idea is to get them used to the corral area, so at least the corral by itself is not a stress factor; if you can't make it a pleasant experience, at least don't make it an unpleasant one.

How many stress factors you can place on a animal before you do serious damage is one of those great unknowns. It has been argued that it's best to do all livestock work at one time and get it over with, rather than make a special roundup for worming, another for weaning and so on. Considering the way most outfits round up and work animals, it's probably better to do everything all at once. On the other hand, if you can eliminate or reduce some of the stress factors through training, proper stockmanship and facilities, then it's better to do only a few jobs at any one time. It is also best to do the really severe jobs such as castration, de-horning and branding when the animal is only a month or so old.

One of the characteristics of IGM is that the behavior of the livestock will change. Because you are offering fresh feed on a regular basis, the animals soon lose their great fear of you; they don't necessarily like you, but at least they don't climb trees to get away from you. At moderate levels of cattle density (35-70 head/acre), their pattern of behavior becomes more herd like and less of an individual thing, however at high densities, or where water and mineral facilities are plentiful and close by, certain otherwise social functions become individual again.

Herds in IGM are often composed of a great many individuals, which in turn split up into sub-herds. There has been some concern express as to the possible negative effects of the size of these herds. In dairy herds there are several examples of an increase in production (around 10%) when herds of over 100 head were split into two sub-herds. However, in each instance there were a number of other factors involved that were causing the problems, such as poor milking management, poor husbandry, long periods on concrete, etc. Anytime you increase the number of animals being worked or handled, there is also an added risk factor that something will go wrong; worker boredom or fatigue, failure to pay attention to details, etc. Remember, when the amount of energy going through a system increases, the amount of control must also increase.

FORAGING:

Cattle will generally begin the day at sun-up, almost all members of the herd will be grazing for about 2 to 3 hours. After this first big graze period, there may be a second big graze period occurring between 3 and 6 PM, or there may not be, depending upon environmental conditions. If the average day temperature is under 59 degrees F, there is little night grazing, if it is over 77 degrees F there is usually one or more night grazing periods. However, wind, moisture and humidity all have some effect. The total number of grazing intervals during a 24 hour day will average about 5, but again this number is subject to wide variations as the total amount of available forage in the pasture will also determine how often animals graze. Generally 8 hours is spent grazing for cattle and 7 hours for sheep; one grazing study on adjacent pastures in Texas found that cattle spent 7.7 hr, sheep 6.6 hr and goats 5.9. In Australia horses were found to graze 2 hr longer than cattle in the winter and 6.5 hours longer in the summer; a horse will graze between 12 to 16 hours a day depending upon feed supply. Horses will usually spend more time grazing at night than cattle or sheep.

Studies in Hawaii conducted on an intensive grazing unit during the spring, found that growing steers have one graze period during the night. The morning graze began at sun-up and the major afternoon period commenced about 3 PM. However there were always some cattle found to be grazing at all hours. In this particular herd of 560 head on 338 acres (divided into 28 paddocks), it was noted that sub-herds were formed. These sub-herds were very fluid, that is individuals joined and quit almost at will, but there did appear to be some permanent members. The size of these sub-herds ranged between 40 to 60 individuals. Watering generally occurred after the morning graze period, with each sub-herd going to water as a more or less distinct unit.

The quality of the diet in the morning is lower than in the afternoon. This is because in the PM the animal is apparently not as hungry and is consequently more picky, doing a better job of selecting. It has also been found that feeding a supplement in the PM gives better results than feeding it in the AM. The reason seems to be that feeding in the AM interrupts the morning graze period, and the supplement becomes a forage substitute. The amount of forage consumed decrease if more than a little supplement is fed; one study found that feeding 8 lbs of oats to steers reduced grazing time by 2 hours. Another trial in Colorado found that feeding small amounts of supplement on good pasture, 1/4 lb of barley per day, resulted in a one to one conversion of the supplement; a similar effect was noted in Australia with 2 lbs of grain on poor quality, but ample amounts of forage. In general the amount of supplement fed should be less than 0.5% of the liveweight of the animal; more than this it becomes a forage substitute.

It has been observed on a wagon wheel IGM unit in Hawaii that the mineral consumption of growing heifers dramatically increase when a portable mineral wagon was moved with the animals and placed about half way along the length of the paddock. This mineral wagon was in addition to the minerals located at the hub. It was surmised that the sub-herds moved as a unit to the center, where the dominant animals drank, then consumed mineral. When the dominants had finished they left, and as they were the leaders, the less dominant animals left with them. The time the sub-herd stayed at the hub was short, and the less dominant did not have time to drink and eat mineral both. However, when the mineral was in the paddock as well as the hub, mineral was available to both dominant and timid animals at their respective leisure. This same increase in consumption was also noted on another IGM wheel unit with mature cows when a portable mineral feeder was added.

Forage intake is dependent upon quality, quantity, the length of the forage being grazed, the mass and the density of the forage. The higher the quality (protein and digestible energy), the more forage is eaten. In general if protein levels fall below 12%, or digestibility much below 60%, forage intake declines. The relation between intake and quality has to do with the rate of passage of material through the rumen (turnover rate), for example stems will remain in the rumen about one third longer than will leaves.

There are three aspects of grazing that animals use to maintain a constant rate of intake; size of bite, number of bites and length of grazing, Figure 6.2 shows this relationship for sheep. The time spent getting a mouthful decreases with increasing forage height. With grass over 3 inches in height (about 3.5 inches for mature cattle), the rate of intake remains the same even with heights of 14 inches. What does change is the number of bites taken and the size of each bite. The denser and/or taller the forage the larger the size bite and the fewer bites taken. For pastures yielding more than 1200 lbs per acre of similar quality, the rate of intake will remain fairly constant. Because of the structure of the mature ruminants jaw (no upper incisors), forage less than 1/2 inch in height for cows and 1/3 inch for sheep can not be eaten, in actual fact forage less than one inch in height is seldom consumed by mature animals, except during drought or severe overstocking.

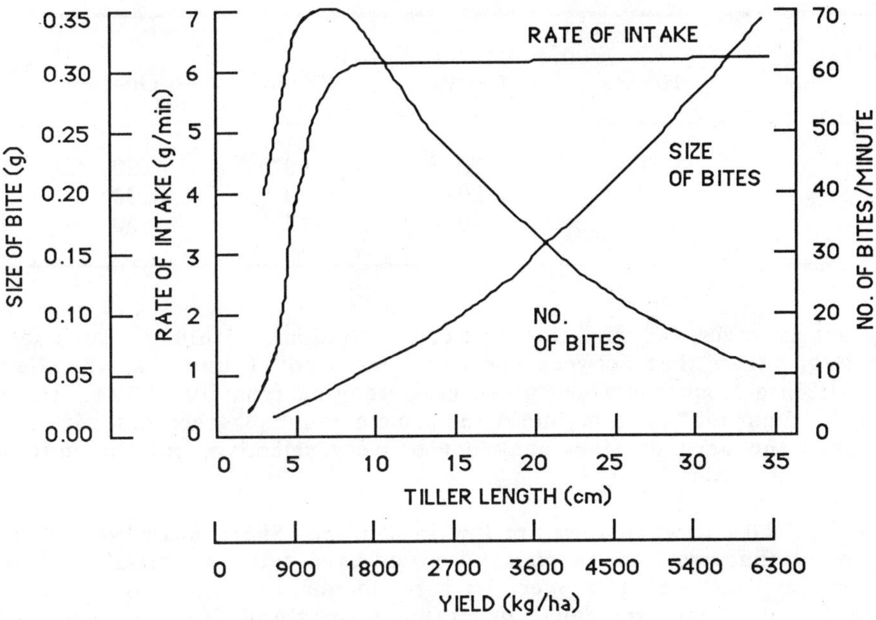

FIGURE 6.2. Relationship between tiller length, number of bites, size of bites and the resultant intake of grazing sheep. (From Arnold and Dudzinski)

Prior experience also determines how much an animal will eat during a day. In one study, sheep with no prior grazing experience had much lower rates of intake than sheep that had been grazing pasture since birth; this difference lasted for 2 years. Animals also learn from their mothers and herd mates as to what is good and what is not. For many foods the animals must be given time to adapt to it. Greenleaf desmodium has many hairs, both on the leaves and stems, inexperienced cattle will not eat this plant unless forced to it; however, once they learn, greenleaf becomes a preferred forage. Inexperienced cattle placed in a new and strange environment will often eat lethal amounts of toxic plants; halogeton and greasewood are two plants that cause this sort of problem in the Intermountain West.

The diet of young calves is different than that of their dams, both as to species selected and plant parts selected. This is due, in part, to the smaller mouth parts of the calf. However, once mature, selection appears to remain the same for any individual animal regardless of age or physiological state. The only difference appears to be that the bigger they are the more they eat, also if they are milking or growing wool, the more they eat. Diet selection however, does not appear to change. There is a definite difference between species of animals in diet selection, but little to none between breeds. Brahman type cattle will do better on a lower quality diet than british breeds, but both do equally well on high quality diets. Different species have different diet preferences, Table 6.1 shows the preference for four species of livestock.

TABLE 6.1. DIET SELECTION OF FOUR CLASSES OF RANGE LIVESTOCK

Type of Plant	percent of diet			
	Horses	Cattle	Sheep	Goats
Grass	90	70	60	20
Weeds	4	20	30	20
Browse	6	10	10	60

The above are averages, the variation between individuals within the same species, may actually be larger than that between species. One study found that the diets of individual sheep within a flock contained green grass ranging from 10 to 70%. However the same individual animal tended to maintain the same amount of green grass in it's diet day after day. Thus the herd or flock may not be very selective, but the individual animal is.

Animals will select salt when they are low in sodium. Sheep have been shown to select parts of a plant that are high in phosphorus, as have cattle. Animals will select green material over dry, and leaf tips over the base portion. Cattle tend to select for sugar content, but not all sugars are equal, generally hexoses and fructose over sucrose, however calves prefer sucrose. Generally animals avoid forages high in tannins and alkaloids. Many mammals will often avoid a food that has made them sick or has resulted in a serious deficiency in the past; how widespread this ability is in livestock is questionable. However, "nutritional wisdom" (the ability to select a diet containing the proper amounts of required nutrients) in domestic animals is probably no longer there, simply because the domestic animal is no longer closely tied to it's original environment. Wild animals in their natural environment no doubt have a certain amount of "nutritional wisdom", as their ability to select a healthy diet has a great deal to do with how well they survive.

The amount of water drunk is a function of temperature and percent moisture in the forage. Most beef cows will require 8 to 10 gallons per day, more if they are milking heavy or pregnant and if the temperature is high. Cattle on high protein diets will require more (up to 26% more) water than if on a low protein diet. Milking dairy cows may require up to 30 gallons per day. Sheep average around one gallon per day, however it can go as high as three gallons; again depending on diet (high salt) and temperature. In intensive environments cattle will drink an average of four times a day, with an individual range of one to six. On the open range cattle may only come to water once a day, or even once every two days in cool weather, depending on distance traveled. However, once at water, they will remain for several hours and drink several times. Sheep drink most of their water before 8:30 AM (between 64-73% of the flock), while cattle drink the biggest share after the first morning graze period; more on water a little farther on.

NUTRITION:

Just about everyone has seen, at one time or another, a table similar to Table 6.2. The traits shown in this Table, are some of the more important animal traits from an economic viewpoint. Genetics, or having the right genes, is important. If you have only scrub genes, then all you can hope for is scrub cattle. While many of the above genetic traits are additive, many are not. Fertility and mothering ability go together, that is you can select for both. But fertility and carcass grade have no additive effect, and in fact may even be slightly negative. In other words if you select for carcass grade you may be slightly decreasing herd fertility; consider the reproductive problems associated with doubled muscled animals.

TABLE 6.2. HERITABILITY OF DIFFERENT ECONOMIC TRAITS IN CATTLE

Economic trait	Heritability (%)	Environment (%)
Fertility (calving interval)	10	90
Conformation score (weaning)	25	75
Weaning weight; Pasture gain	30	70
Birth weight; Mothering ability; Efficiency of gain; Carcass grade; Conformation score (at slaughter).	40	60
Feed lot gain	45	55

Many breeders start out a breeding program with a whole shopping list full of traits. The problem is the more traits you select for, even if they are compatible, the slower progress will be made on any one trait. The pitfalls of a selection program are almost endless. This is why records are important as well as having well thought out goals and objectives.

The item that up until recently has not been emphasized, other than in a feed yard sort of way, has been the environment the animal or those genes have been living in. As a casual glance at Table 6.2 will show, the biggest effect on all of the traits is the environment. Providing the proper environment for those genes to "do their thing" is by far the most important item that can be done to improve herd performance. Proper environment means, good handling procedures and facilities, shelter and most importantly, good nutrition.[6] Without adequate nutrition all the genetics in the world won't help. It makes little sense to pay $5,000 for a bull to get a 10% improvement in fertility, then throw him into an environment that can't even meet your present herds nutritional needs.

The ruminant animal is very good at what it does. It can take a low quality feed (forage) and convert it to meat and milk. It can also take a high quality feed (grain) and convert that to meat and milk. But it doesn't do that near as well as non-ruminant animals can. It takes 7 to 8 pounds of feed grain to obtain one pound of steer, 4 pounds to make a pound of pig, 2 pounds to make one pound of chicken and catfish are almost one to one. Depending on quality, 15 pounds of grass can produce anywhere from a minus weight gain to over 3 pounds of beef; you can't produce very much pig, chicken or catfish from grass. In cattle, grain is being used to mask the effects of sloppy animal

[6] You can have the best "designer jeans" in the world, but if they are not put on a well fed frame, "all's" you got is another pair of baggy pants.

husbandry and management, just as sloshing on fertilizer was used to hide pasture mismanagement in the not too distant past.

Animals require water, energy, proteins, oxygen and an assortment of minerals and vitamins in order to live (maintenance) and produce. If any one of the above is absent the animal will either die or, at best, barely maintain itself. Consequently, there is no one more important than the other. There are some however, that are used in larger amounts than others.

MINERALS:

If any of the necessary minerals are in short supply, the animal will not be able to fulfill it's genetic potential, in short, production will suffer. Minerals required for cattle and sheep production are broken down into two groups; those required in large amounts (major elements) and those required in much smaller amounts (trace elements). Table 6.3. list the essential minerals and the average percent that each should be found in the diet. A more detailed description of each mineral and suggested levels for performance may be found in Appendix 2 for suggested intake levels for major nutrients and Appendix 3 for trace minerals.

Table 6.3. The Essential Minerals.

MAJOR ELEMENTS:	Symbol	percent of DM intake sheep	cattle
Calcium(a,b)	Ca	0.29	0.44
Chlorine	Cl	0.10	0.24
Magnesium(a)	Mg	0.12	0.19
Phosphorus(b)	P	0.20	0.32
Potassium(a)	K	0.36	0.58
Sodium	Na	0.09	0.12
Sulphur*	S	0.15	0.18
TRACE ELEMENTS: parts per million (ppm) of DM diet intake			
Cobalt	Co	0.11	0.09
Copper*	Cu	5.00	7.00-10.00
Iron*	Fe	30.00	40.00
Iodine	I	0.50	0.50
Manganese	Mn	25.00	25.00
Molybdenum*	Mo		
Selenium	Se	0.03	0.03
Zinc	Zn	25.00	25.00

(a) Ca, Mg and K are all involved in grass tetany, if pasture K levels are high and Mg levels low, there may be a potential problem. (b) Ca and P are inter-related, excess Ca interferes with the absorption of P. The Ca/P ratio should lie between 1:1 to 2:1 * Cu, Mo, S, and Fe are all inter-related. For every part of Mo, one more part of Cu is required, if the pasture has high levels of S and/or Fe, higher levels of Cu are required.

SALT:

Before pushing on a word on salt might be in order. It has been claimed by some that supplementation of salt is not needed, while others claim that every adult cow must consume 2 ounces a day or more. All animals require salt; the size, rate of gain and if milking or not determines how much salt or sodium is required. Appendix 2 gives the salt requirements of cattle and sheep. A quick look at that table shows that a 838 lb dairy cow giving 66 lbs of milk daily would require 21.3 grams or 0.75 ounces each day. To satisfy this requirement the forage would have to contain 0.125% sodium. For an average milking beef animal the forage would have to contain 0.11% and if dry 0.10%, younger growing animals would have ample sodium at a forage level of 0.07%.

As mentioned in Chapter 5, the sodium content of most "green" temperate grasses and many temperate legumes are sufficient. Perennial rye grass, orchard grass, white clover and sub clover usually contain sufficient sodium; tall fescue, timothy, alfalfa and red clover often do not. However, the warm season or tropical forages typically contain marginal or low amounts, for example kikuyu and Dallas grass (paspalum) often are in the 0.02-0.06% range. Also important to remember is that once the plant has matured, gone to seed, or dried up and turned brown, sodium is rapidly leached out, in other words, generally the forage is no longer capable of supplying the sodium requirements of livestock. How do you tell for sure? Get a forage analysis.

On the other hand the amount of salt that must be consumed before levels become toxic is quite large; if ample water is available. The 1980 National Research Council committee on mineral tolerances of domestic livestock gave the following as the maximum tolerable levels of salt in the diet: 4% for lactating cows; 9% for non-lactating cattle and sheep; 8% for swine; 2% for poultry; 3% for horses and rabbits. In Texas nearly three pounds of salt were fed to cattle daily without any harmful side effects, other studies have shown no harmful effects on pregnant animals and their young, or on rates of gain by feeding large amounts of salt (lots of good clean water was available at all times).

If the forage is green, more often than not cattle eat salt because they like its taste, rather than because they need it. Much the same way some of us eat ice cream, not because we need it, but because we like it. The question of whether or not to put out salt when the animals are on green forage, in the absence of a forage analysis, is one more of economics and personal conviction rather than need. However, when the forage is brown or dead, salt had best be out there, unless you know for a fact that the forage contains sufficient sodium.

VITAMINS:

There are 15 vitamins that have been identified thus far. A distinction should be made between dietary and metabolic requirements. Some animal species are able to make vitamins; for instance the rumen flora of cattle, sheep and goats are able to make the B and C vitamins. Consequently these vitamins may only be required during the early stages of a young ruminants life, prior to good rumen activity. Other species must have the vitamin in the diet, for example man, apes and guinea pigs have a dietary requirement for vitamin C. Only a very brief description will be given here of a few of the major vitamins.

Vitamin A is essential for the normal maintenance of eyes and epithelial tissues (mucous membranes). It has also been implicated in retained placenta's and birth of dead, uncoordinated, or blind animals. Vitamin A is found in green feeds in the form of carotene which is then converted to the vitamin; for cattle and sheep one milligram (mg) of carotene is equivalent to 400 International Units (IU) of vitamin A.

The B-complex vitamins is a group of vitamins that at one time were all lumped together under vitamin B. In cattle, sheep and goats once the rumen is functioning there is no need for this complex as the rumen flora manufactures it's own. Cobalt is a main ingredient of B 12 (which is made by several rumen organisms); if cobalt is in short supply, not only does it affect production of B 12, but it also messes up the relative numbers of the different types of rumen microbes. The main concern with the B vitamins is with mono-gastric animals, chickens, swine, man, etc. Of interest is that thiamine (vitamin B1) is also manufactured by higher plants.

Vitamin C, or ascorbic acid, is required by all animals, however most can manufacture their own. There is some evidence that supplements to swine diets is beneficial.

Vitamin D is involved with the metabolism of calcium and phosphorus. In young growing animals a shortage of this vitamin can lead to rickets. The clinical signs are; thickening and swelling of the joints (particularly the knee and hock), stiffness and humped backs. The pre-cursor of vitamin D is manufactured by the skin of most animals on exposure to sunlight. This pre-cursor flows onto the hair where animals pick it up either by self grooming (licking) or grooming others.[7] Most green feeds have ample supplies of vitamin D, as does good hays and silages. Whereas grains and grain by-products are practically devoid of vitamin D. Feeding large doses to animals near the end of gestation has been shown to be helpful in controlling milk fever, as long as adequate amounts of calcium and phosphorus are also present. Shortages of this vitamin have also been implicated in cases of grass tetany.

Vitamin E is an antioxidant and facilitates the absorption and storage of vitamin D. A deficiency of vitamin E shows up as a generalized loss of muscle coordination (white muscle disease). Animals walk with a crossing-over of the hind legs. Muscles of the tongue are affected which impairs the young animal's ability to suck. Selenium and vitamin E are inter-related and some problems due to a shortage of vitamin E can be corrected by additions of selenium in the diet. This vitamin is found in green feeds and whole cereal grains.

Vitamin K is necessary in the diet to preserve and enhance the clotting power of blood. Under normal situations the vitamin K requirement is met by natural foodstuffs and the production of the vitamin by intestinal microflora of swine and in the rumen of ruminants. Moldy feeds and some legumes, especially clovers, have been shown to increase the clotting time of blood and result in generalized hemorrhaging. This condition can be corrected with vitamin K supplements or injection.

OXYGEN:

Normally oxygen is not considered a problem, and usually it's not. There are some situations, especially in trucking or transportation, where a shortage of oxygen could be a problem if there is not proper ventilation. Another possibility is the hauling of animals from a low elevation to a high elevation and then working them. They should be given ample time (10 days to 2 weeks) to "condition" themselves to their new environment prior to any hassling. Dusty feedyards, pneumonia, brisket disease, and anything that affects

[7] This appears to be a case where a necessary biological substance has given rise to a whole set of social interactions.

the lungs, reduces the amount of oxygen available to the animal. In these situations the animal should be handled with TLC (tender loving care) to avoid additional stress.

WATER:

There seems to be considerable misunderstanding concerning the role of water in the diet of the grazing ruminant and nowhere is this misunderstanding greater than in the water contained in forages. The amount of water contained in a forage will affect how much additional "free" water the animal will drink. If the forage has an exceptionally high water content of 90%, an animal that eats 20 lbs of dry matter also ingest's 180 pounds, or 22.5 gallons of water. Consequently, if the feed is lush (water content around 85%) the animal may not need to drink much additional water. In New Zealand, during the winter, it is common to graze animals on lush forage and only provide drinking water every 3 to 7 days, this does not appear to lower performance all that much.

Young lush forage has been reported as causing a lowering of intake, this is possibly due to acidosis resulting from a high concentration of one or more chemical substances found in the young plant. The confusion arises because lush forages also have high water contents. However, unless the moisture content starts getting up around the 90 to 95% level (such as aquatic plants like water hyacinth) water does not cause the lowering of intake. The very nature of a ruminant animal requires that it be able to intake, eliminate and re-cycle large amounts of water every day. In an average mature cow the salivary glands will secrete around 40.5 gallons (150 liters) per day, in addition the animal will consume 10 or so gallons of water (milking dairy cattle as high as 30 gallons) each day, all this ends up in the rumen.

Ruminants have a compound stomach consisting of four parts, the rumen, reticulum, omasum and the abomasum. The total volume of all four in a mature cow is around 52 gallons with the rumen accounting for about 40 gallons. The reticulo-rumen portion functions as a fermentation vat in which the bacteria do their thing, but in order for them to function properly the feed (digesta) must be very fluid.[8] The main job of the omasum is to remove the majority of this water in the digesta coming from the reticulo-rumen. Later, down the line, the colon (large intestine) also removes excess water. These two organs are largely responsible for maintaining the water balance of the animal; take it out of the digesta, put it back into the blood so that more saliva can be made so that it can enter the rumen with the forage, so that - etc. The omasum and colon are very good at what they do.

An average cow on pasture (water content around 80%) will void in urine and feces between 7 to 9 gallons of water each day. Water is also lost by respiration and evaporation from the skin, this varies depending on temperature, breed and activity (somewhere around 4-6 gallons per day at a temperature of 65 degrees F and about 6-8 gallons at a temperature of 86 degrees F for a resting Holstein, about half this for a Zebu). If the animal is milking, water is also removed in the milk. The average number of urinations for a normal cow is 9 per day; the more water the animal takes in, the greater the number.

[8] The whole digestion system is based on the continuous production principal and not on the "batch system", which is why the feeding principal "a little but often" gives the best results.

QUANTITY:

Young animals consume more water in proportion to their body weight than do mature, non-lactating animals of the same species; a 33 lb pig will require 5.5 lbs (pints) of water per day, or 17% of its body weight, while a 220 lb pig needs only about 10% of it's body weight in water per day. A 880 lb cow giving 44 lbs (5.5 gal) of milk per day may require over 11 gallons of water per day, whereas a 880 lb dry cow on a maintenance ration may only require a minimum of 4 gallons.

Animals obtain some water from their feed and may obtain most of it from a succulent feed source, as do the wild cattle in the wet mountains of Hawaii. Most green forages have a water content between 65 to 85%; stem dried forages may be around 25 to 35% moisture and hays near 15%. Non-lactating cattle in cool temperatures on lush forage may need to drink only every week or so. Dry sheep native to dry areas get along with watering once every two days, if allowed to remain around the watering point for several hours. However sheep deprived of drinking water for 96 hours showed a decrease in feed utilization. Goats native to tropical areas have a lower water turnover rate than sheep; as much as 11% lower at high temperatures. Because of this, they need to consume less free water than sheep to survive.

In hot temperatures cooling water has shown to increase performance, however the economics do not appear to justify the refrigeration cost. Burying water pipes under ground will keep temperatures down to that of the soil and works especially well if under shade. Pipes should be buried at least 6 inches deep, as the first few inches of the soil, if bare, can be much hotter than air.

Air temperature can have a great effect on the amount of water animals drink. On the average cattle will consume 50% more water during the summer than in the winter. Tables 6.4 and 6.5 show the effect of air temperature on daily water intake for cattle and sheep.

QUALITY:

For some reason almost everybody worries about having enough water, but few concern themselves with the quality of the water on offer; both are equally important. All water contains some dissolved nutrients. The average composition of surface water from 140 different stations in the Western U.S. over 12 years showed all the essential nutrients present; sulfates 136 ppm, calcium 57 ppm, sodium 55 ppm, chloride 478 ppm and magnesium 14 ppm, were present in the greatest amounts. The availability of elements in solution is believed to be at least equal to solid feeds or dry salt mixes. Some nutrients may be present in sufficient amounts to provide up to 25% of the daily recommended intake.

The salt content is one method used to express water quality. By salts, is meant all mineral salts, not just sodium chloride (table salt). Water containing less than 1000 ppm (parts per million) should present no serious problem to any livestock. Water containing up to 5000 ppm may cause temporary diarrhea in livestock not accustomed to it. Water from 5000 to 7000 ppm should not be used for pregnant or lactating animals. Water containing over 7000 ppm of dissolved salts should not be used. Hydroxides in water are more harmful than carbonates which are more harmful than bicarbonates. Animals can consume water of high salinity for a few days if they are then offered low saline water. Abrupt changes from low saline water to high saline water will cause more problems than a gradual change. Table 6.6. lists the safe limits of various toxic substances found in water.

TABLE 6.4. ESTIMATED DAILY WATER INTAKE OF CATTLE AND SHEEP IN GALLONS, AT DIFFERENT TEMPERATURES.

Body weight (lbs)	Air Temperature in Degrees F 50	59	68	86
beef cattle on maintenance ration:	Gallons per day*			
220	1.4	1.6	1.9	2.4
441	2.3	2.7	2.9	3.8
662	3.0	3.5	4.1	5.0
882	3.7	4.2	4.7	7.1
1103	4.4	5.0	6.0	9.2
fattening cattle on dry feed:				
441	4.4	5.1	5.9	8.2
662	6.0	6.8	8.0	11.0
882	7.4	8.4	9.8	13.7
1103	9.7	11.2	13.0	18.2

For a lactating cow take the maintenance amount and add the amount of milk produced daily (average English breeds give 1.5 gals/day, superior milkers 3.0 gals/day) to determine the total amount of water needed.
*The amount of water shown is the combined amount consumed plus the amount contained in the feed.

TABLE 6.5. TOTAL WATER REQUIREMENTS OF SHEEP IN GALLONS, AT DIFFERENT TEMPERATURES

Type of sheep	weight (lbs)	Air Temperature in Degrees F under 50	50 - 68	over 68
			Gallons per day*	
Growing	55	0.6	0.6	0.8
Fattening	88	0.7	0.8	1.0
Early pregnant	100	0.9	1.2	1.4
Late pregnant	100	1.7	2.0	2.5
Early lactating	132	2.6	3.3	4.0
Late pregnant	132	2.2	1.5	1.8

*The water required is the total water contained in the forage plus the water consumed.

TABLE 6.6. RECOMMENDED LIMITS OF CONCENTRATIONS OF SOME POSSIBLE TOXIC SUBSTANCES FOUND IN DRINKING WATER FOR LIVESTOCK.

Material	Safe upper limit (ppm)	Material	Safe upper limit (ppm)
Arsenic	0.2	Mercury	0.001
Cadmium	0.05	Nickel	1.0
Chromium	1.0	Nitrate-N	100.0
Cobalt	1.0	Nitrite-N	10.0
Copper	0.5	Vanadium	1.0
Fluoride	2.0	Zinc	25.0
Lead	0.1		

Fecal contamination, dead animals (mice, rats, etc), algae, dust, grass and other feed stuffs left in the water trough causes animals to drink less. A reduction in water consumption is usually accompanied by a reduction in dry matter intake, thus reducing performance. Water tanks, troughs and reservoirs should be cleaned on a periodic basis. The frequency of such cleaning should be in direct relation to the degree of intensity of the livestock program.

WATER SUPPLY:

In extensive range situations the rule of thumb is no more than 50 animal units per watering point and the length of the trough should be such that 10% of the herd can drink at once. The distance between water points depends to some extent on topography and climate. Various range management textbooks state that water points should be no farther than 2 miles in flat country and 1 mile in rough. Obviously this is seldom practiced. I am not sure how these numbers were arrived at, but most desert operations I know of have water points at about twice these distances or greater. If you're attempting to raise fat British cattle in arid environments, then perhaps you need your water points closer. If you're selecting for an animal that can produce in that environment, then the water points should be farther apart; the economics of placing a well every mile or so seldom works out in extensive environments.

Under moderately intensive operations with paddocks of 10 or more acres, graze periods of one day or more and herd sizes of 200 or greater, the length of the water trough should be 1.5 inches, times the number of animals. For example, to water 200 head a straight trough, 300 inches (25 feet) in length would be sufficient. Troughs should be shallow and only need to be open on one side; a 12-13 inch PVC pipe cut so that the top 1/3 is removed, braced in a cradle, painted to protect it from sunlight, makes an ideal, inexpensive water trough, Figure 6.3. This type of watering requires that there be a high discharge rate. Pipes should be at least 1.5 inch in diameter and provide a minimum discharge of 10 gallons per minute. Due to bovine sociology, seldom do groups larger than 50 to 60 head come to water at any one time, unless forced to. Water should be kept cooler than the air temperature during the summer, but differences greater than 18 degrees F, appear to retard rumen flora.

FIGURE 6.3. WATER TROUGH FOR LARGER PADDOCKS.

Under very intensive strip or break grazing, a 55 gallon drum, cut in half, fitted with a demand valve will suffice, Figure 6.4. The suspected reason why this small unit will work under these conditions, where under more extensive operations the animals would tear it up, is due to the changed social context of animals under very intensive grazing. The social order of cattle dictates that when the herd moves there is a leader and under range conditions or in large paddocks, going to water involves traveling and consequently becomes a social function. However, when water is close by, going to water is no longer a herd function and becomes an individual one. As a consequence, there is little jousting around the water barrel, as individuals approach it one at a time, rather than a whole herd. This also allows timid or less dominant animals to get their fill.

This type of water trough can be easily constructed. The drum is cut so that it can rest on the flat bottom, a demand valve with a float attached by a light chain is connected on the bottom, a quick connect coupler and a garden hose can be used to the water supply. The drum can be easily moved when the animals are, with very little bother. This set up, along with a portable mineral wagon, is ideal for break grazing.

FIGURE 6.4. WATER TROUGH FOR SMALL PADDOCKS - HI DENSITY.

ENERGY:

A plant captures the suns energy and stores it as carbohydrates, fats and protein. These materials are eaten by the animal and digested; when this happens energy is released. The amount of energy released depends on the ability of the digestive system to break down the complex plant parts into simple substances. Digestibility is the measure of how much of a plant is capable of being broken down into these simpler substances by the animal. If an animal eats 10 lbs (dry matter) of a forage and 5 lbs (dry matter) comes out as manure, the digestibility of that particular forage is 50%. When fed to the same animal, different plant species and different parts of the same plant, will have different digestibilities. In fact, parts of the same plant at different stages of maturity will have different digestibilities, higher when young and lower when mature.

Because the rumen has a limited size, plants or feed with a high digestibility are preferred over feedstuffs of low digestibility. For example, say a steer requires 10 lbs of dry matter of 70% digestibility to meet it's energy requirements; if another material of only 35% digestibility were substituted, the steer would then have to eat 20 lbs to get the same amount of energy. What often happens when an animal is placed on feeds of low digestibility is that the rumen is not large enough to hold all the feed required for the animal to meet it's basic energy needs, and the animal loses weight. Thus an animal may literally starve to death while standing belly deep in low quality forage. "Hay belly" is another sign of poor quality forage; however, very young animals that have lost their mothers before the rumen is completely functional will also show this.

In the U.S. the unit of energy is the calorie, which is the amount of heat required to raise the temperature of one gram of water one degree centigrade (C). In human nutrition 1000 of these calories (1 Kilocalorie) equals one diet calorie. Most of the rest of the world uses the Joule. One calorie is equal to 4.184 Joules. The path that energy travels as it moves from the plant to stored energy in the animal is shown in Figure 6.5.

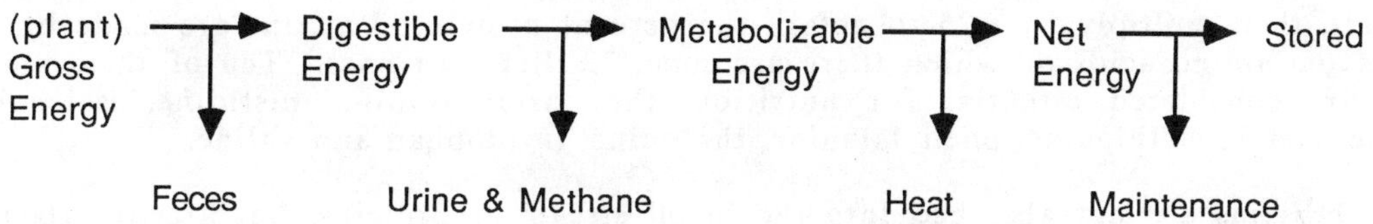

FIGURE 6.5. Energy flow through an animal and the associated energy losses.

Gross energy (GE) is the total amount of energy contained in the feed. The gross energy content of pure fat is about 9.4 Kcal/gram, pure protein and carbohydrate are both about 5.4 Kcal/gram; however the GE content of most grass is about 4.4 Kcal/gram, due to the presence of minerals and other non energy substances. The difference between gross energy and digestible energy (DE) is the energy contained in the feces. Not all the digestible energy is available for the animal's use. Some energy is lost in the urine and some through methane gas formation and subsequent lost when the animal ruminates or belches.[9] Carbon dioxide and hydrogen are also produced in the rumen, but they have no energy value. After the energy of urine and methane have been subtracted, what's left is metabolizable energy (ME). This is the energy that is available to the animal for all the things an animal does. On a maintenance diet the energy lost to urine and methane is about 19% of the digestible energy (thus ME = 0.81 DE). Net energy (NE) is the amount of energy left after metabolizable energy is converted into forms that can be used for maintenance and production. The difference between ME and NE is lost as heat. Net energy is further broken down as the energy available for maintenance (NEM) and the energy available for production (NEP). The amount of energy used for maintenance depends on the metabolic rate for the particular animal and the level of activity; this energy is released as heat. What's left is available for production.

There are various tables out that list the amount of energy required by different classes of animals for different rates of gain. For pasture animals the best work has been done in New Zealand and the United Kingdom, "Sheep and Cattle Nutrition" is

[9] It has been estimated that one fourth of all methane contained in the atmosphere comes from ruminant animals.

recommended (see suggested reading section). For feed lot animals, tables such as the Nutrient Requirements of Beef Cattle, 6th ed. published by the National Academy of Science are available. There are equations out that one can use, but they can get pretty hairy, especially for feed yard animals, better to use a table. Appendix 4 lists two tables for fattening steers; one based on feed yard data and the other the energy requirements for pasture animals. The energy requirements listed in the appendix assumes that the animals are not subject to environmental stresses such as heat, cold, wind, moisture or excessive solar radiation.

PROTEIN:

Proteins are very complex chemical compounds that contain on the average 53% carbon (C), 22% oxygen (O), 16% nitrogen (N), 7% hydrogen (H) and 2% sulfur (S). Sulfur is not present in all proteins but the other elements are. Because the proportion of N is fairly constant in all proteins (16%) it is common practice to determine the amount of N in feeds, then multiply by 6.25 to obtain the percent protein. Proteins are made by the body from amino acids of which there are some 22 different ones. Ten of these amino acids are considered essential for nutrition, they are; arginine, histidine, isoleucine, leucine, lysine, methionine, phenylalanine, threonine, tryptophan and valine.

Proteins are not absorbed into the blood stream as proteins, but are absorbed as amino acids. These amino acids are the result of the digestion of different proteins that takes place in the mid-gut. Consequently it is not necessary to feed the exact protein that the ruminant will require, but is only necessary to feed the right combination of the building blocks (amino acids). Thus if milk is the desired output, the right combination of amino acids to make casein must be supplied to the mammary glands by the blood stream. For young ruminants (before the rumen becomes functional) and monogastrics (horses and swine), the right combination of proteins is important. For the ruminant animal it is not necessary to feed the entire requirement. This is because the ruminant obtains its amino acids from two different sources; the feed it eats and the microbes in the rumen; about 70% of the ruminant protein source comes from microbes.

When the ruminant eats a protein source, most of the protein is degraded in the rumen into various nitrogen fractions. Most is used by microbes and a fair percentage of what is not, is absorbed directly into the blood stream; however, some will escape degradation and arrive in the mid-gut as amino acids, which are then absorbed into the blood stream. Of the proteins degraded in the rumen, the nitrogen content, usually in the form of ammonia, is used by the microbes to make microbial protein; about 80% of all bacterial species in the rumen are able to utilize ammonia. These are the same microbes that transform carbohydrates into fatty acids. These microbes are constantly being moved down the gut along with other material. They are killed in the abomasum by secreted acids and enzymes (mostly hydrochloric acid and pepsin), broken down into the individual amino acids and absorbed the same way the non-degraded protein from the original feed are.

There are two types of microbes of the rumen, bacteria and protozoa; each represented by several different species. Some are capable of digesting only carbohydrates while others digest cellulose and some both; many of the protozoa prey on the bacteria. However the protozoa are more digestible (91%) than the bacteria (74%) and contain more lysine, which makes them slightly more valuable. But all microbial protein is of high value, normally with a better balance of amino acids than most plant protein. In addition to the microbes, a fungus is also present, which aids in the breakdown of cellulose.

Not all nitrogen is capable of being digested, about 3% is tied up in un-digestible plant fractions, primarily lignin. Strictly speaking lignin is not a carbohydrate but a chain of phenylpropane molecules. Lignin is incorporated into the plants cell walls in a lattice like fashion which makes the walls hard; the older a leaf, the harder the walls.[10] Also a certain amount of the forage does not get sufficiently broken down by chewing and thus escapes degradation. This is why fine chopping forages (about 1 inch in length) improves performance.

The modern trend for animals on concentrated rations is to define the various amino acids and/or proteins actually required for the performance desired. For pasture work the term digestible crude protein is used in figuring protein requirements. This is obtained by multiplying the N (in %) content by 6.25 to obtain the crude protein (CP). The crude protein is multiplied by 0.9 then 3 is subtracted: ([CP x 0.9] - 3), this is the amount of protein actually digested. As may be noted the lower the crude protein content, the greater the percentage that is not digested. A forage with a CP of 15% will have a digestible protein of 10.5, but one with a CP of 4 will only have a digestible protein of 0.6%. Many dead forages have a CP of 4 or less, especially tropical forages. For all practical purposes they have no value other than filler.

The crude protein requirements of young animals are high, normally above 13% of the dry matter intake. As the animal grows and ages, the crude protein requirements drop to around 10% of the DM intake. Lactating animals have a higher requirement than do dry animals. If the crude protein content of the feed drops below 7% the animal, regardless of class does not have enough protein to maintain itself. Conversely if the level is very high, above 20%, energy must be used to break down the protein and production may suffer. Thus as the level of protein in the diet increases, so must the energy content.

THE ANIMAL UNIT:

As noted in the section on nutrition, it is not all that simple to describe what a particular animal should eat to obtain a given level of performance. What follows is even less accurate, but none-the-less of considerable use. If the forage quality is high (70% digestibility) a stocker will consume 3 to 3.1% of its liveweight in dry matter per day; if digestibility drops to around 50% the intake will drop to 2% or less. On the average a mature cow will consume about 2.5% of its live weight in dry matter per day. This translates to about 12.5% of its live weight in fresh green material, due to the moisture content (assuming 80% water) in the forage.

An animal unit (AU) is defined as the average amount of dry matter a 1000 lb cow, with or without a calf, will consume in one day; 25 lbs. Without saying it, this definition assumes that the animal is being fed in a manger, and waste's not. Also it ignores the fact that a lactating animal will eat about twice as much as a dry one. The value of the unit is that it allows a rough estimation to be made of the carrying capacity of the pasture or ranch over a period of time. There is a better method (feed budgets) and this will be covered in Chapter 11.

By convention there are 5 sheep units to 1 AU, a goat equals 0.17 AU, a horse 1.25, a bull 1.25, a weaner 0.6 AU. Converting from one species to another has some inherent difficulties, of which failure to recognize the different food habits is the worst. Likewise expecting a weaner to perform on a range suitable only for dry cows will equally lead to

[10] Kinda like arteries in old folks.

performance problems. However, for a quick and dirty estimation of what a given area will do, the AU concept does have a minimum value.

Where the AU approach does work well is in drought calculations where existing feed must be allocated out over time and the primary consideration is not performance, but keeping them alive until forage growth starts again. Allowing 25 lbs of feed per day for a 1000 lb cow as the basis (1 AU), the AU equivalent can be determined by use of the above conventions (crude).

If you have a scientific calculator, you can get the AU's a little closer. A 1000 lb cow raised to the 3/4 power equals 177.8 (use the y*x key; where *=exponent). Take the average weight of the group of animals under consideration, say 650 lb heifers, and raise this weight to the 3/4 (0.75) power; this is equal to 128.7. Now divide this by 177.8, the answer is 0.72. This 0.72 is the AU equivalent of a 650 lb heifer. A bull weighing 1500 lbs would be equal to 1.36 AU. The basis for this approach is derived from the equation for the basal metabolic rate of animals; 0.077 times the weight of the animal in kilograms, and this amount raised to the 0.75 power (0.077W*0.75).

REPRODUCTION AND GENETICS: by An Peischel and Burt Smith

In any livestock operation, reproduction and genetic goals must first be spelled out, if any sort of progress is to be made. For example, in a cow/calf operation, two very important goals that come to mind are: 1. To produce the optimum percent of weaned calves per herd. 2. To produce a quality carcass from each cow's calf. These goals have to be attained if the operation is going to be an economically feasible enterprise. To reach these goals, various objectives have to be outlined and followed in an organized manner as well as making various management decisions along the way.

Reproductive efficiency surfaces in any discussion of the optimum number of weaned calves for a herd. After all, it may not be economical to obtain a 100% weaned calf percentage, just as it is not economical to attempt to kill every fly on the place. To have a cow produce a calf each year of her reproductive life, she must return to first estrus (heat) within 80 days of calving, breed the first heat cycle, be in good physical condition, settle into pregnancy through a good bull or an artificial insemination (AI) program, breed so that her age at first calving is 2 years, and the list goes on.[11]

To produce a quality carcass, one has to first have the weaner (or stocker) to sell. This is reproduction; getting the animal on the ground. Getting the animal to weaning is based upon the maternal ability of the cow, or your ability as a manager. Maternal ability is defined as how well the mother cares for and raises her young to weaning age. In intensive environments, such as dairy operations, the maternal ability of the cow does not count for much; the important thing is to get the cow back on the milk line. Likewise in intensive livestock operations, if the economics work out and if the manager is willing to take over for poor or weak mothers, then maternal ability is not of primary importance; however the ability of the animal to gain weight, grow wool or whatever, is a primary objective and that is the one upon which the animal is culled or retained.

Genetics is the ability to turn out a U.S. Choice carcass at the age of 16 months, or turn out a premium forage fed carcass at 22 months, nutrition and management is making it happen. Scrub genes are scrub genes, all the nutrition and management in the world

[11] AI also means artificial intelligence, which is OK, I guess, since each is working on opposite ends.

won't change what they are; something about not being able to make a silk purse out of a sows ear. On the other hand good genes can't do their thing if nutrition and management are sub-standard.

REPRODUCTION:

The ability to manipulate the nutritional environment is an important criteria in managing the cow-calf herd, especially with emphasis placed on reproduction, since conception rates are influenced by the cows nutritional status. Once the cow has calved, she has to go through uterine involution (get her uterus back in shape), produced milk, put weight gain on her own body and rebreed. All of this takes an extra amount of dietary energy, some protein, phosphorus, vitamin A and trace minerals. The necessary requirements will depend on the age of the cow, the amount of milk that she is producing and her body condition.

At rebreeding, the cow must be in good flesh with a very slight fat covering on the ribs, a little fat over the withers, but none around the tailhead. Cows that are overconditioned (too fat) are hard to keep pregnant, while those that are too thin are hard to rebreed. The cow should therefore be gaining body weight at breeding. This gain should start to take place approximately four weeks before rebreeding; it will also depend on the quality of forage offered the cow. With intensive grazing, the animals that are being flushed can be run through the pastures first, to consume the upper third of the plant which is the most nutritious part (stocker feed).

Cows should be coming into heat approximately 35 days after calving (this is called first heat or estrus). It would be perfect if all the cows could be bred on first heat, but this is seldom the case with herds much larger than one animal. It would however, be ideal to have a 45 or even a 60 day calving period. All the animals would be stressed over a short period of time and so would you, as manager. You would be able to place more emphasis on watching the cows, marking the calves, checking for disease and calving problems, not to mention having a uniform bunch of animals to sell.

It is necessary to get each cow bred within 80 days after calving to stay on a regular yearly cycle; as the average gestation period for the cow is approximately 283 days for European (British) breeds; (294 days for Brahman). The main objective is to get as many cows bred during first heat as possible (usually about 60 percent of the herd) with the rest of the cows settling during the second heat and the remaining few during the third heat period. In the above discussion, we are talking about using bulls to do the work for us. There is another way and that's using AI.

The most important reason for the use of AI is the ability to select superior sires. More than 50% of the improvement that will be made within your herd is from the sire - the improvement will also be made within the shortest period of time. Be sure to have set goals as to the performance qualities (or targets) you are attempting to reach through the program as they will be important when selecting bulls, for example, superior growth of the calves, eg. 450 lbs at 205 days of age.

Don't forget, older mature cows and first calf heifers will be bred at the same time. Therefore, be aware of the associated calving difficulties that can also accompany some of the faster growing bulls (it is sometimes a matter of attempting to make too much change in too short of time). It will take more intensive management at this time as heat detection must be done by a competent individual or an animal that has been prepared for detecting heat (gomer bulls or cows equipped with markers).

Estrus synchronization is another powerful tool to use in combination with AI. Synchronization can group cattle into tighter calving groups and can be used to shorten the breeding season. For example, with first calf replacement heifers it is possible to group them so that you can calve them out is as little as 20 days. First calf heifers should be bred one heat period before the mature cows, (2 to 3 weeks early). This gives the heifers one extra heat period to get cleaned up and get ready for re-breeding for their second calf. The heifer goes through more stress at calving than the cow, this is the first time for everything, plus, she still has to gain weight (grow).

At breeding, the heifer should be approximately 65% of her mature weight for her breed; 650 lbs for English breeds. She also needs to gain approximately one pound per day until calving. At calving, if everything has gone OK, she is about 85 to 90% of her mature weight, milking and recovering from calving stress. She still needs to gain 10 to 15% more body weight and is also expected to rebreed. This is a lot to ask of a heifer, so give her a break, breed her before the mature cows.

These heifers are in need of high quality forage, every bit as good as for steers. Actually as much money can be made with heifers calving as two year old as with steers. For this to happen the heifers must be maintained in a separate grazing unit. Their first calf may need to be weaned early, depending on quality and quantity of forage, remember they must still gain weight and re-breed. They should stay in this unit until they are safely in their second calf. You may need to feed a protein and/or energy supplement.

Shortening the calving season is a natural with an IGM program; a 45 or 60 day season is a realistic goal. A shorter calving season has many advantages including the ability to concentrate on a uniform and salable product, a more tightly managed unit and the ability to effectively select the best replacement animals. A group of calves born within a 60 day period is obviously going to be much more uniform (age at weaning, weaning weight, etc.) Than those born during a four month or worse, a 365 day calving season. In addition your average weaning weight will be higher. Why?

Estrus synchronization takes management, recording of dates and giving injections, but it will help with record keeping and updated information as to the breeding status of the herd. The cleanup bull can then be turned in with these animals. Any AI program normally involves some cleanup bulls, "range janitors". They get what's left, ideally not more than 50 out of 100 cows in the AI breeding program.

By pregnancy checking cattle, more efficient use of available forage can be attained. Identification of open cows allows for optimal marketing and grass utilization. Pregnancy checking cattle at weaning allows for decisions to be made much earlier than if you waited until calving. If the market is high or there are drought conditions the free-loaders can be sold. If there is an abundance of stocker feed, then the open cows can be fattened before going to market. If you don't identify the cattle that are open before calving and carry them over the winter or dry season, not only are you misappropriating valuable forage, but the open ones may only be in moderate flesh come calving and you will have to take a lower price. There are a few things one should check out before rushing into a preg testing program; things like actual costs and expected benefits in your area, as well as possible alternatives.

The bull battery should be based on the age and social status of the bulls (covered in the first part of this chapter) and the serving ability of the animal (covered later). Very simply, bulls should be fertility tested and very possibly libido tested. In breeding, bulls will work best if they're of the same age, you'll get more out of each animal. If you have bulls of different ages, then separate them and rotate them through one or more breeding herds. Give each bull 4 to 5 days then replace with another. Single sire herds are another possibility. On extensive environments more bulls can be used at the same time since they will tend to space themselves out, but you have to watch them. One high desert Nevada ranch had 12 bulls, two of which were Santa Gertrudis; the next year 68% of the calves born were Santa Gertrudis crosses. As for numbers there are a few general rules of thumb that can be considered; unless they have been libido tested and you know they can handle more. For yearling bulls, 15 to 20 head, for older bulls, approximately 30 cows per bull. This is a management decision based upon the environment and conditions persisting on each individual operation.

The bull should be gaining weight at least a month prior to the start of the breeding season. Be sure the grass he is consuming has a high energy content - this is an important time, equal to the flushing of the cows. During the breeding season they may need to be removed to rest and build up reserves. You have to watch them, at this time they would rather be doing other things than eating.

GENETICS:

Bulls offer the fastest method of incorporating new genetic material into a herd; whether it be breeding up an existing breed or getting into a cross breeding program. On the other hand selection can only work with what is already there. However first you must define your production and performance targets. What is this future cow and/or calf going to be expected to do?

If you're selling weanoffs - your income is from the weight gain of the calves; selling stockers - the income comes from the weight gain of the stockers. No matter what way you look at it - money is muscle and vice versa. The calves have to gain!

Environmental factors often hide an animals genetic potential. If two animals are to be compared for selection be sure both animals have been subjected to the same environmental factors at the same stages of their life; the differences are then more likely due to genetic differences and not just environment. Therefore, if one wants to make genetic improvement it will come faster if the animals to be evaluated are all approximately the same age. When selecting the bulls that you produced, whether they are for the bull battery being used to breed the herd, or from a company selling semen - know your production targets and select accordingly. Do not try to select for too many traits because then you will not have quality improvement in any of the traits. Also, be sure that you are selecting traits that are related so that as one trait is improved so is the other. Do not select traits that may work opposite to each other, such as fertility and increased birth weight.

Reproduction, genetics and intensive grazing management go hand in hand. With IGM, control has been gained over the ranching operation (forage and animal production). Therefore, the environment, nutrition, disease and management can all be modified or manipulated to reach maximum production goals. Under IGM, the cattle are controlled, moved, accustomed to people and concentrated. This fits quite well into a genetic improvement program. Tremendous gains can be made fairly quickly by use of AI and a selection program geared to identifying and retaining quality replacement stock. It still takes five generations to set a trait, but there's no reason to take any longer.

SELECTION:

There is only one goal in a commercial livestock enterprise and that is production; maximum sustained yield measured as pounds per acre. Most of this book is directed at ways to manipulate your environment so as to allow the genetics of your animals to express themselves to their full potential. However, production requires a marriage of both genetics and environment and this is where selection comes in. For any selection program to work you must have your objectives well thought out and clearly stated. Just what is it you are trying to accomplish and how soon do you want this to happen? What segment of the market are you trying to produce for?

Selection is a powerful tool. But knowing what to select and what to cull are often the major problem areas. For example, you want to graze your animals on alfalfa. What do you select for? One way of course is put them all out there and mate the survivors. This method is fast and simple, but a little messy and expensive. There is another way, not as fast, but a lot cheaper. After fasting the animals for 16 hours, take a measuring tape and measure the rib girth (around the body at the last rib) and the heart girth (around the body just behind the front legs). Divide the heart girth into the rib girth then plot this on a chart with the heart girth as the horizontal axis and the rib girth as the vertical axis. Do this for every animal in the herd; then draw a straight line so that about half of the points are above the line and half below. Only graze and breed those animals that lie below the line, the farther below the better; do this prior to each breeding season. You will still have to watch for signs of bloat at first, but if you keep at it, in a few generations you'll have a herd that can graze alfalfa with little worry on your part.

An extensive range operation operates under a different set of rules than does an intensive setup; Chapter 11 gets into this in greater detail. For this discussion it is important to realize that the first criteria of a extensive operation is the ability of the animal to survive and reproduce in that environment. Maternal ability in such an environment is an important consideration, whereas in an intensive environment it is not nearly as important. If you are selecting for maternal ability, never creep feed calves or buy any bull or dam that has been creeped as a calf; creep feeding hides maternal ability.

The genes that allow an animal to survive and reproduce in a harsh environment are not the same ones that put on super gains in feed yards. Work in Australia has shown that animals which perform well in harsh environments do not perform equally well in easy environments and vice versa. Also animals selected to perform in very harsh environmental conditions perform better than those not so selected, but only if the conditions remain harsh; when the stress levels are reduced they no longer have superior growth rates. Similar work has also demonstrated that animals bred and raised in easy environments go to pieces when placed in a harsh environment. How many fat and fancy bulls have you bought, turned out on your range and then watched them nearly die?

So what is the best selection strategy for extensive environments? Well, if you never have any good years, then there's no problem, just select those cows that bring in a calf every year and sell your steers as "light" beef. Actually your secondary criteria would be the weight of her calf at weaning, but don't expect those steers of yours to be super feed yard performers. However, if you have about as many good years as bad years your selection program must account for both. What in effect you are attempting to do is to strike a happy medium.[12] You still cull on the basis of bringing in a calf, but your secondary criteria is how much calf that cow produced over at least two years, hopefully over one good and one bad year. In other words you don't cull that cow until you have a record of her performance over more than one year.

In any environment the first and foremost effort should be directed at fertility; you must first have a calf (lamb, kid, etc.) on the ground before anything else can be considered. Culling can be done through pregnancy testing or shortly after birth. Anything that is not pregnant or fails to give birth, goes; no if's, and's or but's. If maternal ability is important, the second culling takes place at weaning. Any animal that fails to wean a young, for any reason, goes; no if's, and's or but's. Livestock production is a business, you can't afford to have one of your basic machines standing around idle all year consuming feed and taking up your time; not when it can be replaced with one that will produce. One commercial herd in Florida started a rigid culling program, coupled to semen testing of their bulls, in three years they improved their percent weaned calf crop from 64% to 92%. Not bad! So what's a one percent increase in weaned calf crop worth? It's worth a 5% increase in average daily gain for every calf weaned.

After fertility and maternal ability comes performance; how well does that animal perform in your environment? The color of the animal should not be of any concern to the commercial cattleman. Performance is how well those animals convert forage to meat, that's where the money is, not in a particular breed or color of the hair. To determine performance a set of scales is required; you can't use your eye for this. By the time your eye has told your that something is wrong, it's too late. Scales can pick up dips or increases in rate of gain while there's still time to do something about it. The new electronic scales are cheap, portable and easy to set up and use; there's no longer any valid reason not to have and use a set of scales.

Measuring performance for a cow/calf outfit requires that the calf be identified to the mother. Simply determining the rate of gain or weaning weights of the calves doesn't mean much unless you are able to trace those superior or inferior calves back to their mothers and cull accordingly. You should know the date the calf was born and a estimate of it's birth weight (most beef breeds newborns weigh about 70-80 lbs). Subtract the birth weight from the weaned weight and divide by the number of days between the two (age of calf), this gives you the average daily gain (ADG). You can use this number to base your culling on, or you can multiply the ADG by the average age of your calves at weaning (in days); this gives an adjusted weight at weaning and you can cull on this basis. There are other procedures such as the 205 day adjusted weaning weight which is a bit more elaborate; check with your local Extension Agent.

One criteria that receives little attention in most cattle operations is selection based on disposition. Tom Lasater, founder of the Beefmaster breed, list disposition as one, if not the number one criteria of selection: "Disposition is judged continually with any noticeably excitable or high-strung animals being periodically eliminated from the herd."[13] The problem of dark cutters is related, not only to how the animal is handled at the slaughter house, but also its disposition. What is disposition worth? One feed yard study (and there have been others) in Florida ranked crossbred heifers according to disposition and found that the docile animals, number 1's, gained 1.66 lbs/day, number 2's gained 1.43 lbs/day, number 3's gained 1.36 lbs/day, and number 4's (very aggressive animals) gained only 1.19 lb/day. That's a difference of 0.47 lbs/day just because of disposition. Those

[12] Did you hear about the midget fortune teller that escaped from jail? The headlines read: "Small, Medium at Large".

[13] It use to be that a high-strung man was just a horse thief that got caught.

rank old "sookies" may be fun to chase and rope, but they have no place on a modern ranch.

Selecting replacement heifers in the ranch of the future will have to be done on the basis of performance and that means: Does she have a live calf and, depending on environment, how well she raised that calf, or how well the calf raised itself? Regardless, you are selecting genes that can perform in a certain environment. In extensive environments it is crucial that the calving season be as short as possible, 60 days, max. A short season allows great flexibility in dealing with the highly variable weather that extensive environments are noted for. On intensive environments it is not as crucial, since you have greater control over the environment that the animal is going to be functioning in.

To get a short calving interval requires rigid culling of all non-producing cows. To get high performance as well, means more rigid culling of the dams of the low performance calves. The main tool you have to work with is selection; genetic improvement through AI or improved bulls is a secondary tool; for the time being at least, embryo transfer should be left to pure bred operations or the people with money, you know, the ones with white picket fences around their outfits. All this culling places a heavy demand on your heifers. You should BREED EVERY HEIFER, except the obvious culls. Your eye is not good enough to see genes, but you can judge the offspring once it's on the ground.[14]

BULLS:

The second item in obtaining a short calving interval is the selection of bulls. A semen test should also be conducted of all bulls entering the herd. Numerous studies in the Western U.S. have demonstrated that fully 20% of all bulls in actual herd use are sterile or sub-standard. The shape of the scrotum also has a bearing of the desirability of a bull. The normal scrotum is sort of bottlenecked, this shape has the best testicular development and also allows for good thermoregulation in response to large temperature variations. A wedge "shaped-scrotum" or "straight-sided", generally generally do not allow for good thermoregulation and more often than not, especially if the scrotum is pointed at the bottom, contains undersize testis. Testicular development is a heritable trait. Herd sires with above average testicular development for their age would be expected to produce sons and daughters that reach puberty earlier, thereby increasing the overall fertility of the herd.

Scrotal circumference is a measure of testicular development and as such is a tool that can be used in selecting bulls. To take a scrotal measurement a flexible tape measure is used; they have designed a special one for the job but any measuring tape will work in a pinch. Unless your bull is very friendly I would suggest that you put him in a chute for this. Palpate the testis gently but firmly down into the scrotum. The scrotal tape is looped over the scrotal neck and tightened loosely, then slid slowly and carefully downward until the largest circumference is reached. Remove the tape by sliding down and off the scrotum; record the circumference.

Scrotal circumference does vary with breed and age of the bull. For British and European beef bulls in good condition a suggested minimum scrotal circumference is: 7.9 inches at 6 to 7 months; 12.7 inches at 12 months; 13.3 inches at 18 months; 13.9 inches at 24 months. These numbers have to be taken with a grain of salt, as there is variation

[14] Cattlemen are too optimistic to trust their eyes. Who else goes through buckets of mountain oysters looking for pearls?

between breeds and different countries. For example one Canadian study found that the average of 663 yearling Herford bulls was 13.8 inches; the average of 162 Angus was 14.2 inches; 194 Simmentals averaged 15.2 inches. Whereas a study in the U.S. with 55 Herfords found the average to be 12.1 inches; 79 Angus averaged 12.9; 28 Simmentals averaged 12.9. There is a lot of information concerning bull selection available at your local University or Extension office, use it.

The bull, as much as the cow, is responsible for obtaining a short calving interval. One item that is beginning to receive attention in the U.S., is libido testing of bulls. Libido is the sexual aggressiveness of bulls; a measure of how many cows can be served in a period of time. There are several methods of conducting a libido test. The simplest is simply to sit on a hill for a couple of days and count the number of mounts each and every bull you have in the pasture makes. This lets you know who is doing the work and who is either a shy breeder or who is dominated by an older bull. It's not perfect and doesn't give you all the answers, but it does give you a basis for selection, especially if the bulls are the same age.

A more detailed test was developed by a Dr. Blockey of Australia and is as follows: Two or three cows are restrained in service crates in a corral (Figure 6.6). These cows do not have to be in heat, 70% of cows will "stand" to the bull if restrained (cows should be replaced after every 20 or so mounts, or sooner if her condition calls for it).

FIGURE 6.6. Bull libido testing corral on Haldron Station, south island of New Zealand. Head stalls for cows are on far side of main corral, the spectator corral on the the other side of the head stalls.

No more than five bulls should be tested together. Any more and there is too much confusion to obtain good records. The person running the test should be familiar to the bulls and feel easy working amongst them. Before the bulls are brought in they are stimulated by being placed where they can see the cows and other bulls mounting (spectator corral). The first group of bulls tested should be tested again later in the day.

Initially the test was run for one hour and the number of services recorded for each bull during this time period. Bulls that scored 2-3 mounts during the hour are normal, bulls that score less than this should be considered for bologna. It has been found by commercial operators in New Zealand that 20 minutes is sufficient if proper preparations are made. These are: Bulls of the same age are tested together; testing is not done during the heat of the day; only bulls that show considerable excitement in the spectator corral are brought in for testing; if the test last longer than one day, conditions must be similar on the succeeding days; the herdsman must ease the bulls into and out of the facilities, no heavy handling. While the bulls are servicing, observation is made of how the bull mounts as well as any other faults, such as; cork-screw penis, broken penis, short protrusion of the penis, plus any other defects that might hinder service in the field.

The value of all this? This type of program coupled with scrotal measurements and semen testing can result in single sire service of one bull to 100 cows. One station on New Zealand's South Island is getting over a 90% calf crop in a 45 day calving season using just 2 bulls per 100 cows. Even if you don't want a short calving interval, if high libido bulls are used on your heifers, you'll see a marked increase in heifers being settled during the first service. This will pay dividends the following year when they start to calve and it will also give them a little longer to put on weight before breeding back.

If you don't use performance records to cull, at least make sure your cows look like cows (feminine like dairy cows) and your bulls look like bulls (masculine like fighting bulls); if either of them look like steers, you're in deep trouble.

SHEEP:

A brief word about sheep. If you are trying to increase the twinning rate (or percent lamb crop) of your flock, don't select the largest or biggest ewe lambs in the flock. If you do, you'll be selecting against twins and selecting for single births. Single lambs are larger at birth and they grow faster than twin lambs, they keep this edge into maturity. The proper way is to separate the twins from the singles, then select the best twin lambs as replacements. That's easy enough to say, but unless you mark the lambs at birth, you'll find it a bit difficult to sort into single, twins or triplets. There are paints and/or dyes that work well on sheep.

Any kind of marking system can be used, but different colors are best with the marks put where you can see them easily. Don't forget to mark those twins or triplets that were raised as a single, as they will be larger. One method in New Zealand called the raddle system marks them this way: Born and reared a single (blue dot on side); born a twin, reared as a single (blue dot on side another on shoulder, same side); born a twin, reared as a twin (red number on side); born a triplet, reared as a twin (green dot on shoulder and number on side); born a triplet, reared as a triplet (green number on side).

If you don't want to mess around with too many triplets, then you'll have to keep your percent lambs born per 100 ewes giving birth, less than 170%. As lamb percentages increase from 100% to 150% the number of double births increase and the number of singles decrease. However over 170%, the number of triplets increase so that at 200%, about 20% of the ewes have singles, 60% twins and 20% triplets, whereas at 170% only 3%

of the ewes will have triplets. The reason is the increased fertility of the ewes, which means more eggs being shed. You think a 200% lamb crop is unrealistic? Not at all, a number of New Zealand farms are this productive, but you won't get it without management.

To select a ewe for mothering ability, pick up her lamb by the tail so she can see it. If she bunts, paws at the lamb, knocks you down; that's a one (1). If she stands within a foot of you and sniffs, a two. A three is when she stands back 5 to 10 feet. If she stands well back and doesn't look too interested, a four. If she runs off, a five. Get rid of all 4's and 5's.

SUGGESTED READING:

An Introduction to Practical Animal Breeding. 1985. C. Dalton. Granada Publishers.

Animal Nutrition, 6th Ed. 1969. L. B. Maynard & J. K. Loosli. McGraw-Hill Book Co.

Beef Cattle Science, 5th Ed. 1976. M. E. Ensminger. The Interstate.

Behavior of Domestic Animals, 2nd Ed. 1969. E. S. E. Hafez. Bailliere, Trindall & Cassell Ltd.

Behavior of Domestic Animals, 1985. B. L. Hart, W.H. Freeman & Co.

Dairy Goat Breeding, Feeding and Management. 1972. Ed. B.C. Colby. American Dairy Goat Association, Box 186, Spindale, N.C. 28160.

Dairy Production from Pasture. 1982. Ed. by K. L. Macmillian & V. K. Taufa. Clark & Matheson Ltd.

Deer Farmers Assoc., Eastern Bay of Plenty, Workshop Seminar, Proc. Oct. 1984. MAF Regional Office, Box 1238, Rotorua, N.Z.

Ethnology of Free-ranging Domestic Animals. 1978. G.W.Arnold & M.L. Dudzinski. Elsevier Scientific Pub. Co.

Farming of Deer. 1982. Ed. D. Yerex. Agricultural Promotion Assoc. Ltd. N.Z.

Goat Farming in New Zealand. 1985. Ed. C. Rumble. Wrightson NMA Ltd. P.O. Box 1895, Wellington, N.Z.

Grazing Lawns: Animals in Herds, Plant Form and Coevolution. 1984. S.J. McNaughton. American Naturalist, vol 124, No.6.

Handbook of Livestock Management Techniques. 1981. R. A. Battaglia & V. B. Mayrose. Burgess

Improved Feeding of Cattle and Sheep. 1981. P. N. Wilson & T. D. A. Brigstocke. Granada.

Improvement of Livestock Production in Warm Climates. 1972. R. E. McDowell. W. H. Freeman & Co.

Livestock Behavior: A Practical Guide. 1984. R. Kilgour & C. Dalton. Methuen.

Mineral Requirements of Grazing Ruminants. 1983. Ed. N. D. Grace. N.Z. Society of Animal Production.

New Concepts of Cattle Growth. 1976. R. T. Berg & R. M. Butterfield. Macarthur Press.

Nutritional Limits to Animal Production from Pastures. 1981. Ed. J.B. Hacker, Proceedings of International Symposium. Commonwealth Agricultural Bureaux, Farnham Royal, Slough SL2 3BN, UK.

Nutrient Requirements of Beef Cattle, 6th Ed. 1984. National Academy of Sciences, Washington D. C.

Sheep and Cattle Nutrition. (No date) J.D.J. Scott, N. Lamont, D.C. Smeaton and S.J. Hudson. Agricultural Research Division, Ministry of Agricultural and Fisheries.

Sheep Production: Vol. I, Breeding and Reproduction. 1982. G.A. Wickham and M.F. McDonald. N.Z. Institute of Ag. Sci.

Sheep and Goat Handbook, Vol. 1. Ed. M.E. Ensimger. Agriservices Foundation.

Stockmans Handbook, 5th. ed. 1978. Ed. M.E. Ensminger. Interstate Printers.

Wortham Lectures in Animal Science. 1965. J. C. Bonsma. Texas A&M Press.

Used with permission of NEA.

Let's get our priorities right.

7 ANIMAL HEALTH

This chapter will address only those aspects of livestock health that have particular relevance to intensive grazing or those practices that, because of their general nature, take on added importance under intensive management.

Under "extensive" management, rough or careless handling of livestock, while not justified, could be lived with, since production targets were low, expectations were equally low and normal frustrations were often taken out on the animals. After all, if average performance was only a half a pound per day, setting them back to zero pounds for a few days, was no big deal. Under intensive management, production targets are high, expectations are high, there is no room for sloppy animal husbandry. When you're shooting for 2 plus pounds per day per animal and maybe 4 to 6 pounds per day per acre, any set-back hurts. There are a number of excellent books that cover various aspects of animal health and techniques of animal handling, some of these are given in the suggested reading section; "The Handbook of Livestock Management Techniques" is highly recommended.

The health and well being of the animal is directly related to its nutritional status; with good nutrition, health problems are minimized. However, anytime animals are grouped together in large herds the risk of disease and other problems increase; as the flow of energy increases, the amount of control required also increases. But, by the same token, the ability to effectively treat these problems is also increased. Since the herd is under control, mass medication is possible (in the water or supplement), individual animals are easily removed for treatment and most importantly, daily observation makes for early detection and treatment. However, before one treats an animal, one should first determine if it is sick. This requires knowing what a healthy animal looks, and/or behaves like.

THE HEALTHY ANIMAL:

Most experienced livestock people have developed a sort of sixth sense of knowing when an animal stops behaving "normally". This comes from being around a lot of animals under a lot of different conditions. The herdsman must know what the normal eating behavior is, how the animal spends its time, what normal manure looks like, the different postures and activities of the animal, the normal animal sounds, he must understand herd behavior and be aware that individual variation in behavior is normal.

In general, anytime an animal does not do what it is expected to do in a given situation, or does not look, or act like it's herdmates, should be cause for further investigation. In other words, a sick animal "sticks out". A few obvious signs are: drooping head and/or ears, lack of alertness, dull eyes, runny nose or eyes, or a very dry nose.

Once an animal has been "seen", it is standard procedure to get it in the chute and "shoot it up" with anything and everything in the vet kit. While this satisfies the need to "do something", it tends to give one a false sense of security. Which is often put to the test the next day when the animal is either better, worse or dead. Other than spending four years at a vet school (and that's no guarantee), there's no quick, sure fire way to arrive at a positive diagnoses of what's wrong. Experience helps, by knowing the kind of problems one has had in the past one can often either confirm or dismiss a particular disease fairly quickly. Sometimes knowing the vital signs, temperature, heartbeat and respiration rate will help; these are given in table 7.1

TABLE 7.1. VITAL SIGNS OF COMMON LIVESTOCK:

Species	Rectal temperature in deg.F range	average	Respiration per minute	Heartbeat per minute	(ave)
Cattle	100.4-102.8	(101.5)	30	40 - 70	(50)
Goats	101.3-103.5	(102.3)	15	70 - 135	(90)
Sheep	100.9-103.8	(102.3)	19	60 - 120	(75)
Horses	99.1-100.8	(100.0)	12	25 - 70	(45)

The respiration and heart beat can be determined either visually, by taking the pulse or with a stethoscope. For taking the temperature a special kind of thermometer is recommended; one that has a loop on the top end so you can tie about 12 inches of string to it. The thermometer is first moistened, then inserted about 3/4 of its length into the rectum and left there for at least 3 minutes. This is where the string comes in. During those 3 minutes that thermometer will move one direction or the other, guaranteed. If you don't stay right there and hold it (pretty messy), it will either end up on the floor in a pile of something or other, or will sort of disappear and a bare arm fishing expedition will be called for. However, if you take the string in one hand, then move the hand forward of the tail head, you have positioned yourself away from the action. Regardless of what happens, all you have to do at the end of 3 minutes is to gently pull on the string and presto, you have your thermometer back; then simply clean it off and read it. If you don't like the idea of holding it, fasten a alligator clip to the string and clip it to the hair. On hot days, or if the animal has been hassled, expect the temperature to be higher than shown in the table.

Now that you know who to treat, being able to treat and treating are next. Being able to treat simply means that the animal is restrained in some fashion, preferably in a squeeze chute, or calf table; treating, gets a little more involved. One word of caution, an animal that is sick or hurt, will, unless in really bad shape, look upon any thing you attempt to do to it as a direct threat. It already feels bad and is probably grouchy; consequently anything you do will likely push it round the bend, so be careful.

GIVING INJECTIONS:

Dose syringes come in all sizes and shapes. For treating one or a few head the disposable syringe is suggested, if routinely treating large numbers the multi-dose gun is used. The needles to use depends on the size of the animal and the type of injection. Needles are rated by gage, the smaller the gage (number) the larger the diameter of the needle; the length of the needle is also important. For most applications a 22 gage, 1 inch or an 18 gage 1 1/2 inch will be adequate. Use needles with a Luer-Lok hub, so that it can be easily screwed on the syringe.

If the dose bottle contains more than one dose, insert a clean large diameter needle into the bottle (18 or 16 gage) and fill with this needle, not the one used in giving the shot. This prevents contamination of the medicine so that the remainder can be used at a later date. Pull back on the plunger of the syringe, drawing air into the cylinder, attach the needle to the syringe and inject into the bottle (this builds up air pressure in the bottle and makes withdrawing the medicine easier). Make sure that only the bevel (tip) of the needle is past the stopper so that all the medicine can be withdrawn. Tip the bottle up so that the bottom is higher than the mouth and withdraw the plunger filling the syringe. Unscrew the needle and replace with the needle used to give the shot. Point the syringe up in the air and depress the plunger slightly to remove excess air in the cylinder and needle.[1] You are now ready to give the injection.

Several types of medication contain suspensions of small solid particles that do not dissolve, but remain suspended in the liquid. Often these tend to settle to the bottom. These medications need to be shaken well to distribute the particles throughout the liquid; comboitic and copper are two examples of suspensions. A good rule is to always shake well and keep all medication out of direct sunlight.

There are several types of injections, but regardless which is used it is important to make sure that the injection site is free of mud and manure. Needles and syringes need to be sterile. If a disposable set is used, keep clean until used, for other kinds boil in water for 30 minutes after using and store in a sterile container.

SUBCUTANEOUS (SC): These are injections made under the skin but on top of the muscle; usually a 16 or 18 gage needle, 3/4 or 1 inch long is used. The side of the neck on horses and cattle and the space under the front leg, flank or abdomen for sheep, goats and swine are the usual places. Depending on the medicine and the size of the animal between 2 - 15 cc's may be given. Lift the skin with the thumb and index finger, insert the needle into this fold; push the needle in until you are holding the fold of skin over the hub of the needle. Lightly pull out the plunger and observe for the appearance of blood, if noted, withdraw the needle slightly before making the injection.

INTERMUSCULAR (IM): These are made directly into a major muscle, the neck muscle on horses and cattle and/or the rear leg muscle (below the tail head) on cattle, sheep and goats. Do not inject more than 10 to 12 cc's at any one site on the larger animals, less on smaller ones, as too much at any one site can cause localized death of the muscle (muscle necrosis). Slap the site several times, prior to inserting the needle, withdraw the plunger slightly and check for blood, if noted, remove the needle a bit to remove from the blood vessel and inject. If more than 12 cc's of medication is required, withdraw the needle part way, redirect and inject again; this way you only penetrate the skin once.

INTRAVENOUS (IV): This type of injection delivers the medicine directly into the blood vessel and requires some knowledge of the animals anatomy. The IV injection is recommended only for the experienced herdsman and not for the novice. An IV injection is dangerous in that if an air bubble is injected into a vein, serious problems can develop, also some animals will have a strong allergic reaction to drugs given in this manner. It does have the advantage of being able to give large doses quickly, such as glucose or Cal-Mag solutions, or for certain drugs that are irritating to muscle tissue.

INTRADERMAL (ID): These are made with a 25 gage, 3/4 inch needle, and are inserted between the layers of the skin. Only 1 to 2 cc's are given at a site. These are particularly useful when slow absorption is desired.

[1] The reason that you remove all air from the syringe and needle is that should an air bubble be injected into a vein, bad things may happen to the heart when that air bubble reaches it; like a heart attack.

INTRAPERITONEAL (IP): Given directly into the peritoneal cavity. The left flank is the normal site on ruminants; it is not recommended for horses. Care must be taken to avoid placing the needle in an intestine. Introduce the needle slowly, this allows the intestine to move away from the needle, rather than penetrate. The (IP) route gets the drug to the blood system faster than the IM or SC route, but slower than the IV.

INTRAMAMMARY (IMI): This type injection is given by inserting a cannula (milk tube) into the teat canal and injecting the drug directly in the udder. It is most important to clean and disinfect the teat opening before inserting a sterile milk tube. After injection, massage the udder to disperse the drug. This treatment is normally for mastitis.

SUBCONJUNCTIVAL: These injections are in the mucous membrane that lines the inner surface of the eyelid; the conjunctiva. A small diameter needle is used, 22 or higher gage. Normally done in combination with treatment of pinkeye.

OTHER METHODS OF ADMINISTERING DRUGS:

TOPICAL: A topical application is one in which the medication is applied directly to the wound, skin, mucous membranes of the eye, ear, or nasal passages. The medicines are available as ointments, water solutions, powers or aerosols.

BALLING GUNS: These are used to give boluses, capsules and/or tablets. The instrument is placed on top of the tongue, just back of the highest portion, when the head is held in the normal position. Be careful not to ram the gun into the throat as a rupture or other injury to the throat may result. The plunger is then pushed to deposit the pill onto the back of the tongue.

DRENCHING: A bottle with a long neck is best, if you have any old milker inflations, attach to the bottle to give the needed length. A large coke bottle (16 oz) placed directly in the mouth is another method for animals in the 500-600 lb range. For young or small animals a basting syringe, or a large multi-dose syringe can be used. Stand at the right side of the animal, reach over the head with your left hand, place your thumb into the animal's mouth (between the incisors and molars) and let your fingers fall under the lower jaw. Lift the head until the nose is raised slightly with reference to the ground. Your left hip should be placed just back of the animal's head, to help in holding the animal. The neck of the bottle, or syringe, should be placed into the right side of the mouth and on top of the tongue. Allow the animal sufficient time to swallow the contents. Large animals should be restrained in a chute. This is a good way to get food into stubborn new born calves.

STOMACH TUBE: Take a flexible tube and lubricate it with saliva from the animal. Mineral oil is not recommended, as should the tube enter the lungs the mineral oil, being a foreign body, could lead to complications such as pneumonia. Place the tube over the tongue and gently pass it to the back of the mouth, keep gentle pressure on the tube and slowly pass it into the esophagus and into the stomach. The trachae and the esophagus lay directly underneath the skin on the front of the neck. Place your hand on the front of the neck and you should be able to feel the tube passing into the esophagus. To find out if you have it in the stomach or lungs, place your ear at the other end of the tube and listen; if you hear air passing through, try it again. Often, but not always, the animal will cough if the tube enters the trachea. It is important that you are absolutely sure the tube is in the stomach and not the lungs. Medication delivered into the lungs will often cause pneumonia to develop and will usually result in death or one very sick animal. If you are satisfied that the tube is in the stomach, put a funnel on the open end and pour in the medication or milk replacer. This is the same procedure used to relieve bloat. Stomach tubes for horses should go through a nostril; needless to say this should only be tried by an experienced person.

FEEDING: The medication is mixed in with a supplement, mineral mix or water. There are some obvious problems with this approach; the first being, the amount of medication to mix in with the feed. Two things have to be known, how much medication is needed per animal and how much of the feed is the animal going to consume? Once these items are known, or more likely guessed at, mixing the ration is simple. The next biggest question, how do you know how much medication each animal got? The answer is, unless you stand there and feed each individual animal, you don't. Since the social order of the herd often dictates who is going to get the lions share of the goodies; goodies are not a very good way to get medication into the entire herd. The only way that using a very palatable "carrier" for the medication works is if it can be placed in front of the entire herd almost simultaneously. The other alternative is to place it in the mineral or salt, or mix it in the water. This is assuming that the medication is able to be mixed. The water route is probably the best approach, since they all drink at one time or another. But whatever way the medication is applied, be aware of the fact that some animals will get more than they need and others will get less.

DRUGS:

Drugs are classified as either pharmaceuticals or biologicals. Pharmaceuticals are used for the treatment of the disease, while biologicals (vaccines) are use to prevent disease.

PHARMACEUTICALS:

None of the pharmaceuticals are effective against any of the true viruses, yeast, or fungi. Also, bacteria can develop resistance to a particular pharmaceutical; for this reason reliance should not be placed on just one "good ol medicine". Pharmaceuticals should only be used for the purpose stated on the label, the proper dosage applied, usually based on the weight of the animal and only used for a good reason. Many drugs have required withdrawal times; the time from application of the drug to the time the animal is slaughtered or milked. A whole lot of unpleasant legal problems can develop if withdrawal times are not observed. It is very important that producers be aware of what drugs are used, as not only is the FDA getting tougher about residuals in animal products, the same thing is happening in the European Common Market as well as other countries. Pharmaceuticals are of the following types:

ANTIBIOTICS: These are substances produced by various species of microorganisms that suppress growth of other microorganisms. Bacitracin is recommended for use on surface wounds as an ointment, or to treat mastitis. Dihydrosteptomycin is given intermuscular (IM), for pneumonia and generalized infections. Dihydrosteptomycin is a short acting drug and must be administered every 8 hours; it is often used in combination with penicillin. Penicillin is available in several forms and can be applied directly on the wound or IM; it is effective against several different types of infection. Tetracyclines have a wide range of application, but do their best when the disease organisms are rapidly multiplying. Tetracyclines are irritating and should not be used on sensitive tissues. Neomycin is a broad-spectrum antibiotic and is used widely in topical applications.

SULFA COMPOUNDS: Sulfa's are technically an antibiotic and have a broad-spectrum action. The ones commonly used are sulfanilamide, sulfamethazine, sulfathiazole, sulfamerazine and sulfapyridine. Two or more sulfa's are often used in combination. They are effective against most types of bacterial infections which cause diarrhea, pneumonia and mastitis. Sulfa's have a few problems, bacteria do develop resistance, they have a long withdrawal time and prolonged use may result in toxicity problems, resulting from kidney mal-function, unless a LOT of water is given to the animal.

NITROFURANS: An antibiotic that is a broad-spectrum drug whose primary action is to inhibit bacterial growth. They are of most use in treating intestinal diseases. They are also used in topical formulations on the skin, eye, ear and genital areas. An occasional animal may be allergic, but in general they are not very toxic. However, they have been banned for some uses by the FDA as there are tissue residual problems; check before using.

STEROIDS: Steroids such as cortisone are used in combination with other drugs to reduce inflammation. However, they lower the bodies own ability to fight disease and develop immunity. Steroids should only be used under supervision of your veterinarian[2] as their indiscriminate use may allow a mild infection to become serious.

HORMONES: These substances are produced by the endocrine glands of the body and their action and function in the body is very complex; consequently they should be used only after consultation with your veterinarian. Oxytocin, for milk let down and the drugs used for heat synchronization are two types of hormones.

BIOLOGICALS:

Biologicals are effective in preventing disease because they trigger the antigen-antibody reaction in the animals body. They do this by causing the body's own defense mechanism to react by producing antibodies against the particular disease being vaccinated for. Biologicals should be REFRIGERATED and KEPT OUT OF DIRECT SUNLIGHT. The bottles containing the biologicals should be wrapped around with a piece of aluminum foil when they are used Never use needles or syringes that have been sterilized by chemical disinfectants, as these chemicals can kill the organisms in live vaccines. Use boiling water or steam to sterilize. Never mix a biological with any other substance, unless stated on the label. Don't try to save money by using biologicals after the expiration date. There's a reason for that date, use of old vaccines will not give the same level of protection, like a lot of things, they get old and worn out just sitting around.

BACTERIN: A bacterin is a killed culture of bacteria. It normally requires several doses (boosters), given over time, to attain adequate protection from many bacterial infections. Bacterins offer excellent protection against certain diseases and organisms; blackleg, leptospirosis, streptococcus, salmonella, pasterrella, bordetella and staphylococcus. There are two types of bacterins, autogenous and stock. A stock bacterin is made from a bacterial culture maintained in a laboratory. An autogenous bacterin is one that is made from the organism that was isolated at the ranch where the disease occurred. An autogenous bacterin gives better protection than a stock bacterin, as it is tailored made, but then it cost more. Bacterins are often mixed to give multiple protection; an example is 3-way for blackleg, malignant edema and pasturellosis. however, recent articles have appeared about the Pasturella part of the vaccine. These

[2] Some Vet's don't believe in acupuncture, they would rather stick you with the bill.

studies have shown that killed Pasturella baccine not only does not protect against Pasturella pneumonia, but that the disease can be worse in a vaccinated animal than in a non-vaccinated one. If you have Pasturella problems it may be worthwhile to check out a modified-live Pasturella vaccine that has just come on the market.

LIVE BACTERIAL SPORE VACCINES: These vaccines are effective against bacteria that have the ability to form spores to carry themselves over bad environmental conditions. Anthrax is one example.

LIVE BACTERIAL VACCINES OF REDUCED VIRULENCE: These are made from certain strains of bacteria which are less virulent (not as capable of causing disease), but are still capable of stimulating the immune system to produce antibodies. These vaccines generally give better protection than killed cultures.

KILLED VIRUS: These are similar to bacterins in that the viral organism is killed and then used as antigens. Because the virus is dead there is no danger of getting the disease. However, boosters must be given at regular intervals; equine encephalomyelitis vaccine is of this type.

MODIFIED LIVE VIRUS: These vaccines are produced by growing the virus in such a manner that its ability to produce the disease is reduced, while still maintaining the ability to turn on the bodies immune system. Infectious bovine rhinotracheitis (IBR) and equine rhinopneumonitis are two examples of where a modified live virus is used; a killed virus is also available for these two diseases. More and more research is showing that under certain conditions a modified live virus actually causes the disease. Several states have laws that modified vaccines can only be used by, or on the order of, a veterinarian.

ANTISERUM: This is a blood serum containing specific antibodies against a disease. The antiserum is obtained from an animal that has been immunized against the disease, the serum is then used to treat an infected animal.

ANTITOXINS: Antitoxins are antiserums that contain antibodies to bacterial toxins; tetanus is an antitoxin.

PARASITES:

The decision to treat for parasites is best left to the individual operator. If you know how much parasites are costing, the economics of treatment can be worked out. Some of the new parasite drugs that have just come on the market appear to be "wonder" drugs, that is until you put the dollars to it. There is really only one way to tell if a parasite program pays on your outfit and that's to run a trial and find out. This is not all that difficult to do. For stocker animals select a group of at least 20 head, tag them with numbered tags, weigh and treat them. Do the same with another group of 20, tag and weigh them, but don't treat. Wait 60 to 90 days and weigh them again. Calculate the cost of the treatment (that's both the medication and labor for working the animals) and figure the amount of gain, the price each additional pound is worth, then make your decision. If you need some assistance with the program, contact your local county agent or veterinarian.

WORMING:

To worm, or not to worm, that is the question. Then after you've answered that one, to worm at branding or to wait and worm at weaning, then what about your mature animals and how many times should worming take place? How can an animal build up resistance if you kill all the worms? Can you select replacements that are resistance to worms? If you start a worming program, can you ever get out of it and if not, can you reduce the number of times you worm during a year? If you get into intensive grazing management can you stop worming?

The answer to the above is like our foreign policy, yes, no and sometimes maybe. Typical symptoms of worms are rough hair coat, diarrhea and poor performance; however, checking for eggs in the manure is the only positive identification. If you have worms and no selection program, then worm, if you don't have a selection program and no worms, then don't worm. If you have a selection program and worms, then worm the little guys and your market animals. Worming at branding or weaning gets the young off to a good start, as resistance appears to develop with age. Resistance to worms can be built up through a long term selection program. Studies have found that this sort of resistance confers no special advantage when the worm load is removed. Also resistance appears to be closely associated with the environment the animal is in. Change the environment and the animal get re-infected, this has been found to be the case with Coccidiosis in feed yard animals.

As to how often, that depends on the level of infestation and the relative economics. In Hawaii, with 40 cent market animals, a study was made that used over 600 animals from 12 herds on 10 different ranches. The trial compared the new, all-purpose parasite injectable, to whatever combination of wormer and external parasite control medication the ranch was presently using. The all-purpose injectable was generally found to give better results than the other combinations, but it was not always economical. The greatest economic returns were when the animals were going onto poor forage. When there was plenty of high quality forage the all-purpose injectable was not economical. Other combinations of wormer and external parasite control were economical although the return per animal was less than when moving onto poor forage. It appeared that on high quality forage there was enough for the worms, animals and everybody else. The economic results of this study indicated that the all-purpose injectable should be used when the animals are going into a stress period, other combinations gave a greater dollar return at other times. Unless there is a high infestation of worms, worming twice a year is normally sufficient. A new sustained release wormer has just been introduced that claims to control worms for a 90 day period, this might have certain advantages in extensive operations; however, there might be some problems with residuals and withdrawal times.

It is possible to reduce the level of infestation through a comprehensive worming program, but the worming program must be tailored to the environment of the ranch and to a lessor extent, the life cycle of the different species of worm, as each is different. Middle stomach worms can be greatly reduced by worming during periods when the outside environment is unfavorable to the worm, this normally occurs during periods of very dry (drought) or during very cold conditions. These worms sort of "over-winter" bad times in the animal; so this is when you can really do a number on them. By worming at regular intervals and every animal that enters the pasture, a "safe" pasture can be be developed. But be aware that just one cow pie from that infected animal of the neighbors that jumped the fence, can blow a "safe" pasture.

It has been stated that rotational grazing will eliminate the need for worming by breaking the life cycle of the worm, not so. Simply rotating animals around different paddocks, will help in the control, but will not eliminate the need for worming and in fact, may make it essential that a worming program be in place. This is for three reasons: 1. Different species of worms have different lengths of life cycles. 2. Some eggs will always survive, regardless of how short the grass is grazed or how much ultra-violet light they are exposed to. After all, they make it over-winter in temperate regions and through some pretty long hot, dry periods in the wet-dry tropics. 3. You are concentrating a large number of animals in a small area and if the period of rotation happens to coincide with a particular worms life cycle, you can have a large infestation, fast. However, herds in multi-paddocks units, with a well designed worming and selection program, will eventually reduce the use of a wormer to that of a yearly maintenance application.

OTHER PARASITES:

In wet areas, especially in warm climates, liver flukes have always been a problem.[3] The new liver fluke medication works on the tropical as well as the temperate liver fluke. However, trials in Hawaii have not shown any significant increase in liveweight gain, apparently the liver can suffer a lot of damage and still function reasonably well. The livers were still condemned at slaughter, even though all flukes were dead. It takes a long time for a liver to get over the effects of flukes, if it ever does. In order to have a healthy liver the medication would have to be given every 8 to 10 weeks during the life of the animal. Economics again, the cost of the medication and a "$5.00 liver" does not justify it's use. If you are attempting to obtain a "safe" pasture for some reason or another, then fluking every animal two days before entering and 6 to 8 weeks later, plus going after the larvae with copper sulfate in all water entering the pasture, plus introducing a parasitic wasp for the snail host, might, just might, do it.

Control of other parasites should be on an "as needed" basis. In wheel designs or water centers, a dust bag or "oiler" can be placed so that all animals must pass under it. During fly season use one of the commercial insecticide impregnated ear tags. If the whole herd is tagged, only one tag per head need to be used. If you have a tick problem, two tags should be used and bulls, due to their size should be double tagged. It has been demonstrated that severe fly problems can reduce daily gains by 1/3 of a pound per day. The tag should be placed in the fleshy part of the ear, not in a vein or through the cartilaginous portion; also try not to put the tag through the tattoo, if present. Rotate the type of insecticide used, this will reduce the chance of building up pest resistance.

DISEASES:

There are any number of diseases that affect livestock, and the reader is encouraged to find out more about the diagnosis and treatment from the suggested reading list. The best treatment of disease is prevention. This is especially true when animals are run at high densities.

PREVENTION:

All animals are continually being challenged by exposure to various disease organisms. Whether or not they actually catch the disease depends on the number or organism they come in contact with, the frequency of contact and the over all health of

---

[3] A near adult fluke has even been found in the liver of a 220 day old fetus.

the animal. The greater the stress placed on an animal, the fewer disease organism required to make the animal sick.

WATER: Water should be clean and ample amounts available. The area leading up to the water trough should be dry; they should not have to swim across a bog hole to get a drink. Open reservoirs are fine if only one or two head drink from them daily; they are totally inadequate for watering a herd. Open reservoirs should be fenced off and the water brought out and placed in a trough (a siphon on the downstream side). There are any number of disease organisms that thrive in dirty or stagnant water. Leptospirosis, footrot, Bovine Virus Diarrhea (BVD), Red Water, Tuberculosis, to name a few. If you wouldn't drink the water, don't make your animals drink it.

FIGURE 7.1. In some environments it will be necessary to bring in gravel to maintain the area around the water and mineral as dry as possible. When working with high densities of animals an ounce of prevention is worth a ton of cure.

FEED: Dirty or moldy feed, if it doesn't make them sick, or kill them, will at least cause lower performance (decrease weight gains).

POISONOUS PLANTS: With a few exceptions, if there is an adequate supply of quality forage available, animals will not eat poisonous plants. See Chapter 5. for more on poisonous plants.

PASTURE: When one starts reading about things that can kill or reduce production, everything becomes suspect. Forages are no exception. Almost every major forage has been, at one time or another, implicated in animal deaths or poor performance; such as ryegrass staggers, fescue toxicity, and of course bloat. There are a couple of others that should be mentioned in the context of intensive grazing.

1. Forage Poisoning (Cerebrocortical Necrosis), usually seen in feedyards, but does occur in pastures. Known to occur on a high carbohydrate diet. Symptoms are; sudden death, sick animals are excitable, uncoordinated, and have impaired vision, animals often go into convulsions upon driving. Certain bacteria in the rumen produce a substance that degrades thiamin (B1) and causes a deficiency in the animal. IM or IV injections of thiamine are suggested.

2. Bovine Pulmonary Emphysema also known as; panters, cow asthma, fog fever, lungers, summer pneumonia, green grass poisoning or grunts. This disease is noticed when animals are abruptly changed from dry mature feed to lush green feed. The disease is characterized by rapid and labored breathing (the animal forces air from the lungs and may grunt with each breath), may also breath through and froth at the mouth. Temperature generally remains normal and the appetite is good. Cause is believed to be associated with high levels of tryptophan in the rapidly growing forage. Atropine and antihistamines are useful in treatment. The prevention is obvious, don't make sudden changes when going from poor to good feed (there are other reasons for not making abrupt changes in feeds that are covered in Chapter 11).

VACCINATIONS: A well thought out vaccination program based on diseases that are prevalent in your area (in consultation with your veterinarian) is a must for any intensive grazing program. By in large the cost of the vaccine is small and can be administered during a regular animal working period. In many cases the deaths resulting from the diseases is not the big cost item, it's the loss of production (lbs/day gain) of those that didn't die, that really hurts; for example, BVD and Leptospirosis.

Calves can be vaccinated at an early age (1-3 days) and they will establish an active immunity. Older recommendations had you waiting in the belief that colostrum antibodies interfered with the foreign protein in a vaccine. Current research is showing that this in not the case, calves are born with an active immune system and colostrum does not interfere. Ranches practicing early calfhood vaccination (1-3 days) have reported increased survival rate.

POST-MORTEM: Posting an animal is messy, smelly and generally down right disagreeable, but do it anyway. Anytime an animal dies you need to know why; it could be the start of a raging epidemic, or an isolated case, you don't know. Posting an animal may or may not positively identify the cause of death, but it can give some leads. It is best to get your vet to post the animal, but that's not always possible. When there is no vet, you should do it. Have your vet show you the procedure and generally what organs to look at, most livestock vet books will tell you what a particular disease does to the different organs, some even have pictures.

TREATMENT:

BLOAT: So much has been written on bloat, that it seems a pity to write any more. Just one thing, speed is important, whatever remedy you choose to use, do it fast. If the animal has gone down, forget the stomach tube or other approaches; you have about 30 seconds to a minute to get that pressure off the heart and lungs. Use a trocar if you have it in your hand, if not, a pocket knife, and open up the rumen. Go in from the left flank, high up, about a hand length in front of the left hip bone, don't worry too much about the niceties of your surgical ability, if she dies, it doesn't matter; if she lives, you can sew her back up at your leisure. In sewing back up, sew up the rumen first, then the hide. Use a whip stitch (over stitch, ask your wife to show you) on the rumen. The hide heals fastest if the cut edges are butted together and stitched; use a curved needle and a pair of needle-nosed pliers. The best stitch to use here is the single stitch (in, under, out and tie off), placed each stitch about a fifth, or a quarter inch apart. If you don't have surgical thread, use a cotton thread for the rumen (since these stitches can't be removed, the body must absorb them, so use a natural fiber). For the hide, use a stout, wide thread or string, such as nylon; the stitches need to be removed in about 10 days. It's also a good idea to treat for peritonitis (sulfa in the wound and a shot of combiotic), as that will be the major problem resulting from your surgery.

GRASS TETANY: The cause of grass tetany is a low level of magnesium in the blood stream. How this happens is not entirely understood. It is known that very high levels of potassium in the forage can set the stage for tetany, if the forage is also low in calcium and magnesium the chance of tetany increase. Lack of vitamin D is also suspected as being one of the triggers (more tetany occurs on cloudy days than on sunny ones). Having a calcium-magnesium mineral in front of the animals at all times, especially during periods of rapid forage growth is the best prevention. Treatment consist of an IV (best), or IP (intraperitoneal) injection of a calcium-magnesium (Cal-Mag) solution. If the animal is down, a speedy treatment is called for. While the Cal-Mag is being given have somebody tie up the animals feet as recovery is very rapid, often before the entire bottle of solution can be administered IV. She is also going to be madder than hell (like what happens when you pull a cow out of a bog hole) and most likely will try to take it out on you or anybody else that gets in her way. Another treatment, not as effective as an IV but simpler, is to take a couple handfuls of epsom salts and place them up the rectum. The magnesium is absorbed fairly quickly through the rectal walls.

DISEASE: If you notice a sick animal, immediately isolate it from the herd and begin treatment; keep it isolated in a hospital pen until well. If you don't know what the problem is, call a vet, but still isolated the animal. If you can't find a vet, take the animals temperature and if it's high, give it a shot of a broad spectrum antibiotic. If it doesn't have a fever, you might still give it a shot of a antibiotic (to keep any secondary infection under control), while you read your vet book. Remember, antibiotics don't do a thing for viral diseases, but they do help control secondary infections.

FIRST AID KIT: Every ranch should have a supply of items that are used to treat sick or injured animals. Table 7.2 is presented as a suggestion for the minimum items. Remember to keep them as clean and sterile as possible.

CARE OF THE NEWBORN:

Generally mother knows best, so let her. Don't get in the way and if you have a nervous mother, stay out of sight; additional stress to the mother is not needed at this time. However, there are times when either the mother can not, or will not take care of the newborn, or in certain operations, such as dairies, where management practices dictate that the newborn must be cared for away from the mother.

TABLE 7.2. FIRST AID KIT FOR RUMINANTS.

HARDWARE	SOFTWARE
good knife, scalpel and blades	mineral oil
pan and/or bucket for water	dextrose, electrolyte and
thermometer	Cal-Mag solutions
syringe and needles	disinfectant
IV equipment (optional)	antiseptic ointments
needle and surgical thread	cotton, paper towels and
trocar	gauze
balling gun	pink eye spray or power
dose syringe (small and large)	penicillin-streptomycin,
tweezers (forceps)	tetracyclines, etc.
stomach tube	(oral & injectable)
scissors	mastitis antibiotic tubes
needle nose pliers	tincture of iodine
bone saw	insecticide and/or repellent
ear tagger	lubricant (petroleum jelly)
COW/CALF	wound spray
de-horner	sulfa tablets and powder
emasculators	rubber or plastic gloves
calf puller and chains	ONE GOOD VET BOOK

Free the newborn of fetal membranes by rubbing vigorously with a clean rag, clean the nose of fluids and check to see if breathing. If not tickle the nose, rub vigorously, shake up and down and sideways, give artificial respiration. Tie, clip and dip navel cord in iodine. See that newborn gets the first milk "colostrum" during first 2 hours after birth. Give immunization, vitamin and mineral injections if needed. Make sure that the newborn is kept dry and free from drafts. Dry bedding is a must.4

If the newborn is chilled, temperature below 102 degrees F for lambs and 101 degrees for calves, warm them with electric blankets, infra-red lamp, and/or a warm bath. If their temperature is below 98 degrees F, give a intraperitoneal injection of about 2 cc's for lambs and 10 cc's for calves of a 20% glucose solution. A shot of penicillin is also suggested as well, to avoid pneumonia. Colostrum (best) or milk (last resort) should be fed as soon as the animal sits up and takes notice of its surroundings, a little brandy mixed in also helps. If the animal shows little signs of life after 30 minutes, pass a stomach tube and feed a table spoon of warm glucose to lambs, about 4 table spoons for calves; repeat every hour. Keep warm. It can take up to 6 hours for chilled animals to respond.

COLOSTRUM:

A very important item; make sure the newborn gets adequate colostrum. Make sure the colostrum is from the first milking, as each milking dilutes the colostrum by half. It has been known since the 1920's that over 90% of calves will die if they do not receive

[4] If you're not sure the bedding is dry enough, sit on it for 5 minutes - if your pants get wet, change them and the bedding.

colostrum. Timing is all important; colostrum must be in the stomach within 12 hours of birth. The reason is that at birth the walls of the stomach and small intestine are very porous; the large molecules (called immuno globules) that make up the active part of colostrum can pass through and into the blood stream where they then trigger the calves own immune system to develop protection. After about 12 hours these large pores shrink so that the immuno globules can no longer pass through. Calves fed colostrum after this time do not develop immunity to the numerous diseases that they are exposed to and usually die or become "poor doer's"; does not grow well, gets sick a lot, and is late maturing. The amount of colostrum is equally important as if the calf does not get enough, it will also likely develop into a "poor doer". Two quarts of colostrum is usually sufficient. In a survey of 20 Arizona dairies it was found that between 3 to 15% of the calves born, died before they were weaned. Another dairy survey found that 42% of the calves born did not nurse during the first 12 hours, they either all died or were "poor doer's". That first 12 hours; important.

Colostrum can be collected and frozen for later use. When needed, simply thaw at room temperature or place container in a pan of luke warm water, DO NOT HEAT in an oven or a microwave (heat destroys the antibodies).[5] Colostrum from your own herd is best, from other beef herds next best and from dairies last. As a last resort colostrum can be used from other species, not too good but better than nothing. Blood serum; bleed dam or other animal of same species, let blood clot, pour off serum (clear liquid) and feed; again, not as good but, if that's all you got ---.

Milk replacers MUST BE TAILORED to the specific species of animal. The best milk replacers are made primarily from milk by-products, not from soybeans. The number of feedings, the concentration of replacer to water and the amount of time a calf is allowed to drink are all important. If nipple feeding, slow is the word; if bucket feeding only place a small amount of the total ration in the bucket at a time. Gulping closes the esophageal groove and allows the milk to go to the rumen and not to the abomasum where it is suppose to go. Milk that enters the rumen becomes rancid and may cause diarrhea, dehydration and a lot of other problems. The idea is to by-pass the rumen; so feed in such a way as to keep the neck in the same position as when actually suckling dam. Calves also need "extra" water. They don't actually produce enough metabolic water and, depending on protein source and concentration in replacer, need to dilute their urine before excretion from kidneys. Always keep electrolytes on hand for use if needed. The composition of various milk from some different mammals, along with different milk replacers and feeding schedules for different species are given in Appendix 5.

GRAFTING A YOUNG TO A DIFFERENT MOTHER:

Grafting a orphan on to a different mother can be a frustrating ordeal, but if accomplished saves a lot of bucket feeding. Smell is the major factor that the mother uses to identify her own offspring, sight is second in importance. The mother labels her young by licking, with her milk and of course birth fluids. Attempting to mask the orphans smell generally is not too successful. Best results have been obtained when the young is washed then rubbed down with the mothers birth fluids, if not available her milk and/or saliva. Skinning the dead young and tying the hide on the orphan often works, but make sure that the head and anal areas are left on the hide. Cutting off a leg of a panty hose then rubbing it over the mother's own young then placing over the orphan often works with sheep. Spraying perfume or talcum powder on the young and into the mothers nose probably only works on those mothers that would have taken the newborn without any

[5] The first natural antibody was the ant-eater.

aid other than temporary confinement. Confining the animal with her own plus another sometimes works, best if the other is first covered with the mothers birth fluids. In some African nomadic tribes blowing air into the vagina is used to induce acceptance of orphan calves. Modern researchers have found that blowing up a balloon in the vagina of a ewe 2 to 3 hours after birth, gives good results.

RAISING THE CALF IN CONFINEMENT:

Anytime animals are placed in close proximity sanitation becomes a major concern; this is especially true for baby calves. A common practice is to isolate calves into individual pens. This is not a recommended practice, unless the calves are being raised as veal. There are several reasons for keeping calves together, the first is economics, the second is that calves raised in isolation have a hard time adjusting when placed into a herd, they need the social interaction of the herd at an early age to develop properly.

The economics of a large pen is better than that of several small ones. There is less labor and material cost involved. If an automatic milker is used, the savings are increased. However, like most things, there has to be better management in a large unit and the unit must be run at near capacity for these savings to be realized.

There are several designs out for movable calf pens, which all have some drawbacks, usually ease of moving, as they need to be moved daily. If raising calves is your business, consider a slotted floor pen with an automatic milker. In New Zealand, running 20 to 40 calves on slotted floors, fed with an automatic feeder, using milk replacer, obtaining 2 lbs a day gain is routine. However, sanitation is watched very closely; pens cleaned and disinfected daily as well as all milk handling equipment. Sick calves and poor doer's are removed as soon as detected.

CARE AT BRANDING:

A calf table is strongly suggested anytime a young animal is to be worked. Roping and dragging them up to the fire is fine if your objective is recreation (roping and rodeoing) or on a "dude" ranch. If your objective is to make money, then you can not afford to have 10 to 15 cowboys running around the corral roping and flanking calves. The idea is to do it with as little labor as possible and to keep the stress at a minimum level. Be wary of "free" labor, especially from your "town-ie" friends and relatives, it usually ends up costing you a lot more than you think. It is a rare "town-ie", regardless of how well intended, that knows livestock and has your financial interest at heart. If you must use them, give them a job, like tending fire, or refilling the vaccine guns, etc., but keep them away from the animals.[6]

There are numerous chores to do at branding time; the obvious one is branding. If many animals are to be worked, a fire brand is best. Freeze branding is time consuming, requires special equipment and often does less than a perfect job; it is of importance where the animal or hide has special value. For a good fire brand, the iron should be heated until it is a dull red in color. Time of application will depend on thickness and moisture content of the hair. A proper brand will generally require 10 to 20 seconds to apply and will have the color of rawhide upon removal of the iron. Thick wet hair should be shaved off, as if it is not, a blotched brand will likely result. The brand should be a clean design, fancy figures with lots of corners come out looking like you used a frying pan. If the brand has a right angle corner, the corner of the branding

[6] Remember, you're suppose to be the professional, if any old "town-ie" can do the job, who needs you.

iron should be notched about 1/4 inch in depth. The brand should be constructed of soft iron or stainless steel (better), about 1/8 of an inch in thickness.

CASTRATION:

There are many ways to castrate a calf; and even more wrong ways. Calves should be young when castrated, preferably less than 3 months of age. Male animals over 500 lbs should not be castrated, but left intact. A recent study found that during the feed yard finishing phase, males castrated over 500 lbs gained less, were less efficient converters, had lighter carcasses and less marbling than groups castrated at lighter weights. Bulls left intact gained almost a pound a day more, were more efficient feed converters and were more economical than those castrated over 500 lbs.

The death loss of bull calves due to improper castration (usually resulting from gangrene) is one of the great secrets of the cattle industry, but what data is available indicates that it can go as high as 5%. Improper sanitation and poor technique being the major causes. This does not include the loss of production that is also going on. The weight loss from the stresses at branding are substantial; the animal will often require over a month just to get back to the weight it had at branding. The loss of a calf at branding, through stupid handling, will probably be blamed on the cow; a lot of good mothers go down the road, when it ought to be somebody else.

SURGICAL TECHNIQUES: Any surgery is major, there is no such thing as minor surgery; anytime the body is opened up, the chance of infection is greatly increased. Regardless of the method used, it is important to have a special castration kit made up. This should contain a pan or bucket, a clean rag, a bottle of disinfectant to mix with water (Nolvasen solution is a good all round disinfectant), a knife (a high quality paring knife with one side sharp, the back side dull), emasculator, tweezers, some clamps and applicator (Versiclips or Ligaclips, these are expensive but good), a needle and surgical thread and a antiseptic power (one that contains furacin works well). Clean the scrotum and the surrounding area with a disinfectant solution; regardless of your choice of instruments, cleanliness is a must. After completion, apply the antiseptic power, make sure that some power gets up into the scrotum; if is is fly season use a fly repellent. Regardless of the type of surgical castration used, calves should be returned to their mother and watched closely for bleeding. After a couple of days, any calves showing signs of swelling, that are stiff legged, or look hung-over, should be brought in and checked.

When castrating surgically, the tunic (tunica vaginalis), which is the thin membrane that covers each testicle, should also be removed. If the tunic's are left, they can aggravate swelling, infection and possibly build up into a mass of scar tissue. In all castrations you must be on the look out for testicles that are retained in the abdomen (cryptorchidism). If you have a cryptorchid you can: 1. Ignore it, maybe it won't descend. 2. Surgically go in after it (dangerous). 3. Have your vet do it next time he's around (better). If the testicle remains in the body cavity the animal will be sterile, but will show the sexual characteristics of a bull.

If large bulls, over 500 lbs, are to be castrated the bleeding must be control by using an emasculator (not too good), tying off the cord and accompanying blood vessels with surgical thread, or clamps (better). If the bleeding is not controlled a large blood clot will form inside the animal. This makes an ideal medium for bacterial growth, which can act as a wick for infections to other parts of the body resulting in serious complications.

Castration is faster and with less animal stress if done on the animal while it is standing, preferably in a chute. The operator works from behind the animal, either through the chute or by opening a special door. The testicles tend to hang down after the scrotum is cut, rather than snap back up into the abdomen like they often do if the animal is laying on it's side. If you have never tried it this way, try it - you'll like it.

There are two types of knives, the Newberry knife and either a scalpel or a special castration knife (high quality paring knife); NOT the one hanging on your belt. Whatever instruments are to be used, should be sterile, keep them in a pan with a disinfectant solution covering them (the solution should be changed every 15 calves). The Newberry and/or longitudinal method of castration will not be covered here; the reader is referred to the book by Battaglia and Mayrose in the suggested reading section.

Make an incision across the scrotum, take off one fourth to one third of the scrotum (a sharp knife is recommended); be careful not to cut the testicles or the large vein (saphenous vein) that runs down the inside of the bull's legs. The testicles are pulled out of the scrotum and the cord stretched. The cord can be scraped with the back side of the knife (or use a separate dull knife), or cut with emasculators. If emasculators are used, pressure should be maintained for at least 30 seconds on each cord. Apply a disinfectant to the area when finished and a fly repellent if needed. If bleeding has not stopped within 20 minutes, the scrotum should be sewed shut and a combiotic given, remove the stitches prior to turning the calf out (an hour or more may be required). Continue to watch closely for several days.

NON-SURGICAL: The elastrator is a very strong rubber band that is applied around the scrotum between the testicles and the body wall. This technique works best on young calves, under a month of age, or lambs. Advantages; quick, bloodless, 100% effective when properly performed. Disadvantages; predisposes the animal to tetanus infections, when the bottom atrophies and sloughs off; tetanus shots should be given when used, especially calves.

The burdizzo (emascuatome) is an instrument which crushes the vein and artery through the wall of the scrotum. Without blood supply the testicle dies. Frequently this is called bloodless castration. Actually there is lots of bleeding, but it's contained inside the scrotum where it can't be seen. Advantages; can be used any time of year without concern of an open wound or parasites. Disadvantages; reduced rate of gain, technique less than 100% successful, even in experienced hands.

CHEMICAL CASTRATION: Recently a product was developed which is injected directly into the testicles. A non-painful swelling develops which destroys the testicles. After several weeks the swelling subside's and fat will fill the scrotum. Advantages; bloodless, no reduced rate of gain. Disadvantages; expensive, should be done on a young animal, technique must be learned and mistakes can be made with animals remaining fertile, may be difficult to tell a steer from a young bull with just a casual glance from horseback.

SOME OLDIES BUT GOODIES: Teeth; used on lambs, however this technique is gradually giving way to other methods. Advantages; very macho, does not require packing around a knife. Disadvantages; reduced rate of gain, increased rate of infection, various degrees of blood loss, occasional hoof marks on operators face.

The two brick technique is used on camels in Saudi Arabia and was explained to a vet class by an East Indian veterinarian. You hold a brick in each hand and very rapidly slap them together over the testicles. When asked "Doesn't that hurt?" The vet replied, "Not if you keep your thumbs out of the way".

DEHORNING:

Use polled animals. If not, dehorn when young, the younger the better. One of the arguments for a short (60 or less days) calving season is that all this messy work can be done while the animal is young, resulting in less stress and a faster return to normal gains. A five or six month old calf may take over a month to get back on the gain after getting the works at branding, whereas one less than a month old may only be set back a week or so. Use a paste (poor), spoon or hot iron de-horner on young, up to 4 or 5 month, animals. The electric de-horning irons are fine if there an electrical outlet handy, if not a special iron with a copper head that fits over the horn is used. Hot iron de-horners have the advantage in that they are bloodless and fairly fast. Obstetrical wire also works well on smaller horns. On larger animals, saws or nippers must be used. A pair of tweezers to crimp the ends of severed blood vessels will help control heavy bleeding (and they do bleed). Again, de-horning is surgery and cleanliness is a must. If you don't dehorn when the animal is young, there are few reasons to dehorn when it's old; tipping the horns maybe all that's needed. On range cattle an argument can be made for leaving horns on breeding animals for protection of young.

OTHER:

I has not been the intent of this chapter to turn you into instant vet's. What has been described thus far has been, for the most part, routine herd management practices. Treatment beyond this level gets into the special domain of the veterinarian. However, from a realistic point of view, the average rancher or farmer is not going to call a vet everytime an animal gets sick or hurt, he will either treat it himself or simply hope for the best. Not too many years ago it was fairly common for the majority of vet work to be done by the rancher; cancer eyes, spaying, prolapse of uterus, vagina or rectum, lancing abscesses, treating and sewing up wounds and wire cuts, etc. Today most people that do a lot of this type of "advanced" vet work on their own, do so for a couple of reasons: They enjoy doing it, they take pride in their ability and of course it's often cheaper. However, if you don't enjoy and take pride in doing it, the chances are it will no longer be cheaper. Sanitation, attention to detail, reasonable speed and follow up are the key ingredients in determining the success rate. There are several good books that describe various treatments for those that wish to get into this subject deeper.

One of the many things coming down the veterinary pike that promises to be interesting, even though a few years ago sounded like something out of the dark ages, is veterinary acupuncture. Several Universities are beginning to take a hard look at acupuncture and actually have staff trained in acupuncture on board doing research. Of course the Chinese have been doing this sort of thing for centuries, mostly on humans. Western people have tended to shrug acupuncture off as being psychological, but when they started doing it on animals the psychological argument becomes a little thin. It is doubtful that acupuncture will ever be used for the routine jobs, such as castration and de-horning. Where it seems to have its greatest value is in treatment of those chronic cases, such as lameness or nervous disorders. When used for anesthesia, its primary use appears to lie in major operations, such as caesarean sections. Speaking of anesthesia, there is an electrical device on the market that attaches to the nose and below the tail head of the animal and immobilizes it while you do, whatever it is you want to do to the animal. It does immobilize the animal. However, it is doubtful that it acts as an anesthesia or blocks any pain; if fact just having that thing hooked up to the nose and rectum would appear to be painful enough, even without any electricity.

SUGGESTED READING:

Animal Health: Livestock and Pets. 1984 Yearbook of Agriculture, USDA, Washington Gov't. Printing Office.

Beef Cattle Science. 1976. M.E. Ensminger. Interstate.

Dehorning, Castrating, Branding and Marking. 1971. Farmers Bull. No. 2141. U.S. Government Printing Office, Washington D.C. 20402

Extension Goat Handbook. 1984. Ed. F.W. Haenlein & D.L. Ace. U.S. Printing Office.

Handbook of Livestock Management Techniques. 1981. R.A. Battaglia & V.B. Mayrose. Burgess. (highly recommended reading).

Hawaii's Poisonous Plants. 1979. R.E. Baldwin. Petroglyph Press.

Horses Health Simplified. 1979. Peter Rossdale. Arco Pub. Inc.

Internal Parasites of Sheep. 1982. A.D.Ross (ed.) Pub. by Lincoln College, Canterbury, N.Z.

Livestock Health Encyclopedia. 1968. R. Seiden (Ed. W.Gough), Springer Pub. Co.

Merk Veterinary Manual, 5th ed. 1979 Merk and Co.

Poisonous Plants in New Zealand. 1977. H.E. Connor. E.C. Keating, Gov't. Printer. (Good text on poisonous plants in temperate regions.)

Rearing Farm Animals. 1983. The New Zealand Farmer. Contains information on 10 different newborn farm animals. NZ Farmer, Box 56-064 Dominion Rd. Auckland. ($2NZ)

Veterinary (Animal) Acupuncture Anesthesia. 1980. S. Young-cai, C. Ci-lin and P. Hong-ze. Agriservices Foundation. 648 West Sierra Ave. Clovis, Calif. 93612.

Veterinary Handbook for Cattlemen. 1972. J.W. Bailey, 4th ed. Springer Pub. Co.

Used with permission of NEA.

You need to know all of what's going on, not just what you see.

8 BASIC ECOLOGY

Ecology is the study of all the factors that affect a biological system, or as it is usually termed an ecosystem. The study of all systems, man and nature, and how they interact is best handled by systems ecology. In this approach you not only try to understand the basics of whatever it is you are studying (let it be a ranch), but how the environment (both natural and man made) acts on the ranch and what can be expected by increasing or decreasing different factors (rainfall, market prices, disease, etc.). In addition you attempt to understand how the ranch affects, not only the community around it, but how the community, because of the ranch, affects the county, the county the state, and most importantly the way the different segments affect each other (feedback loops).

There has been a considerable effort of late to define a system as being rigid and a method as having considerable flexibility, whereas accordingly to the dictionary definitions, the reverse is closer to the truth. This has resulted in some confusion. According to the dictionary there are any number of meanings for system, but the number one definition tends to be something along the following: "An orderly combination or arrangement, as of parts or elements, into a whole; specifically, such combination according to some rational principle. Method on the other hand is defined as: "A general or established way or order of doing anything, or the means or manner by which it is presented or taught." Most secondary definitions tend to use systematic in the definitions of method. The point being, the argument over system or method is a tempest in a teapot. This book will use system according to the common definition of the word, which is is given above and implies flexibility.[1]

Any farm or ranch system does not operate in isolation, it is dependent upon other outside systems for its ultimate well being. In the past it was fashionable to study pieces of a system, with the idea that by knowing how all the pieces worked, one would somehow or other be able to construct the whole. This led to a whole generation of specialist, but few generalist.[2] In the United States the end result was that the livestock industry became pre-occupied with genetics and the application of large amounts of energy to overwhelm the system, such as grain feeding, fertilizers and mechanical brush clearing. When things didn't turn out so well, the tendency was to do more of the same, only bigger and better. Scientists, farmers, ranchers and just about the whole country got so wrapped up in single purpose technology that many of the basics got trampled underfoot. In todays modern livestock industry one can not be, just a specialist. It's not enough to know everything there is to know about the workings of a ranch; you must look beyond the boundary fence and see how the ranch can affect the outside world. You must look to the consumer, for if you can influence them, you can change the ways they in turn influence you; its called positive feed-back and it works both ways.

The word "holistic" has been tossed around quite a bit of late as a word to express this approach. It doesn't quite measure up. Holistic means all things that affect, say a ranch; it does not mean how the ranch affects all things. The word holistic and the concept of holism, like intensive grazing, has been around awhile. Holistic was first introduced by A. J. Lotka in 1922 in a paper titled: "Contribution to the Energetics of Evolution". A South African Statesman, Jan C. Smuts developed a philosophy of "Holism" which he introduced in 1926 in a paper titled: "Holism and Evolution".

---

[1] See Webster's 3rd International Dictionary, unabridged.
[2] A specialist is one that knows more and more about less and less and a generalist is one who knows less and less about more and more.

MORE ON ENERGY:

On Earth all energy, with the exception of nuclear, comes, or came at one time, from the sun. Actually sunlight is nuclear energy; stars like our sun, are in reality gigantic hydrogen bombs. The light that is given off, sunlight, is a waste product as far as the sun is concerned. For life on Earth, without sunlight (energy), there is nothing. Without control of energy there is only random luck. With increasing control comes a greater chance of a successful existence. This is why every living organism, or biological community, spends a great deal of effort to obtain control over as much energy as possible. This control is in two forms, the first, especially for young systems is to capture more and more of the available energy. The second, but no less important, is the reduction of energy losses due to operation and maintenance; the knowing when to drop a structure (organism or enterprise) when it can no longer pay its way (shedding leaves). This second becomes more evident in mature systems.[3]

FIGURE 8.1. DEPRECIATION IS AN ENERGY COST. It takes energy just to keep a structure, such as a building, leaves, legs or even the whole organism from deteriorating. Maintenance is a fact of life.

---

[3] A young bull said to an old bull: "Lets run over to that pasture and get us a couple of heifers!" The old sage replied: "Lets walk and get 'em all."

Energy does not cycle around a system as minerals do, it just sort of flows through, some of it gets stored for later use, all of it changes form, most is lost as heat. As a result, a successful organism or community must either have a big storage capacity to get past the bad times, or be diversified so that some parts (enterprises) are bringing in some energy at all times, or be able to hibernate, or die and leave seeds to germinate when the good times come back.

Because the earth is tilted on it's axis the sun appears to move higher in the sky during summer than winter in the temperate regions. This results in a higher energy flow striking the ground during the summer months. In tropical regions such as the equator this tilting only results in the sun being north for half the year and south the other half. However, there are still shifts in energy flows due to climatic factors, the wet-dry season and droughts for example. As a result, in most places in the world, energy must not only be used on a day-to-day basis for maintenance, repair and building new structures, but must also be stored for later use. The organism, community, or system that can capture, use and store more of this energy than its neighbors has a much greater chance of outsurviving them.

In oceanic environments such as a coral reef, the seasonal fluctuations of temperature are dampen, the daily variations all but eliminated. Waves and tides move food and nutrients into the reef and remove waste products. A stable, steady state ecosystem is quickly established. A similar situation also exists in the rumen; the temperature is constant, food and nutrients arrive in a predictable fashion and waste products removed. The rumen has a faster response time than a coral reef, change the diet and a stable, steady state ecosystem is established in about 14 days.

Each of the above systems are dependent on food and nutrients arriving from somewhere else. Each system has a benign environment, one is very productive and the other has the potential to be. Both can withstand severe shocks, if of the kind that there is a system memory for, eg. hurricanes for one, droughts for the other, both can be wiped out by shocks for which there is no system memory; sudden temperature drop or importation of a poison, muddy water from a dredging operation will do it for a reef, nitrate fertilizer for the rumen.

THE CONCEPT OF ECOSYSTEM CLIMAX:

Any system can be defined by four components; the substrate, the magnitude of the various energy flows into and out of the substrate, the periodicity of those flows and time. If the first three of these are not changed and sufficient time is allowed, the system will eventually arrive at equilibrium. At this point the components of the system, the plants and animals, will remain more or less the same, as long as the three components do not vary. The process of arriving at this equilibrium position is called succession.

SUBSTRATE: The substrate is the amount and texture of simple inorganic compounds available and the medium in which they are to be found; terrestrial, aquatic, atmospheric, etc. The elements that are required for optimal plant growth are: Carbon (C), oxygen (O), hydrogen (H); these are the "Big Three" and they account for about 90% of the total dry matter of plants. The other 13 (Chapter 4) make up the rest. Water is unique and no living system can function without it.[4] Water not only contains energy, it is a necessary part of all substrates and it is also a substrate itself (oceans, lakes, rivers, etc.). The substrate determines what kind of system is possible. This is so obvious that

[4] There are only two liquids that occur naturally on Earth, water is by far the most abundant, the other one is mercury.

it's importance is often overlooked. You can do certain things with sand in central Florida, entirely different things if that sand is at the bottom of the ocean and different still if it's in a river bed.

MAGNITUDE: How much energy is available for use. A deep cave has a lot of substrate but precious little energy entering. There is no sunlight and unless there is an underground river bringing in organic material little organization or biological activity is possible. A coral atoll has a lot of energy in the form of wind, waves and the sun, but precious little substrate, consequently not much can develop. While the sun is the ultimate source of all energy, how that energy arrives, the form that it takes can be many and varied. Any farm or ranch has numerous energy flows into and out of it; in addition to sunlight there are the animals, fertilizers, fuels, machinery, herbicides, electricity, etc. The magnitude of each energy flow and the sum total of all the flows will determine the potential climax of the system; desert or rain forest, town or city. The greater the magnitude, the greater the potential, if a suitable substrate is available; however magnitude and substrate are not enough.

PERIODICITY: The periodicity (does the energy come on a regular basis or all at once) of the flows determines how close to its potential the system can get. If the energy arrives as a linear function with little variation (a little bit every hour or day), the system will attain a high level complexity eg. tropical rain forest. If the energy input is periodic, seasonal or daily, without too large a variation between the highs and lows, the system can still attain a high degree of complexity, for example temperate forest and pastoral agriculture. However if large bursts of energy arrive in a random manner the system will be of a low level of complexity, not a lot of structure; a surf line on a beach is an example.

TIME: Most people consider time a constant, sort of flowing along like a river; it's not. Oh it's constant for clocks, but they are physical things and have little relationship to biology. Time depends on the magnitude of energy; too much or too little and time can be very short, or very long. While true time varies depending on amounts of energy, apparent time depends on who is doing the observing. A day is a whole lifetime for some organisms, for those living for 100 years only 0.0027% of their life; but a minute spent in a dentist chair can seem like hours. Thus the clock time required for an ecosystem to reach climax, or to begin to heal itself after an injury, depends on how much energy is flowing into the system. For example a clearing made by a bulldozer in the high desert (fragile environment) may require a year to start the healing process, whereas the same clearing in a jungle may start healing itself within a week. Similarly a insect may live out its life in a day, a bristle cone pine in 5000 years; while the clock time is greatly different, the biological time was the same, one life span.

The substrate, the magnitude of the energy flows, the periodicity of those flows and time determine just what that mature system is going to look like. As long as all are held constant the system will end up in the same. Change any one of the four and you will end up with something that looks and behaves differently.

In the real world it is obvious that all these components change, sometimes very dramatically. Time, of course, is always changing, but also the other ones; just look at the different weather patterns from one year to the next. As a result the system never reaches a true climax from which no further change is made. In reality it just sort of hovers around an average climax, plants and animals enjoying the easy winters and in turn being thinned out during the hard ones. But even this average climax is changing as the earth leaves one ice age and heads for another one. If there is one thing in this universe that does not change, it is change itself. Change is the universal constant.

For any given latitude, or elevation (assuming a reasonable type soil), where a system will end up depends to a very large extent on the amount and frequency of available moisture. Figure 8.2 shows the relationship between moisture, the amount of effective rainfall and basic soil type and how these affect the type of climax that can be expected. Woody species also increase as the moisture increases.

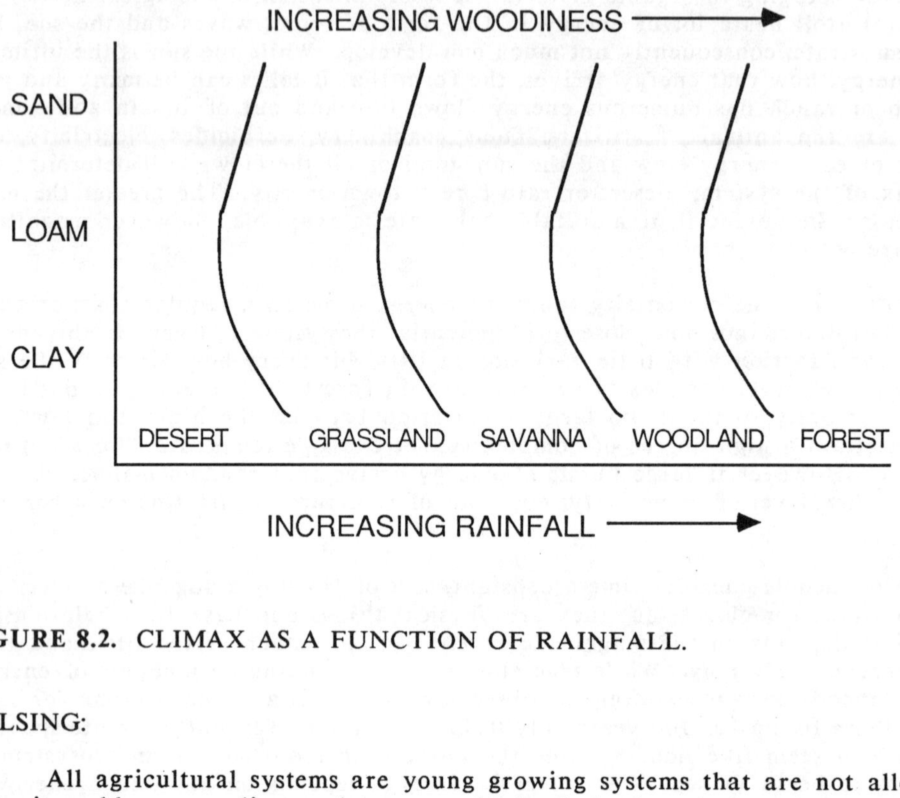

FIGURE 8.2. CLIMAX AS A FUNCTION OF RAINFALL.

PULSING:

All agricultural systems are young growing systems that are not allowed to go to maturity, old age or climax, they are pulsed. It is these pulses that keeps them young growing systems, just like tidal marshes. Greater net production (energy) can be obtained from a pulsed system than from one that just sort of flows along, like some rain forest (Chapter 1). Man is interested in the net energy of a system, he has to be, as that is what he uses to stay alive, or in the case of agriculture, sells. In figure 8.3 two systems are shown. One is an agriculture system where the yields are removed at regular harvest intervals, the other a natural, unpulsed system where the yield is used up slowly by respiration. The natural system may have a one time yield (standing crop) higher than the ag system, but the total yield, over time, will be lower.

The larger the potential energy content of the system, the more rapid the pulses, or the more severe they must be in order to bring the system back to the initial starting point. Paddy rice agriculture is one example of fairly rapid pulses, slash and burn agriculture practiced by various tribes in tropical regions is an example of a severe pulse. Either way, you're dealing with large energy flows and it doesn't take long to mess up if you don't stay on top of it.

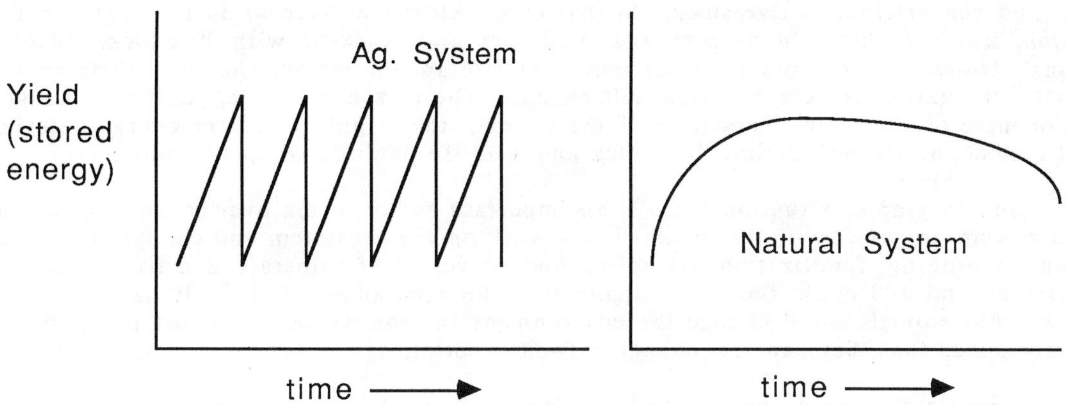

FIGURE 8.3. ENERGY PULSED SYSTEMS.

For example, say you wish to establish and maintain a grassland environment in a woodland. In a high energy environment (high rainfall, high sunlight area) a lot of effort (energy) must be spent to first clear the area, then a great deal must be spent just to maintain it. There is so much energy available that if you are unable to utilize it to its full potential, nature will. The surplus energy will be utilize and you will be faced with continual invasion of woody plants and other undesirables. You can handle the invaders at the time and point of landing, which requires little energy, but constant vigilance, or you can wait until they are well established, at which point a great deal of energy is required. The first case can be handled by a couple of ways; mob grazing the appropriate animal on the area to eat the seedlings, mowing the area for hay, silage or just for maintenance, or by timing a controlled burn to kill the seedlings. The second case requires a bulldozer, root plow, chainsaws, and/or herbicide applications.

Or you wish to establish a grassland in a low energy environment, such as a desert (lots of low quality sunlight, but little water), you are also faced with a high energy cost to become established. The major cost is to obtain sufficient water to allow the grass to grow. This may be little more than the construction of spreaders to move water out of existing stream beds for distribution over larger areas, or pitting range areas so that grass may grow in the bottom of the pits. The other extreme is the building of dams or drilling of wells for irrigation.

Anytime you attempt to establish a different "climax" in an area that is geared for one that uses higher or lower energy inputs, there is a major energy cost associated with establishment, as well as a higher cost for maintenance. In the case of a grassland in a desert the energy inputs are additive, you are in effect pushing the system into a higher energy category. In the case of the woodland, you are having to cancel the natural energies, in effect pushing the system into a lower category. There is an energy cost, which can be directly translated into time and/or money, that is associated with these changes.

Ok, so you take an existing "natural" grassland that is being grazed by wild animals and convert that to live stock. In that case: "All's" you have to do is to fence off a section, kill off the wild grazers and predators and stock it with livestock. Right? Wrong! Unless you, or your livestock can duplicate and/or replace the work done on the area by the native grazers, the area will change. The reason is simple, you have changed one or more of the basic components of the system; the magnitude of the energy entering that system and its periodicity. Livestock and wildlife don't do the same things.

In any grazing situation there is an important factor, often overlooked, that of animal power; the ability of the animal to do work on the ecosystem and change it. In addition to pruning, fertilization, trampling and recycling of minerals, a mature cow will release around 100 cubic feet of methane into the atmosphere daily.[5] It has long been known that animals could change the environment for the worse. It is just beginning to be recognized that they can also change it for the better.

The secret is of course, control. A farmer does not open the gate and turn in his combines to randomly wander all over his wheat field when he's ready to harvest. Nor does he allow those combines to harvest that field until the whole field is ready for harvest; he doesn't let them selectively graze (harvest). When the field is ready for harvesting, the combines enter and in a matter of hours the field is clean. The field is pulsed with energy, it's pulsed again at plowing, at planting, weeding, herbicide application, etc. All crop production is the same.

The only agricultural crop that this major pulsing action has been reduced and in some cases all but eliminated is in livestock grazing. Selling or removing animals in the fall and increasing numbers through birth in the spring is not the kind of pulsing that increases yields. This type of pulsing is essentially the adjustment of numbers to the available forage (set stocking) and is more in tune to steady state with low net energy returns, rather than the high net returns required by modern man.

All grazing systems are in fact, multi-pulsed, but all these pulses are small, micro-pulses; a bite here, a urination there, etc. If these pulses are controlled and timed, then a large pulse can be developed; much like taking a lot of little waves and making a big one out of them. Whether the pulses are macro or mini depends on the level of intensity, or more exactly, the level of control. Chapter 1 talked about the animal as a bundle of energy walking around the field expending so many kilocalories per day. If the animals are spread out thinly, as in continuous grazing, we have a micro-pulsed system; the energy is spread too thin to be effective. However, if we group 100 cows together, animal energy is now concentrated; we have the same potential to do work as a barrel of crude oil. In fact we now have the same potential as any macro-pulsed agricultural system, but with a re-newable energy source.

---

[5] That's enough to make a person think about hanging "a no smoking" signs over their corrals.

ENERGY IN GENETICS:

An animal or plant that exists in the wild must channel some of its' energies into simply being able to survive there. A desert plant may develop large storage roots to conserve water; or only open its stomata to take up oxygen during the night when the temperatures are cooler, thus avoiding excessive water loss. Animals may develop highly refined kidneys to extract all the water possible out of the urine before voiding. Then there are all of the many different behavioral patterns that go into eating, or not being eaten. All of these adapted strategies require energy to operate and maintain. These energies, while necessary for the survival of the plant or animal, are not available for harvest.

Man, in his infinite wisdom, has taken the jungle fowl and turned it into an egg laying machine, the cow into an udder with four legs, plants that produce only a few seeds in the wild, into 200 bushel/acre corn. But at what cost? The leghorn, the holstein and corn can not long survive in the wild. In fact they have a hard time surviving anywhere unless man is right there to supply them with what they need, when they need it. The higher the productivity of crops or livestock, the greater the supplementary energy inputs that man must invest, either in the form of time, labor, material and/or money. The reason is that those survival traits that allowed the animal or plant to make it in the wild have been bred out of them in the quest for increased production. Man now has to supply those things that the animal can no longer supply itself.[6]

Where do all these supplementary inputs come from? Not from the local feed store, gas station or other suppliers, they only expend energy storing, transporting and selling it. If you follow the trail far enough you'll find that it all comes from fossil fuel; oil, gas and coal. Oil and coal make fertilizer, electricity and, in fact power this whole civilized world. Without fossil fuels we would not have tractors, pickups, etc. and most of the food we eat. It is only by subsidizing modern crops and animals with fossil fuel that the U.S. has been able to attain those exceptionally high production figures. The modern plant is no more efficient at converting sunlight, or the animal in converting plant material into different energy forms than were their ancestors thousands of years ago. The progress that has been made, remarkable as it is, has been based on a selection process that included humans as providing the additional energy supplements.

This frenetic search for increased production is fine, but only where man is able to provide a controlled, or at least a semi-controlled environment. This desire to work in controlled surroundings is understandable; nobody likes to see years of work ruined by a flood or drought. This is one reason why so much research has been done in feed yards, the environment is at least semi-controlled. This desire also explains why the majority of animal research today is best described as "analytical" and only has economic relevance in controlled (high energy) environments.

Those traits that allow an animal to survive and produce under range conditions are of little use in the feed yard and in fact are often counter productive. Consider maternal ability, of extreme importance in "extensive" range areas but of no value in a dairy, feed yard or in any "intensive" system where man is willing to take care of the newborn if the mother can't or won't. Yet that "poor" mother will will be allowed to breed again, as fertility (the ability to give birth) is important, whereas the maternal behavior (ability to care for the young) may or may not be, depending upon environment.

[6] It must have taken someone with enormous courage to discover that snails were edible.

There is some hope that gene splicing will allow increases in basic efficiencies to be realized. While some researchers are working in this area the majority are busily designing new products or varieties that, when used will increase production; but only if the proper energy subsidies are included. For strictly economic reasons this means in environments that are being, or are capable of being operated intensely. It is a cruel hoax when we take our subsidized plants and animals and hold them out to the third world as their salvation. Our technology has relevance only in other developed countries where fossil fuel inputs are readily available.

CYCLES:

When a substance is said to be cycling, it does not mean that a particular atom will eventually return to the exact location it left some time ago.[7] Cycling refers to the major pathways that the various minerals follow, over time, to get back into the same form it had. For example, where does a carbon atom go and how long does it take, after the plant dies that the atom was originally in, to get picked up by another plant. One way to think of a cycle is, from here to there and back again. Some cycles are relatively short and others are measured in millions of years. Turnover rate is a measure of how long it takes to complete one cycle. All minerals cycle, only energy does not.

WATER CYCLE:

Energy from the sun causes water to evaporate (raised to a higher energy state), this rises, forms clouds (releases energy) and returns to earth as rain or snow. The water is absorbed by the soil or runs off as the case may be. If absorbed, it is either evaporated back to the atmosphere, or filters down and becomes ground water, where it may be picked up by plants and returned to the atmosphere by a process called transpiration. From a plants point of view, the amount of rain that falls is irrelevant, it is the amount that gets down to its roots that counts. This is the effective precipitation. Effective precipitation is dependent, not so much on how much rain fell, but how fast the rain comes down and if it penetrates into the soil. Topography also plays a big role, the steeper the slope the faster the run-off, the less gets into the soil. How fast the soil can absorb rain is called the infiltration rate. This in turn depends on the amount of organic material present on the surface and in the top soil layer, the soil texture, how much moisture is already in the soil and the presence or absence of a soil cap.

A soil cap is a relatively hard surface layer, which prevents or impedes moisture from penetrating (Chapter 4). This cap forms as a result of the vibration of the upper soil particles due to rain and wind. It is most prevalent in semi-arid environments rather than wet or moist ones. The reason is that organic matter (litter) on surface forms dams and traps water, holds it and allows it to sink in. Also organic material gets embedded in top soil layers and forms channels for water to flow into. Anything that breaks up this cap enhances the rate of infiltration. Considerable investment has been made in designing machines to break up this cap, animals can also do it and the cost is not near as great.

Evaporation is the amount of water that is lost by abiotic (not biological) mechanisms (sun, temperature, wind, color of substrate, etc.), recorded as pan evaporation. Transpiration is the loss due to plant processes, which includes respiration and translocation of nutrients and photosynthate. Evapotranspiration is the total water lost to the atmosphere from all mechanisms and can be in excess of an inch of water per day, especially in windy, dry areas. Another loss of water to the system is by respiration and

[7] Nor does it mean that it's out for a bike ride with it's favorite molecule.

evaporation of water from animals, which can be surprisingly large. On cool days (65 deg. F) a average size holstein cow will lose 17 quarts of water per day; this will increase to 25 quarts on hot days (86 deg. F); this is in addition to water lost in urine and dung.

MINERAL CYCLES:

Each mineral cycles a bit differently than others, but all cycle. The solid minerals have as their base the cycle of the earth. The minerals are picked up by plants, then by animals and back to the plants. Eventually erosion occurs and the minerals flow into the ocean and deposited on one of the various tectonic plates which cover the earth. These plates are in constant motion and are continually pressing up against one another, some sliding over and others under. The ones that go over often become mountains and the minerals that were deposited on it begins its journey all over again. The ones that slide under are in turn melted and the minerals mixed with other material; finally they are expelled as surface lava or intrusions into existing material and the process repeated. These processes take many millions of years and are of little immediate help to the rancher.

Of more immediate concern is how the common minerals move through the ecosystem. Since phosphorus (P) is usually in short supply, this will be considered. The movement of P in the soil is slow (most soils fix P) consequently leaching losses are low. Most of the P loss is due to erosion, both wind and water and harvesting. Counter balancing this loss are the inputs to the system. Phosphorus arriving from the atmosphere ranges from almost zero to well over 10 lbs per acre per year (Chapter 4). Phosphorus is also released from bedrocks through weathering, the actual rate is a function of moisture levels, organic materials, and temperature. In a grazing system P also enters from the mineral supplements placed before the grazing animals.

On average, a mature cow will consume 2.2 ounces of mineral per day for a total of 50.2 lbs/year. Of all the minerals consumed by the animal, over 90% will be returned to the pasture. This means that for a typical trace mineral mix, 4.5 lbs of P and Ca, 1.4 lbs of Mg, 0.7 lbs of K and S will be returned to the pasture. Not a great deal, but more than enough to make up for the harvest lost.

The average percent mineral composition of a whole cow is: Ca=1.33%, P=0.74%, Na=0.16%, K=0.19%, Cl=0.11%, Mg=0.041%, S=0.15%. The phosphorus loss of a 450 lb calf going to market would be 3.33 lbs. Figuring a 85% calf crop the P lost to market would be 2.83 lbs/year. This give a surplus of 0.5 lbs/year, which is enough to cover a slightly high erosion loss of 5 tons of soil per acre per year for a stocking rate of one cow to the acre.

Trees act as nutrient pumps, picking up deep minerals that have leached past grass roots, and recycles them back to the surface in the form of leaves and branches. Turnover rate is the average time it takes for one molecule to cycle. Reduce the turnover rate by one half and production is doubled. Introduction of dung beetles into Australia has been estimated to have increased production by 25%, simply by increasing the turnover rate of critical minerals. Nitrogen in a temperate humid environment will take anywhere from 2 to 3 years to move from the forage through litter, decomposition and back to forage (this will be much less in humid tropical conditions). But, when a grazing ruminant is added to the system the cycle drops dramatically to around 2 to 3 weeks; however, the losses due to volatilization and leaching are greater.

PRODUCTION AND ASSOCIATED THINGS:

Production is a quantity, not a rate. A rate is a quantity per unit time, like gallons per minute. However, productivity is a rate, for example, average daily gain (ADG) is a measure of productivity, so many pounds per day. Standing crop is occasionally used to measure production. But standing crop is a one time measurement and only measures what is there, right now. Only in systems where organisms are large and the turnover rates relatively long, can good estimates be made of productivity from measurements of standing crops, for example in areas that have very short growing seasons, such as mountain meadows. Yield is another term that gets used a lot in agriculture, it is similar to standing crop, but generally denotes total production at the end of the year.

There are several factors that are involved in measuring production. Gross production is the direct conversion of sunlight into plant material, like gross dollars, the cost of getting it is not included. In an ecosystem the cost of obtaining this plant material is the respiration (work) required. What is left after the respiration cost have been removed is called net production. Different units can be used to measure these items, such as pounds or gallons, but since work (respiration) is involved, it is best to use energy terms for all (Chapter 1). Table 8.1 shows the gross production (GP), respiration (R) and net production (NP) of several different ecosystems.

TABLE 8.1. ANNUAL PRODUCTION AND RESPIRATION OF GROWTH TYPE AND STEADY-STATE ECOSYSTEMS (Kcal/m2/year)

	U.S. alfalfa field	English pine farm	N.Y. oak forest	*Fla. silver springs	P.Rico rain forest	N.Y. coastal sound
GP	24400	12200	11500	20800	45000	5700
Rp	9200	4700	6400	12000	32000	3200
NPp	15200	7500	5000	8800	13000	2500
Ra	800	4600	3000	6800	13000	2500
TNP	14400	2900	2000	2000	+/-	+/-

Rp = The respiration done by plants in the system. Ra = The respiration done by the animals in the system. TNP = The total net production of the system.
* Silver springs is a very large fresh water spring in central Florida.
Adapted from *Systems Ecology* by H.T. Odum

A tropical rain forest has the largest gross production of any natural ecosystem, however there is little to no total net production. For one adult to survive in a rain forest it requires about one square mile per year (640 acres), however in a paddy rice culture using only solar energy, one acre will do it.

As a general principle the gross productivity of cultured ecosystems does not exceed that found in nature, however the energy available for man's use often does. Man's success in adapting a natural system to his use has resulted from the importation of energy subsidies. Much of this energy goes towards doing the maintenance work that before had to be generated by the original system. Consequently, when outside energies are introduced the original species, quite often, can not adapt since they continue to duplicate the work effort by providing energy that is now being done by the outside subsidies; as a result there is no saving and those species are replaced by species that are better adapted to the new subsidies. For example how long would indian rice grass persist if the area it was growing in received two inches of water every week during the growing season? Not long, it would soon be crowded out by plants that are better adapted to high moisture conditions.[8]

Nature maximizes for gross production, humans for net. It is important in this day and age of environmental concerns to recognize this difference.

THE LIMITING FACTOR:

Productivities can be increased by adding what is missing or in short supply, the limiting factor. A limiting factor is that element, that if added to the system would increase its gross production. Justus Liebig first expressed this concept in 1840, and it has come to be known as "Liebig's Law of the Minimum". Strictly speaking it is only applicable to steady-state systems, ie. systems in which inflows and outflows are in equilibrium.

In most systems, if the limiting factor is added, yields will increase in direct proportion to the amount added. However, if another unit of the same material is added, then another, the point is finally reached where the yield response will decline and eventually the point is reached where further added material becomes toxic to the plant. When yields start to decline the limiting factor has shifted to another nutrient. Figure 8.4 shows a typical yield curve for yield versus fertilizer added.

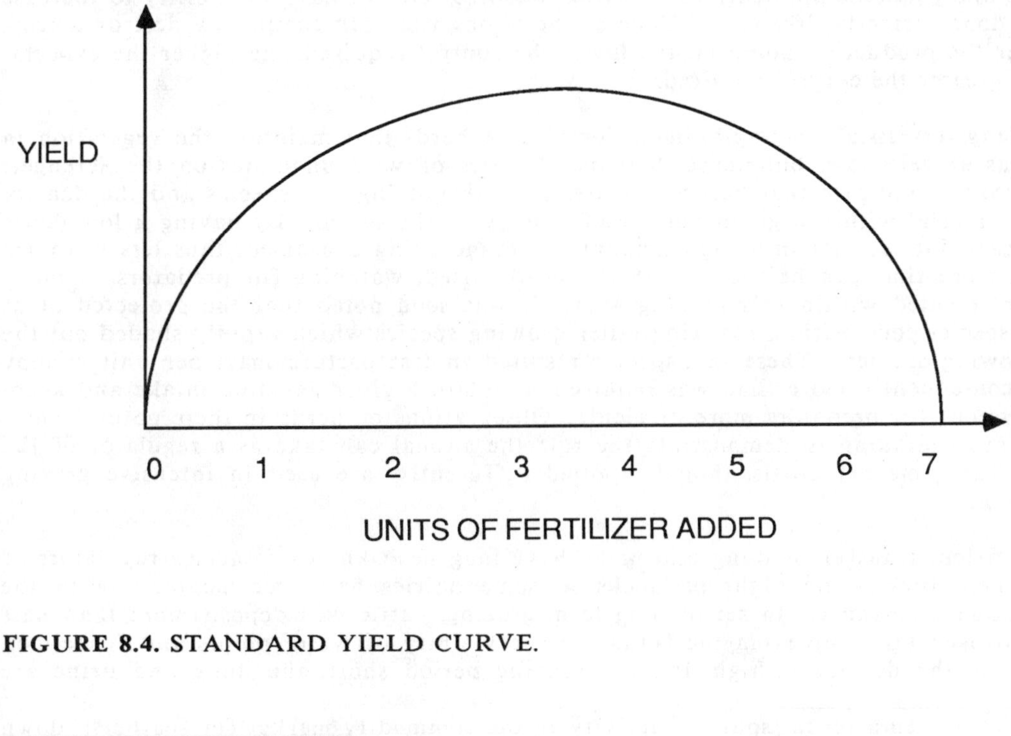

FIGURE 8.4. STANDARD YIELD CURVE.

8 Some things are like blisters, they don't show up until the work's done.

Light is often a limiting factor, not only for plants but for animals as well. It is well known that chickens will lay more eggs and grow better if given 16 or so hours of light during a 24 hour period.[9] Castrated sheep (wethers) as well intact as male and female sheep grew faster if given 16 hours of light daily than if given only 8 hours. Heifers exposed to 16 hours of light daily reached puberty 1 month earlier than those that received only 8 hours. Also heifers that had more light during the winter had average daily gains that were 10 to 15% greater than controls; milk cows also turned in 6 to 13% higher milk yields. The amount of light did not effect steers, but did bulls, which suggest that gonads are necessary for light induced growth effects in cattle. The extra light causes animals to eat more and become more efficient at converting feed to weight. The extra weight is not fat, the percent protein in the animals exposed to high light was higher than in the controls.

It is not too likely that the economics of hanging lights out in the pastures will work out. But by knowing that better performance can be obtained during periods of high light (summer) may allow an advantage to be taken, such as when to really pour the coal to replacement heifers. The same is true of many limiting factors, it may not be economical to do anything about it under traditional management, but once control is obtained, the situation may be entirely different.

THE ROLE OF THE ANIMAL:

A regulator is an agent that because of its behavior, changes or modifies the energy flows in an ecosystem. There are many different regulators, in fact any energy source can become a regulator, if applied at the right place at the right time. Fertilizer, herbicide and pesticide applications, burning, weeding, etc. all have the ability to increase yields if done correctly, likewise if done at the wrong time can completely destroy a crop. The lower the production potential the lower the control required, the higher the expectations the greater the control required.

Many terrestrial grazing animals form dense herds that maintain the vegetation in their areas at very low, but dense, heights. Studies of wild ungulates on the Serengeti plain in Africa indicate that herds regulate, through grazing, the species and the density of plant material within a given volume of canopy in the sward. By having a low dense canopy, each bite results in a high amount of forage being consumed, thus less bites are required, more time can be spent, with the head raised, watching for predators. Fenced plots were erected within this grazing area. It was soon noted that the protected areas changed species composition, favoring taller growing species which rapidly shaded out the lower growing species. These tall species resulted in less pasture mass per unit canopy volume, consequently more time was required to obtain a given level of intake and keeping an eye out for predators more difficult. Observations of herds in their natural environment are beginning to demonstrate the role the animal can take as a regulator of the system. The same principals, though applied differently, are used in intensive grazing management.

Nutrient transfer of dung and urine have long been known to act as regulators of plant communities. The night paddocks in some dairies have become toxic with the accumulation of manure. In set or long term grazing, cattle will deposit more than half of the nitrogen taken up from the forage in the vicinity of water points and fencelines. However, if the density is high and the grazing period short, the dung and urine are

[9] There has a been a recent spurt of activity in the commodity market for feathers; down is up.

scattered throughout the paddock. If electric fences are used, dung and urine are not deposited next to the fence. In Hawaii on lava rock, where roads have been bulldozed through to allow movement of livestock, the only place grass grows is along the trails made by cattle. The organic material in the dung makes an otherwise inhospitable environment for plants at least tolerable.

FIGURE 8.5. MANAGEMENT MAKES A DIFFERENCE. It would often be nice to say that it rained more on one side of the fence than on the other, but the best one can say is that the fence was so tight that the grass couldn't crawl through.

The most conspicuous animals with two notable exceptions, man[10] and large grazing herds, are the least important energetically in acting as regulators of a system. Birds contribute almost nothing to production and consume little, individual or small groups of mammals contribute a small amount and the above ground arthropods contribute a greater amount. Above ground consumption represents only 3-8% of the total consumption in typical natural grassland and above ground production represents less than 34% of the total biomass production (both roots and shoots).

[10] We say man, but that also includes women, as man embraces women.

There is a lot more going on below ground than on top, from the standpoint of both producers and consumers. The biomass of the below ground consumers will often exceed, by several magnitudes, that of the above ground grazers on a per area unit basis.

Nematodes appear to be the major controllers of productivity in grasslands. Forty-six to 67% of root and crown consumption is due to nematodes They also account for 43 to 88% of below ground predation. One trial used a nematicide and increased above ground herbage weight by 59%. Table 8.2 shows the relative amounts of different consumers in a continuous grazed temperate grassland.

TABLE 8.2. RELATIVE BIOMASS OF DIFFERENT CONSUMERS.

	ABOVEGROUND	BELOWGROUND
Herbivores		
arthropods	2.6	15.8
cattle	3.2	-
nematodes	-	13.7
Carnivores	0.2	5.5
Saprophages		
arthropods	-	2.2
nematodes	-	19.4

RANGELAND:

The ranges of the world have been deteriorating; one only has to go back and read the accounts of the early explorers to understand that. The early Spanish as they made their way up the Rio Grand River in the U.S., or the early accounts in Australia, or in Africa, all mention the abundance of "sweet" grass, belly deep to a horse in many records. What's there now? In most cases, precious little "sweet" grass, often as not no grass at all; the area has been taken over by woody plants, annuals, or other un-desirables. The common denominator in all of this has been two-fold: The reduction, or in some cases elimination of the wild graziers and the introduction of domestic livestock. Many of the environmentalist have seized upon the introduction of livestock and ignored the removal of the "natural" graziers as the major cause of deterioration.

The reasons, are of course, a bit more complicated, than the simple "slogans" that the more militant environmentalist would have us believe.

The ranges developed from the interactions of climate (including rainfall), soils, fire, vegetation and the grazing animals. Remove, or change one of those and you have changed the entire system; it all goes back to energy inputs as already discussed. Further confounding the picture is the naive belief that continuous grazing, as practiced by the majority of the worlds ranchers, is natural. Continuous grazing, as practiced, bears little resemblance to the "natural" grazing practiced in Nature.

Overlooked is the fact that under natural conditions, the dominant controllers of the environment were animals that formed large herds composed of thousands of individuals. These herds were constantly on the move; they rarely returned to the same area in successive years. The sedentary graziers, those that did not migrate in huge herds, were thinly spread over the available resource. The effects of their continual presence were overcome by the pulsing (energy inputs), of the system by the large herds and/or fire. It was this pattern of use that made the majority of the World's rangelands.

Attempting to duplicate nature with paddocks of a few hundred acres and a rag-tag bunch of animals scattered all over, is just not the same, especially so when fire is also suppressed.

Environmentalists express concerns about disrupting the natural order of things. This concern is not only being directed towards the introduction of intensive grazing programs into many areas of the world, but also a continuing effort to get livestock off the public domain. Such as in the Intermountain West, where large herds of grazing animals were not present until the introduction of livestock, a couple of hundred of years ago. Australia is another area where hooved animals were recently introduced and are now in competition with the pad footed marsupials (kangaroos and such). What environmentalists have, in general, failed to understand is that range deterioration, which has been well documented on both areas, will continue unless the grazing animal is used (under control) to established a new stable system. Either do this or remove all humans and all human introductions and re-established the original community. This is, of course, possible in a laboratory context, but impossible in the world of humans.

The only way one can return to the past is to make the conditions exactly as they were then. Even if it were possible to somehow or other physically eliminate all exotic, or introduced plants, insects and animals, without also destroying what was left of the original inhabitants, politically it could not happen, some group would raise the rallying cry of: "Save the Beetle", or fish, or mustang, or some other introduced species, and the whole process would grind to a halt. Meanwhile, Mother Nature, continues on her way, letting some species go extinct and introducing new ones; it must be admitted that man often assist in the former, but seldom in the latter.

ECOLOGICAL TRENDS:

The waste products of our fossil fuel economy is changing our environment; whether for good or bad is difficult to say at this time. The concentration of carbon dioxide in the atmosphere has almost doubled during this century; the injection of other gases as well as dust has raised serious questions as to ability of the earth to cleanse itself. The overall effects may be to increase the global temperature of the earth, the greenhouse effect; there is some evidence that this is happening. The increase of carbon dioxide would give the C3 plants a greater advantage than the C4 plants; higher concentrations of carbon dioxide increases production, especially in C3 plants. Since most of the major food crops are C3 plants, and the worst weeds are C4 plants, the net result may be an increased world food production. But what the warming trend will do to existing weather patterns and climates, is largely unknown. If this wasn't bad enough, escaped freon gas is destroying the ozone layer. This allows a greater amount of ultra-violet light to reach the earth, which in addition to causing more skin cancers, will speed up the mutation rate plus other thing that can only be guessed at.

Added to this is the fact that the worlds population is expected to double every 30 or so years; which if nothing else will increase the production of waste products, thus speeding up the above mentioned trends, not to mention the doubling in food consumption. Technology may arrive on its white horse and save us yet again, with another source of cheap energy. People have been predicting doomsday ever since when; thus far they have been wrong. However, the simple fact remains, any ecosystem can only hold so many consumers. How many depends not just on the amount of energy flowing into the system, but how well that energy is used. If we are to obtain the maximum benefits from our energy sources, we must manage well.

SUGGESTED READING:

Cows, Pigs, Wars and Witches: The Riddle of Culture. 1975. Marvin Harris. Vintage Books.

Environment and Plant Ecology. 1975. J. R. Etherington. John Wiley & Sons.

Fundamentals of Ecology. 1971. E. P. Odum. W.B. Saunders Co.

Rangeland Management. 1975. H. F. Heady. McGraw-Hill Book Co.

Systems Ecology: An Introduction. 1983. H. T. Odum. John Wiley & Sons.

Used with permission of NEA.

New ways call for new techniques.

9 ELECTRIC FENCING AND DESIGN

Fencing, herding, tethering and distribution of water and minerals, are methods historically used by stockmen to control livestock movement. All are being used in the world today, fencing more so in the developed countries and herding in the less developed. While all control methods can be used in IGM, fencing allows for the greatest return per man hour spent.

A fence presents the animal with a barrier; how effective a barrier, depends on the type of fence and how bad the animal wants through. A fence to be 100% effective must prevent whatever animal is under consideration from passing through it; however the expense involved in building and maintaining such a barrier, seldom justifies the cost. Most fences are not expected to be 100% effective; a few are, such as prisons and the fences around lions and tigers and such, but even these leak on occasion. Livestock fences aim to turn animals about 95% of the time, in most cases this is sufficient, with IGM, more than sufficient.

Up until fairly recently, it was believed that for a fence to be effective, it had to present a physical barrier; most early fences were of this type, sturdily built rock walls or wooden poles and/or planks. The introduction of the barb wire fence changed, not only the face of the American West, but also initiated a change in the way people thought about fences. The barb wire fence not only presents a physical barrier but also relies on a psychological one as well; the animal often, but not always, gets hurt when it attempts to cross through or over. Because the pain is not always there with each fence encounter the physical barrier is still the major deterrent. However, the damage to hides and udders from barb wire, has been estimated to be over $110 million yearly in the Australia-Asia region alone.

The modern electric fence presents almost no physical barrier, but instead relies almost entirely on the fact that the animal gets hurt (shocked) whenever it touches the fence. Because the fence is psychological, rather than physical, the animal must be TRAINED to the electric fence before control is attained. Without such training, the fence will leak animals and there will be some fence repair to do every day. However, if every time the animal touches the fence it gets shocked, it won't take long (overnight for some, 48 hours for most, two weeks for the really hard cases) before they learn to have a healthy respect for the fence.[1] An occasional animal will always go through, regardless of training or length of exposure; a trip down the road will usually resolve this problem. A comment sometimes heard in this context is: "Well, electric fences may be fine for you, but I have this cow that will go through any fence." When ranch policy is being dictated by one cow, there are some really serious problems with management that need to be corrected before worrying about IGM, or what type of fence to use.

Today's electric fence is often termed "Permanent Power Fence" to distinguish it from the types that were commonly available before the mid 70's. While intensive grazing was in use long before the first electric fence was ever made; it was the development of the Power Fence that gave rise to the considerable economies and ease of use that were impossible to obtain with the early electric or non-electric fences. The material cost of an electric "Power" fence will be 1/4 to 1/3 that of a standard four barb wire fence; the saving in construction cost are even greater. Table 9.1 shows the results of a 1959 Hawaiian study which compared the cost, in man and vehicle hours, of constructing several types of fence; the figures use for the electric fence are based on construction techniques presently used in Hawaii.

[1] The only exception is man, for some reason he never learns.

TABLE 9.1. COST IN MAN AND VEHICLE HOURS OF CONSTRUCTION OF VARIOUS TYPES OF LIVESTOCK FENCE IN HAWAII:

Type of fence	Man hours per mile	Vehicle hours per mile
Lava rock (3 ft.high, two row)	2814	998
Barb wire (four strand)	324	124
Net wire (48 in. high)	241	92
Electric (three strand)	16	4

SOME ELECTRICAL BASICS:

The working principals of electricity are kind of similar to that of water. Both use some sort of conduit to get from one place to another, pipe for water, wires for electricity. Both can be stored, tanks for water, capacitors for electrons. Neither can be compressed. Increase the water pressure at one end of a pipe (like adding a pint of water), the pressure (equal to a pint of water) appears at the other end; add one electron at the beginning of a wire (increase voltage), the increase in voltage appears at the other end. However, unlike water, if the wire is cut, electrons don't fall out. But, lay trash or other conductors on the line and you will get the electrical equivalent to a broken pipe, a short (the electrical pressure goes to zero).

From a practical viewpoint electricity has two major components; amperage and voltage. Amperage is the flow of electrons (like gallons per hour) and voltage is the pressure or force exerted on those electrons (like pounds per square inch). Amperage (amps) and voltage (volts) are related to each other by the amount of friction or resistance (ohms) in the system. This relationship is Ohms Law: R=V/A, volts (V) divided by amps (A) equal ohms (R). One remaining unit is power. Power is energy delivered per unit of time, (a rate). The power of the electrical current is measured in watts (W) and is calculated by multiplying amps times volts (W=AV). In electrical appliances, the per second is usually not stated, but that's what it means, watts per second. The electric company charges you on watts per second, or kilowatts per hour. Other units of power are: Joules, Calories, BTU's and horsepower; 1 Joule = 1 watt per second, 1 calorie = 4.186 Joules (1 Kcal = 4186 Joules or 3.97 BTU's); 1 horsepower = 776 watts, or the heat required to raise the temperature of 1 pound of water 0.7 degrees Fahrenheit per second).[2]

All electrical components have a resistance to the flow of current through them. The greater the resistance, the higher the voltage required to make current flow. Copper is an excellent conductor, a cubic centimeter of copper has a resistance of 0.000,000,1 ohms; a cubic centimeter of glass, which is a very poor conductor, but an excellent insulator, has a resistance of 1,000,000,000.0 ohms. Insulating materials, such as glass and

[2] I'm not sure, but they must of used a scrawny horse when they defined horsepower, as a good draft horse can develop over 27 horsepower.

most plastics, have very high resistance to current flow and can be handled without fear of shock (unless wet). The total resistance in an electric fence system depends on the number, size and lengths of the wires and how well the connections are made (a loose connection can be 5000 ohms). The resistance that the energizer "sees" as it "looks" down the fence, is additive, that is, the resistance of each component is added together, the sum of the parts making up the overall resistance of the system (the higher the resistance the more power required). For most small permanent fences, resistance is not a big worry, as the modern units have a lot of power; but as the size of the system increases, resistance becomes an increasing concern.

A short is not resistance, it is the lack of resistance, a dead short has no resistance, zero ohms. A dead short, such as a copper bar connected from the ground wire to the hot wire, will shunt all the current down the copper bar to the ground wire. The fence will be alive (current flowing), up to the point of the short and dead (no current), from thereon. Trash, vegetation, wires touching other wires or steel post are shorts, but not dead shorts.[3] Each one shunts part of the current away from the fence; enough of them and the effect is the same as a dead short.

In order for current to flow, there must be a complete circuit, that is, electrons must be able to get back to where they started from. This is usually accomplished with wires, current flowing out from the source on one wire and back on another. In two or more wire electric fences, one or more of the wires is a negative or ground (-) and the other wires are positive (+) or hot. Current is viewed as flowing out on the hot wires and back on the ground. However, in the electric fence the hot and ground wire are not connected together. The fence's power unit merely places a potential on the hot wire, current does not flow until an animal touches both the hot and the ground wire. When this happens, current flows, and the animal is shocked. This type of set up is called an earth return wire system (current is returned to the power unit by one or more wires).

Many people think of electricity as only flowing through wires or metal however, it can also flow through the earth. For the earth to be a good conductor it must be moist or wet and the soil particles should be small; rocks or lava are not good conductors, nor is dry dirt. When the earth is used as a conductor, it is called an all alive system (no ground return wires, all wires "hot"). The power unit places a potential on the hot wire(s), when the animal touches it, current flows from the hot wire through the animal and returns to the power unit by way of the earth. In moist areas where the soil is damp for most of the year, single wire fence systems are often used. Where the soil dries out for a portion of the year a ground wire return system (earth return) is recommended. Another way to tell which to use is if your forage is growing all year long, you can probably use an all alive (hot) system.

COMPONENTS:

Each separate component should last for the expected life of the entire fence. In other words, it should all fall apart at the same time. The difference in cost between the most expensive item and the cheapest, does not affect the total cost of the fence all that much. The modern electric fence is an inexpensive fence, don't cheapen it by using inferior materials. With quality components, a well constructed electric fence will last 2 to 3 times longer than a conventional barb wire one. What follows is not meant to be a detailed description of all the different materials and methods used in electric fence construction, new components and methods of using them are changing rapidly; what was

[3] It is tempting to call them short-shorts, but those are something else again.

good by last years standards is only adequate by this years. The reader is urged to get all the information he can lay his hands on; the suggested reading section is a start, commercial sources another.

THE ENERGIZER:

The energizer is the heart of the system. Today's energizers are a far cry from the old type "charger" that were in common use just a few years ago. Modern units are capable of powering tens of miles of wire compared to the less than a mile of yesterdays "chargers". The new units are also able to "power through" a considerable amount of trash on the fence, that shorted out the older chargers. The main reason for this difference lies in the ability of the energizer to deliver a high voltage (5000 volts) pulse for a very short period of time (0.0003 seconds). Basically the electrons are stored in a capacitor, the bigger the capacitor the more electrons are stored, the capacitor fires (kinda like a dam bursting) and down the wire they go. The pulse, as it leaves the energizer looks something like that shown in Figure 9.1a, the shape will change the farther down the wire it goes, towards the end it may look like Figure 9.1b. The actual pulse shape will also vary, depending upon the design of the particular energizer and the manufacturer. Since it is not just the voltage that's involved but the total energy delivered, you can have higher voltage, but lower amps, or vice versa; however, for safety reasons there can only be so much power in that pulse.

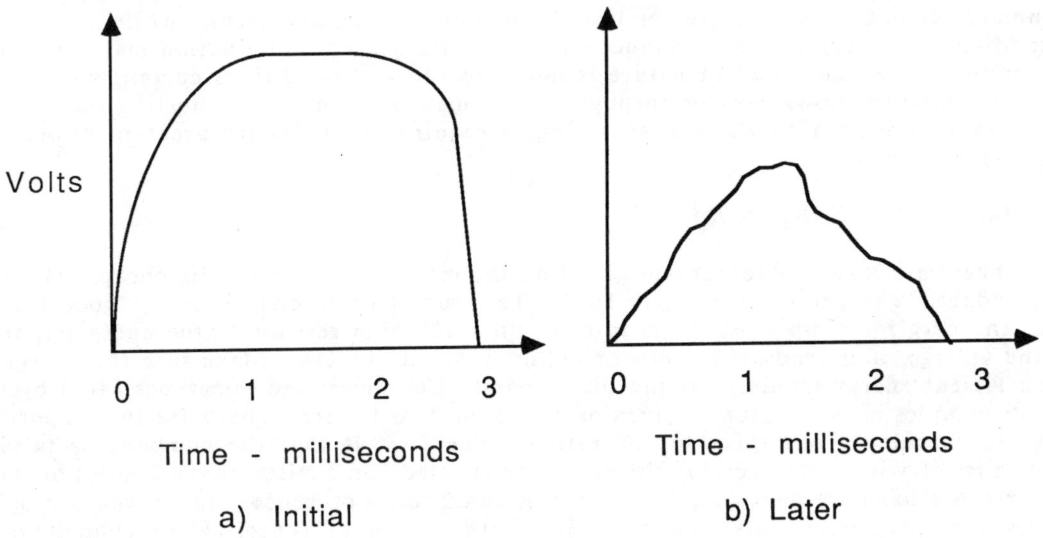

FIGURE 9.1. THE ELECTRIC PULSE FROM AN ENERGIZER.

Most energizers deliver a pulse every second, however some models can vary the pulsing frequency so that one is delivered more often (normally used in training animals). This pulsing action allows for an important safety factor; the wire can be released between shocks. While 0.3 milliseconds does not seem long, by the time the pulse has ended, the first part has traveled 30 kilometers (18.6 miles) down the wire. The shape of the pulse, plotted as voltage over time, changes as it travels down the wire. In general the height, or voltage, will decrease the farther along the fence (greater resistance), Figure 9.1b. However, at some points the voltage may actually be higher, due to the additive effect of the reflected voltage coming back up the wire. This effect can be liken to water waves in a bathtub. A wave strikes the side of the tub and is reflected back; when the reflected wave hits an incoming wave, it builds in height, peaks. This effect sometimes causes concern when testing the voltage along a fence, unless the reason is known.

The amount of energy, or power of an energizer is measured in Joules (United Kingdom) or Kilocalories (United States). One Joule (J) equals 1 watt for 1 second; for conversion purposes, 1 calorie equals 4.186 Joules, 1000 calories equals 1 Kilocalories (Kcal). The number of Joules an energizer is capable of delivering, determines the "kick", or shocking power and how far that shock can be felt down the fence.

Energizers come in all shapes and sizes, but are of two general types, battery powered (direct current, DC) and mains (operating off of house current, which is alternating current, AC). Battery powered energizers are usually rated from 0.05 Joules, small strip grazer units that can handle a half mile or so of wire, to units rated at 0.9 or so Joules, capable of handling up to 9 miles. Mains units can have a rating from 1 to 20 Joules, with the larger units powering up to 50 miles or more of wire.

It is difficult to rate energizers as to the number of miles of wire they can handle. This difficulty is due to the different environments the energizer will be operating in. In wet humid regions there is a greater loss of current than in dry areas. While 100% pure water (distilled water) does not conduct electricity, the least contamination makes water a fair conductor; water found in nature is not pure. In wet conditions current will travel over the insulators, down post or through porous materials, in effect shorting the system. In wet environments a higher energy rating is required than in dry areas to power the same length of wire.

PURCHASING AN ENERGIZER:

There are many different energizers on the market, some are quite cheap. There is an old adage: "You get what you pay for."[4] This runs true in energizers. A good test to see if an energizer is up to snuff is to hook up a 100 ohm resistor to the energizer, then test the voltage, if it reads 1500 volts or better, it should be OK. Make sure the energizer has sufficient energy (Joules) for the job at hand. Energizers are sometimes rated by the number of miles of wire, usually given as 1 to 5, or 1 to 15, etc. There are two important things to consider about this type of rating. The first it that the company is talking about miles of wire, not fence. If the energizer is rated for 2 miles, that's 2 miles of wire. If your fence has only one wire, then you can run 2 miles of fence; but if you are using 2 wires, the energizer is only good for a little over 1 mile of fence. The second item, as mentioned before, is that the energy requirements in wet environments are higher than in dry ones. It is usually good practice to purchase a higher energy rating than actually needed; the additional cost is not all that great. What usually happens is that is once the energizer is in place, more fence (in the form of additional sub-division) is added than was originally planned.

Mains units are preferred over battery powered ones for permanent fence systems; temporary or strip grazing units are often battery operated. Not only are mains cheaper

[4] If you want nice clean fresh oats, they'll cost so much a pound. If you're satisfied with oats that have been through the horse, they come a little cheaper.

to purchase, on a per Joule basis, but they are cheaper to operate and easier to maintain (no batteries to worry about). However, in remote locations a battery unit may be called for. In these situations a small wind generator or a solar panel may be used in conjunction with a battery unit. The amount of electricity that a solar panel can generate, depends on the amount of sunshine available and the number and size of the photovoltaic cells. In cloudy areas a larger panel will be required than in high sunshine areas. The wind generator depends more on the duration of the wind rather than its strength, consequently should only be used where there is fairly constant wind. Both the generator and the solar panel should be equipped with a load regulator to prevent overloading of the battery. The best batteries to use are industrial type batteries. These have a longer life and can be recharged back to full power after draining, something many car batteries can not do.

Assistance in sizing energizer units may often be obtained from your local Soil Conservation Service office, or your County Agent. If they are unable to help, you will have to rely on the dealer and/or what your neighbors are using.

SAFETY:

The safety aspect of electric fences is occasionally raised. This is to be expected, as the shock hurts; it's suppose to. As mentioned before, the pulsing action allows one time to get away from the wire. If 5000 volts were continuously applied to the wire, there would be cause for concern. In the past many "home made" electric fences consisted of the 115 volt house current passed through a light bulb then hooked directly to the fence; this lash-up can kill. To date, no human has been killed by a modern energizer. A very few animals have been killed, but in every instance that the authors are aware of, the animal became entangled in the fence, was lying on its back in a moist ditch and remained that way for a considerable period of time. The fact that the animal was on its back (which by itself can kill a ruminant) and the continual shocks, eventually did it in.

Never use more than one energizer on a fence. All energizers in the U.S. should be UL approved. It is also a good policy to place signs on or near the energizer, stating that this is an electric fence. Check your local State or county laws, sometimes there are additional requirements to meet.

LIGHTNING: In areas where there are lightning strikes, damage to energizers can be minimized by the use of lightning diverters (make sure the diverter has a choke). The best precaution is to unhook the unit from both the fence and the house during electrical storms. In high lightning areas, having a ground wire as the top wire of the fence, will help; but only if the wire is well grounded.

GROUNDING:

Over 80% of all electric fence problems can be traced to poor grounding. Without satisfactory grounding or earthing, the fence simply won't work correctly. It requires about 20 feet of metal rod to make a satisfactory ground. This can be a 20 foot galvanized steel stake, 3/4 to 1 inch in diameter, driven into a permanently damp site, or laid in a deep trench. If it is not possible to drive a 20 foot stake into the ground,[5] then use three or more stakes, about 15 feet apart, connected together by non-corrosive electrical clamps and 10 gage wire, or two or more 12.5 gage wires twisted together. If the soil is very wet and of fine texture, a shorter earth stake can be used. In earth return systems

---

[5] Just finding a 20 foot high place to stand, steady enough to swing a hammer, is going to be your first problem.

additional ground rods (5 or 6 feet in length) will need to be placed every 1500 feet in dry environments and 3000 feet in higher rainfall areas. If steel post are used, grounding to them also helps.

In dry areas, try to place the ground rod in soil that is always damp. If a large diameter water pipe is used as a ground rod, pour water into the pipe every time the unit is checked. Another possibility is to use a non-conducting drip tube hooked up to a water supply. In areas where this is not feasible, drill or dig 3 or 4 holes along the fence at least 40 feet apart. Make each hole at least 3 inches in diameter and 4 feet deep. Fill these holes with a wet slurry of 2 parts Bentonite and 1 part course salt. Place a ground rod in the center of each hole and connect to the ground wire of the fence with suitable clamps. Because of the corrosive nature of salt, stainless steel ground rods should be used. Spread an inch or so of course sand over each area to reduce evaporation. A ten-fold improvement in earthing has been obtained by this method.

Do not place the ground rods closer than 10 feet (preferably 40 feet), to any other electrical ground, or to water pipes.[6] Do not place the ground stake near milking barns, near metal water pipes or other metal items entering the barn. Cow are very sensitive to even a few volts delivered to the udder or mouth by the milking machine or water pipes; they can detect as little as one half volt.

The ground system should be tested once a year, during the dry period. Short the fence out with several steel stakes placed on the hot wires and the earth (the soil should be damp, if not, wet it), do this at least 350 yards from the energizer. Use enough stakes to lower the fence voltage to at least 1000 volts (the lower the better). Connect one lead of the voltmeter to the energizer earth terminal and the other to the ground (damp) at least 3 feet from the energizer's earth rod. The reading should be zero or at most 200 volts. A quickly method, but not as good, is to attach one lead of the voltmeter to a metal rod, then push this rod into damp earth at least 12 inches, attach the other lead to the ground wire on the fence, the reading should be zero. If you get a positive reading, add more permanent ground rods until the reading is acceptable.

When using the small strip grazing energizers, grounding can be a major problem, since it may be moved as often as the fence. Make a special ground rod (12 inch long), attach a terminal or clamp for the ground wire of the energizer. Carry a small water plastic jar (like those used by bikers and joggers), stick the rod in the ground and give it a squirt of water. If there's a steel post or an old net or barb wire fence nearby, attach another ground wire to that. Use the ground system of the mains energizer only as a last resort; a leak or short in the mains system can cause problems with the smaller unit.

WIRE:

If the energizer is the heart of the system, then wires are its veins. Wire comes in different sizes and is made of different material. The best electric fence wire is 12.5 gage, triple galvanized, hi-tensile steel wire, in most situations this will last twice as long as conventional barb wire of similar gage. For those areas near the ocean, volcanos, or other corrosive sources there is an aluminum coated hi-tensile steel wire that is reported to last two or three times as long as the triple galvanized wire. The temptation is often great to skimp by using less than class three galvanized wire, or wire of a smaller diameter, or even worse barb wire. The risk of becoming entangled in barb wire by both humans and animals and taking repeated shocks, isn't worth it, not to mention the liability aspects. Don't make an inexpensive system cheap.

[6] Someone could end up taking a very "hot" shower.

The resistance of steel wire depends on the gauge (diameter) and the length; the larger the diameter, the less resistance, the longer the wire, the greater the resistance. Table 9.2 gives an idea of this relationship.

TABLE 9.2. ELECTRICAL RESISTANCE OF HI-TENSILE FENCE WIRE AS A FUNCTION OF DIAMETER (GAUGE)

Diameter (mm)	Gauge	Resistance (ohms/Km)	(ohms/mile)
4.0	8	14	22.5
3.15	10	22	35.4
2.5	12.5	35	56.4
2.0	14	54	87.0
1.6	16	85	136.9

An energizer that is capable of charging 3 miles of 16 gage wire can charge 7.25 miles of 12.5 gage. If more than one hot wire is used these should be connected together at the beginning and at the end of the fence. Two wires connected in this manner have considerably less resistance than if not connected. Multiple hot wires should be connected together when passing under or over gates.

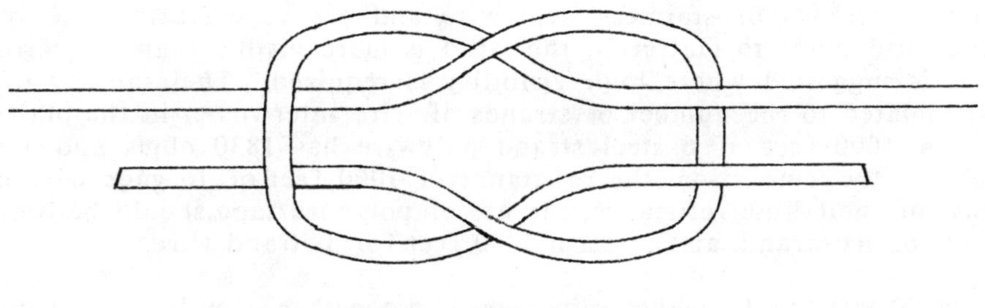

FIGURE 9.2. ONE TYPE OF SPLICE FOR HI-TENSILE WIRE.

Splicing the hi-tensile wire may be accomplished with a reef or figure eight knot or with a nico-press tool, don't use a barb wire type splice (Figure 9.2 shows one type of splice for Hi-tensile wire, a figure eight knot is another good one to use). If a lot of fencing is to be done, it is cheaper and faster to learn how to use just your hands to tie knots, break wire and tie off insulators. For one thing if you depend on a small hand tool, it always seems to be at the other end of the fence. Breaking wire by hand is not all that difficult. The key elements are to make a handle, make sure the spot where you want the wire to break makes a right angle and when you make your move, do it fast. Your dealer can usually show you how to break wire and make ties by hand.

Hi-tensile wire should never be kinked, dragged over rocky ground, or dropped on rough surfaces. Anytime a break is made in the galvanization, the life of the wire is shorten. Wire should be stored where it will dry out quickly after rain and away from fertilizers, acids, lime or other corrosive materials. In splicing a wire, it is best to turn off the energizer first. If not, connect a large diameter wire (copper is best for this) to the ground wire and then to the hot wire(s); this shorts out the system from that point on.[7] The fence can then be worked on without shocking results; don't forget to remove the short when you're finished.

LEAD-IN WIRES: Use 10 gage insulated wire wherever there is a chance of shorting out, otherwise regular fence wire can be used. When passing wire under gates, 10 gage insulated wire is required. Never use a lead-in or under gate wire that is made of different material (such as copper) than the main fence wire. When different materials are used in electrical connections, electrolysis results. Electrolysis may be thought of as very rapid corrosion; spraying or covering with tar is no help. The problems resulting from electrolysis are not worth whatever benefits that may result from using a different composition wire. Clamps should normally be used, however if the hook up is temporary, commercial flexible connectors can be used, or you can make your own by taking a couple of quick wraps; remember this is temporary.

TEMPORARY WIRE:

The wire used for strip or break grazing is usually 16 gage metal wire, or polywire and/or polytape. Polywire and tape are made from an ultraviolet resistant plastic cord interwoven with thin strands of stainless steel wire and are very flexible. Poly tape, because it is wider and tends to flutter in the wind is more visible than polywire; for this reason its use is suggested where high visibility is required. Their ability to carry current is directly related to the number of strands of wire interwoven in the plastic, the more the better. A 1000 feet of 6 steel strand polywire has 1830 ohms and 3 strand polywire, 3050 ohms. By comparison, the resistance of 1000 feet of 16 gage wire is only 110 ohms. Because of their high resistance, the use of polywire/tape should be limited to less than 2000 feet for six strand, and under a 1000 feet for 3 strand wire.

TIPS ON POLYWIRE: To make splices, or join together, simply tie a good knot. When hooking up to a power source, make sure more than one of the thin strands come in contact, if not, the strand may burn out. On long spans, over 500 feet, best results are obtained if the energizer is placed in the center of the span. If the metal strands are removed from about 6 inches at one end, that end can then be tied to another fence or whatever without use of a special insulator; it also makes winding up the wire easier as a bulky insulator is not dragging in the vegetation. When using a hot off-set wire or hot fence wire as a power source, tie an insulator to the end of the polywire, hold on to the

[7] If you hook up to the hot wire first, you will quickly understand why you should hook up to the ground first; also, when you disconnect, unhook the hot wire first, same reason.

insulator and pass it over the hot wire and wrap it around several times; this makes a good connection and can be easily removed by giving the insulator a flip to unwrap it, Figure 9.4 shows this type of quick connection made to a single strand of galvanized wire. Remove the polywire from the temporary post before winding up on a reel, this increases the life of the wire. Do not put a lot of tension on the wire as the wire strands may break but not the plastic; there would be no power beyond the break. Remember, no matter how long you use polywire, it is not a permanent fence.

FIGURE 9.3. QUICK CONNECT FOR TEMPORARY FENCE.

SOME NO-NO's:

Never run a wire directly underneath hi-voltage lines. Wires running parallel to hi-voltage lines pick up voltage, though a process known as induction (transformer effect), this can be a sizable jolt. If you must cross hi-voltage lines do so at right angles to the lines. If you must run parallel, then be at least 100 yards away. The same applies for telephone lines, the same effect, induction, but for different reasons. The pulse may interfere with the phone conservation, or if the energizer is grounded to close to a phone ground, interference can also result. Interference with a radio is seldom a problem, unless the fence is arching somewhere near a radio antenna, usually within 10 or so feet. Only have one energizer connected to a fence at one time. Never use polywire to fly kites, that wire is made to conduct electricity.

POSTS:

Almost any post can be used in electric fencing, even trees; but, most conduct electricity and insulators must be used. Some post, such as plastic, fiberglass and insultimber (a high density hardwood made from selected species of Eucalyptus), have very high electrical resistance and can be used without insulators. The major problems with fiberglass and plastic post are: When used on a ridge (down pressure on the post) tend to bury themselves. When used in a gully (up pressure) tend to pull out. This can be partially corrected by drilling a hole in the bottom of the post and inserting about a 6 inch length of fence wire in it; the post is then driven in, the wire acts as sort of an anchor. If problems continue, either change to a steel or wooden post or, if in a gully make a proper anchor. Some brands of fiberglass post can not take much of a compression loading (they shatter), so don't use on ridges. Many plastic and fiberglass post do not withstand ultraviolet light very well and tend to deteriorate with time, similar to PVC pipe. Hopefully this problem will soon be corrected. Fence wire must be attached to all posts so that it runs free, only end posts and spacers have wire tied off tightly. NOTE: Post for electric fencing should be set at right angles to the ground surface, not on a plumb line as is often done with conventional fences. The reason is that if a post is set on a plumb line the top of the post may only be couple of feet from the surface of the ground on steep slopes; unless you use a very long post.

BRACES: Because the tension on the wire is not high, ideally just enough to keep them from sagging, 150 to 200 lbs maximum, massive end braces (either H or N braces) are not required for one or two wire fences; a deadman will usually suffice. If more than two wires are used then a regular H or N brace is suggested.[8] To construct a deadman brace, first dig a hole 4 feet deep and place the end post (this post should be 8 feet long and have a diameter between 4 to 6 inches). After the post has been tamped in, dig a 4 foot long, shallow trench about 5 to 6 inches deep, next to the post and at right angles to the fence. In this trench place a 4 inch diameter log or cut-off post, make sure it is touching the end post, cover and tamp and that's it. A good idea whenever an end post is required is to twist about 3 strands of 5 foot long fence wire together and staple to the bottom of the end post (before you bury it), run the wire up along the side of the post so that a foot is left above ground level and connect it to the ground wire. This greatly helps in obtaining a good ground. Different braces are illustrated in Appendix 6.

BATTENS: A batten is a short piece (35 or so inches), of fiberglass, plastic or insultimber stake connected to the fence; it may be used as either a SPACER or a DROPPER. A spacer is used between post to maintain wire spacing and to help stabilize the fence; it may rest on, but does not penetrate the surface of the ground. Spacers are tied tightly to the fence wire, the wire does not run free. On the other hand fence wire going past a dropper must always run free as the dropper is tied to the earth. A dropper is used to pull the fence down in a gully or low area. A steel or insultimber stake (with a hole in the top end) is driven into the ground at about a 45 degree angle and acts as an anchor. A batten is then tied to the hole in the stake and the fence pulled down to the desired height. A rock with a length of fence wire tied around it and buried can also be used as the anchor. Battens, for use with anchors, have been made out of old PVC pipe or even black plastic pipe. The pipe is cut to the desired length and grooves cut into it with a table saw (angle the cuts at about 45 degrees), at the proper spacing. The major problem

[8] If goats are being run, either remove all N braces or electrify them; goats will climb up the brace pole and jump out.

is the length of life, the PVC gets brittle with age and the black plastic tends to flow a bit (paint on the PVC will help some). Figure 9.4 shows a home made dropper from PVC pipe.

FIGURE 9.4. HOMEMADE DROPPER FROM PVC PIPE.

TEMPORARY POST: Temporary post are used in strip and break grazing. They may be made from almost any material; some require insulators, some don't. Several commercial types are available (Figure 9.3 shows one type) or you can make your own, however, the economics are questionable. But if you must, points to remember are; a temporary post, must go into the ground a sufficient depth so that it doesn't fall over, must be easily placed in the ground and must be easily pulled out, you usually end up carrying a lot of them around, so they must be light, yet sturdy. A single piece of reinforcing rod with an insulator mounted on it works on soft, level ground. On hard ground, if an upside down L is welded to the bottom of the rod, this provides a place to put your foot to push in, it also helps in stability. Reinforcing rod is a conductor, so care is required in erecting the fence.

OTHERS:

INSULATORS: A successful insulator does not conduct electricity, should have a minimum of 3/4 to an inch spacing between the wire and support, has a smooth surface that does not collect dust or any other conducting material, is strong enough to support the wire with both horizontal and vertical forces acting on it, is able to be attached to the post, is able to withstands temperature extremes and is not degraded by sunlight (ultraviolet). Insulators are made out of porcelain, glass and any number of plastics. Porcelain insulators should be used at any point where the wire places a heavy strain, especially on braces and corners. Check the porcelain for cracks, if present they will collect dust and moisture and cause shorts. The best insulators for the T type steel post have been those that require holes to be drilled in the post, clamp on type have not worked very well. There are however, some new insulators on the market the tie securely to the steel post which appear to work much better than the earlier clamp-on types.

SWITCHES: The basic idea of a switch is to be able to turn off the current to a section of fence without getting shocked in return. The older versions with a porcelain base, a brass throw and a bit of plastic on the end, did not do the job very well, especially when raining. A good switch should also easily indicate if the current is on or off. Switches should be liberally used in fence construction as they greatly aid in trouble shooting. They also keep the power requirements of the energizer low; not all of the fence must be on at all times, only that portion around the livestock.

OFFSETS: Often called stand-offs or out-riggers, they come in several different designs, with porcelain or plastic insulators. Offsets are used to remove animal pressure from new fences or to rejuvenate terminally ill fences. They attach to the old fence and a hot wire connected to the insulators; the old fence can often serve as the ground. A single offset wire should be placed at two thirds to three fourths the height of the animal to be controlled. Offsets can add many years of life to a new non-electric fence by removing the animal pressure on the fence (no more rubbing against, or sticking the head through). Their use on old, rusty fences is not recommended as a permanent solution, as the old wire often breaks and shorts out against the offset. Also, the cost of the offsets and wire, along with the problems associated with a run down fence generally make it advisable to replace the old fence with an electric one.[9]

MISCELLANEOUS: Strainers; there are several types available, some require special tools, some don't. Tension springs; for maintaining tension on short spans or, where there is heavy snow loading. Ties; used with insultimber post and spacers. Spring loaded gates; both hand opened and drive through. Strip grazing reels. Joint and ground clamps. Automatic energizer turn offs; these connect to the energizer and with a special gadget, the energizer can be turned on or off anywhere along the fence. These are good when a lot of fence repair is expected, for example when new stock, not trained, are introduced to a paddock. Flood gate controller; used in conjunction with watergates and limits the amount of current that flows to the watergate during a flash flood thereby maintaining the fence in an effective state. Also used when there is excessive forage growth that causes shorting of the fence. Signal lights that attach to the fence that blink if the fence is operating satisfactory, usually only visible at night.

TOOLS: The basic tools for electric fencing are: 1. Voltmeter. There are several different models available, digital read out, meter read out, or neon light. If a voltmeter is not handy and you wish to find out if the fence is hot, take a long piece of grass, hold one end in your hand and place the other end on the wire; slowly advance it towards the wire until a slight shock is felt (remember the pulse only comes once a second, so don't

[9] There is a great temptation to make do with an old fence, one keeps making do, until all one has left is do-do.

advance it too fast).[10] 2. Wire pliers. 3. Heavy screwdriver. 4. Electric fence tool. 5. Crescent wrench. 6. Post drivers. 7. Shovel. 8. Pry bar and/or tamper. 9. Strainer handle. 10. Pay out spinners, a MUST if your fence measures more than a few yards.

THE ANIMAL:

Before one can possibly hope to control an animal with a fence, one must first know something about the animal one is attempting to control, not to mention the reasons an animal attempts to go through a fence. The reasons why animals attempt to penetrate fences depends on the species of animal, circumstances and if the fence is "newly constructed", or one that has been around for some time. However, a brief summary would be: 1. Curiosity; the "grass is always greener". 2. Nutritious and/or favorite food source. 3. Fence lays on traditional migration routes. 4. Accidental wanderings. 5. Mothers following young. 6. Momentum of large herds. 7. Drought. 8. Harassment by predators, hunters or poachers.

In addition to knowing why an animal may attempt to penetrate a fence, one also needs to know how the animal intends to go about doing it. Table 9.3 shows the general preference of animals when attempting to get past a conventional wire fence. The first preference of the majority of animals is to push through the fence, only snakes, zebras and kangaroos prefer to push under as a first choice. Grizzly bears seem simply to like to tear them up, especially if there is hive of honey bees on the other side. Some herd animals, such as the Wildebeest, apparently use a random milling approach, where, after a lot of shuffling and jostling around, an individual, or small group, finally gets pushed through. In addition there are those animals that are both determined and potential learners, such as elephants, primates and poachers. Consequently there is no one fence design that is going to work for all species of animal, under all conditions, for all time. However, generally speaking livestock are the easiest animals to control and predators and game animals the most difficult. A coyote fence will control livestock as well as many game animals, but not a determined poacher. By knowing where and how the major thrust is expected to come, hopefully gives one an edge. For most animals this means at least one hot wire erected about two thirds the height of the mature animal. For game animals it may also mean an aluminum spinner to attract their attention and arouse their curiosity so that they will cautiously investigate the fence. For poachers and rustlers an alarm that tells when the fence has been cut, or power levels have dropped.

How the animal will react to its first shock has been found to depend not only on the STRENGTH of the shock but also the species of animal and where that first shock is delivered; nose, ears, body, etc. Generally if the shock is to the nose, the animal will back up, if the animal gets part of its head through, it will often as not go forward. For example coyotes generally will touch the wire with their nose and/or whiskers and will snap back upon shock. However one coyote seemed to sense the pulse and moved under the wire during the off phase, two others upon receiving the shock farther up the head, went forward through the too widely spaced wires. Stock dogs that have been used to passing through non-electric fences do not seem to be able to sense the electric pulse, they will continue to pass through until they get "nailed". Some dogs get quite neurotic; one, after the area around the headquarters was electrified, would not leave the area unless riding in the back of a pickup. On the other hand opossums will curl around the wire and if touching both hot and ground will die from repeated shocks. Very small animals,

[10] Another method, which can be used only once per person, is to hold a friend's hand and grab the wire; the friend (or ex- friend) will take the shock.

mice and rats, may be killed by the use of high powered "mains" energizers, in which case they learn very little, or a great deal depending on ones point of view.

TABLE 9.3. METHODS PREFERRED BY DIFFERENT ANIMALS IN ATTEMPTING TO GET PAST A FENCE.

	push over	push under	push through	dig under	climb over	jump over
Cattle	2		1			
Sheep			1			
Goats		2	1		3	
Pigs		2	1	3		
Deer			1			2
Geese			1			
Snakes		1				
Kangaroos		1	2			3
Rabbits		2	1	3		
Opossums			1			
Dogs		2	1		3	
Cats			1		2	
Foxes			1	3	2	

1 = first choice; 2 = second choice; 3 = third choice
Table adapted from R. Lund and S. DeSilva in Gallagher 2nd World Wildlife Power Fencing Seminar.

The electric fence is very passive, there is nothing very threatening about it, animals rarely attack the fence after receiving the first shock. However, many animals will make quite a display around the fence after first contact, digging the ground, pawing, bellowing, shaking the head, etc. They will then cautiously approach and test it again, they may do this several times, one rhinoceros made 15 contacts with a fence in three successive nights, the next contact occurred three weeks later. Horses will usually take the first contact on the nose, whirl and run to the center of the paddock and stand there "all alert" for a short period of time. On the second contact they will usually snort and back off a short distance, then continue doing whatever they were doing.

Elephants will always first touch the fence with their trunk. After the first shock, they will test the fence at repeated intervals, checking each wire and post, one bull even tried to push over a fence by backing into it. In one report, it was noted that after the introductory phase, whenever a herd of elephants came to an electric fence, only a leader would test it, upon finding it "hot", would move away with the rest of the herd.[11]

In designing fences for wild animals, finding what works and does not work is not so much using available theory (quite often there is simply little to no information about the animal) as it is trial and error; the first electric elephant fences are a case in point. The first fences, consisting of a slightly taller version of a three wire cattle fence, were

[11] One thing about elephants, it doesn't take many to make a dozen.

placed along oil palm plantations. They worked for a little while, however the elephants learned that the post were not hot, so they used their trunks, pulled up the post and went through. The builders then electrified the post; it worked for a little while. However, the elephants soon found that they could take trees and other large objects and throw them on the fence, knock it down and go through. So the builders moved any material the elephants could pick up and drop a couple hundred yards away. This worked for a little while, but the bull elephants found that they could throw the baby elephants into the fence, knock it down and go through. The builders gave up, figuring the elephants won, however the problem turned out to be self correcting. It seems as though baby elephants are also pretty smart, they don't really enjoy getting thrown into a hot fence, so they sort of disappear when there's an electric fence near-by.

TRAINING:

The strength of that first shock will often determine how the animal is going to react to the fence in the future. If the pulse is weak, the whole effort is likely doomed to failure. If the pulse is strong, the message gets through the toughest hide to the smallest brain and the lessons have begun. Do not use low power energizers, or ones that can be easily shorted out. Never erect a fence and fail to electrify it as soon as the wire is up. Those first contacts must be met with a strong shock and that shock must occur each and every time contact is made; this is especially important during the introductory or training phase.

Learning in animals does not occur merely by chance. If an action by an animal results in something pleasant occurring, then the chance of that animal doing that same action again becomes greater; likewise if the animal is hurt or frighten by an action, then the chance of the animal doing that particular action again is less. Learning and training are closely related; the animal must learn if it is to be trained. However, training does not produce new ways of responding to a situation; the animal can't be trained to do things that it is not capable of doing. All training can do is to reinforce some desirable behavior that the animal already is capable of doing. Thus opossums will continue to curl around the wire after receiving a shock. Training in this instance will depend upon not killing the animal by designing a fence so that the animal can get off after the shock, but at the same time allowing significant punishment to be delivered so that the lesson is learned to avoid the fence in the future.

All animals, domestic or wild, can be trained to respect an electric fence; however, all require time and space to quietly discover the major disadvantage of fussing with an electric fence on their own. This initial contact period must be done under conditions as near natural as possible, stress should be kept to a minimum. An animal is not going to learn much respect for a fence, electric or otherwise, if it is being chased by hunters, "Ya-Hooed" by rodeo cowboys, or has just had it's rear end tenderized with a "hot-shot".

LIVESTOCK:

Livestock are easier to train to electric fences than are many wild animals. Livestock are generally calmer,[12] are not quite as alert as are animals in the wild and the operator has a greater degree of control over them. Also, with intensive grazing management one of the major reasons for going through a fence is often not there, adequate nutrition is available at all times. If the animals are on a fast rotation program, they know that fresh forage will be given to them shortly, thus the need to test the fence is reduced.

---

[12] Cattle are not born wild, man makes them that way through stupid handling.

The easiest way to train livestock is to take an existing small, well fenced, paddock (net wire is best) and place an offset wire about two thirds the height of the animals to be trained. Make sure that the energizer is on full power and on fast speed (if there is a variable pulse control). Test the power delivered to the training fence with a voltmeter, a reading of 4000 to 5000 volts is suggested. Quietly introduce the animals to the paddock, then leave them alone. A minimum time of 12 hours is required, most animals will be fully trained in 48 hours. Another good way is to simply build a short length of fence near a common gathering area, such as around water.

If a special training area is not available, select a level paddock that has more than one wire, preferably three or more, tie some ribbons on the hot wires so that it will flutter in the wind. Animals are curious and if not under stress will investigate any thing "strange". The idea is to get them to approach the fence slowly. It also helps to leave the gates open, these become recognized as "safe" areas and makes future moves easier.

If you are building a new fence in an existing pasture with animals in it, make sure the newly constructed portion is turned on before securing for the night. First impressions are important. If that first encounter does not result in a hard shock, the animal will continue to challenge long afterwards. The animal may even turn into one of those that is willing to take a weak shock just to get to the other side. This often happens when a poor quality energizer is used. If stock are properly trained to electric fencing they will have respect for the fence even if the power goes off. Animals that have grown up with electric fencing have remained in paddocks with the power off for several weeks. Animals newly trained, or where there is a strong temptation to get through the fence, a few days. Where the moves are frequent stock seldom test a fence.

TABLE 9.4. DISTANCE CATTLE REACT TO DIFFERENT TYPES OF ELECTRIC FENCING MATERIAL.

Wire type, color, etc.	Average distance seen (feet)	range (feet)	Average final stopping distance (ft)
New orange tape	25.9	15-50	8.5
Old orange tape	19.0	13-26	6.2
New black-yellow twist	19.7	15-23	5.9
Old black-yellow twist (grey)	13.1	10-15	7.5
Old orange twist (yellow)	12.8	15-26	5.9
Old orange-black twist	17.1	10-33	7.5
No. 10 galvanized wire	24.9	6-82	6.6
Baling twine (green)	30.2	10-66	12.5

Table adapted from R. Kilgour in Gallagher 2nd World Wildlife Power Fencing Seminar.

The more visible the fence the easier it will be seen, this has to do not only with the physical size of the fence, wire and post, but also to its color. Individual cows and small groups of five were tested on their awareness and their stopping distance to several types of fencing material and color. Table 9.4 shows the average distance different electric fence wire and tapes are recognized and the final stopping distance for cattle. The brighter the material the sooner it is seen and generally the farther back the animal stops. New polytapes and wires should be used for training rather than old mud soiled ones. When galvanized wire is used, make it visible with ribbons or spinners.

Tropical breeds of cattle sometimes jump fences, but seldom do so if trained with an offset on a high fence. Persistent jumpers may have to be culled; you don't let one animal dictate ranch policy. If you have some really hard cases, take a tin can lid, punch a hole in it, make a short hook out of wire, dip in molasses and hang on a hot wire. This will do two things; the animal will learn respect for the fence and it will probably never eat molasses again.

Sheep, because of the insulating properties of wool, will require a tighter fence than cattle. You want the first contact to be on the nose, not the back of the head. If the shock is delivered to the back of the head the tendency is often to go forward rather than back up.

It has been noted that during thunder storms stock grazing in wheel designs will not bother the radial fences, unless they are newly trained animals, or have just been placed in the unit and are not familiar with the layout. Typically they will gather at the hub (they are used to going there), bunched up through fear. They often cause considerable damage to the center, unless the hub is electrified; but even that is no guarantee. Stock grazed in block designs will usually go to the center of the paddock and mill about there.

Dogs can be trained to jump into pickups, they can also be trained to jump over electric fences. Use the command "jump", or the same command you use to get them to jump into your pickup, and "no" if they try to crawl under the fence. It's a good idea to try this first with the power off. If during the day, you shout "no" every time the dog happens to receive an accidental shock, this will reinforce the "no" command; which may be of some help in future situations.

WILDLIFE:

If first impressions are important with livestock, they are doubly so with wildlife; that first contact must be a memorable event. The whole idea of wildlife training is to get that animal to investigate the fence in a slow and cautious manner, so if at all possible, that all important first shock is delivered to the nose. There are several ways to help bring this about. The first and most obvious one is that the fence must be visible. This can be accomplished by using lots of wire and post, offsets, making visual fences in front of the main fence, and/or tying things (hangers) on the wire that flutter in the wind, such as strings, ribbons or aluminum twisters or spinners. It has been found for African wildlife that hangers colored white are the best deterrent or at least stand out better than yellow or silver hangers.

The electric power fence should be designed for the animal that it is meant to control, rather than attempt to erect an "all purpose" fence. Some animals may need to be let through the fence. It should also be understood that most wildlife fences will leak some. A 100% animal proof fence is, in most cases, not economically feasible. It is entirely possible to build a "Berlin Wall", but it's hard to justify it. Even chainlink fences are torn down by grizzly bears trying to get to hives of honey.

Each animal has different behavior patterns, that should be considered. Bears and beekeepers have been going at it for years. In addition to shooting and trapping, beekeepers have adopted a number of unusual and bizarre methods of protecting their hives, such as placing on tall platforms with overhangs, in old school buses, chain link cages, and lately electric fencing. The use of flexnet (essentially a net fence made out of plastic and interwoven stainless steel wire) has been found fairly successful. Often a woven wire fence (chicken wire) is laid down, outside the hive enclosure and grounded to make sure the bear makes good electrical contact when he touches the hot wire.

The fence should be placed such that the animal, when first coming upon it, can have a chance to recognize and explore the fence on its own. Animals in the middle of a stampede learn very little about events that are going on along the mobs edge, or even what triggered the stampede in the first place. Often a trial fence near a water hole or favorite resting or grazing area alerts the animals that there is a "new critter" in the woods. This way when the animal gets shocked it can get away from the fence and sort of think things over. The fence should stand out from its background and be identified as something special as the eyesight of some animals is poorer than others, such as bears and giraffe's. If you don't make the fence visible to them, at least make it strong enough so that if they stumble and fall into it they won't knock it down.

One basic rule that is overlooked more often than not; the fence must be effective when the animal is. In other words if the animal operates at night, it does little good to check the fence out during the day. Check it out at night, when dew or other night things may be causing shorts.

The issue of animal rights and welfare concerning the electric fence has not, at least at this time, become a blurring issue. By any welfare criteria the electric fence must be rated humane. The animal is being punished when it does a certain act. However, this is no different than any other "natural" experience any wild animal is liable to have. Animals learn by trial and error, only if it is punished does it learn to leave certain "bad" things alone. It's better to shock than shoot; like old age, it's not too bad, considering the alternatives.

THE FENCE:

Fences are made in straight lines for two reasons: 1. Aesthetics. Somehow or other we have come to believe that straight fences look better than crooked ones. 2. Economy. A straight line is still the shortest distance between two points. From the point of view of most animals, it doesn't matter, since a straight line is completely alien to them. Many ruminants have eyes that are constructed such that they give an image similar to a fisheye lens on a camera, also straight lines do not exist in nature; a couple of the reasons why curved corrals and chutes work better than straight ones.

Chances are, when you build your first electric fence, it will look very similar to one of your barb wire fences; straight as an arrow, post every 16.5 feet, 3 or 4 wires, H braces on each end, etc. The second electric fence will look a little different, by the time you have 10 or so miles under your belt, your electric fence will finally begin to look like an electric fence should; a fence that is a psychological barrier, not a physical one. There are two aspects to a psychological fence, one is the training of the animal and the other is your training; both take time.

In general, an electric fence should be straight, as long as it is easy to build straight; but, if you need to zig or zag to miss some obstacle, then do so. The low tension (just enough to keep the wires from sagging, 150 to 200 lbs Max.), on the fence allows plenty of leeway. However, keep any change in direction less than 30 degrees from the original, greater than this and the post will have to be braced. The distance between post and battens (spacers or droppers), will vary depending on terrain; on level ground, post should be about 44 yards apart with 3 spacers between. Post should always be placed at right angles to the ground surface and not on a plumb line. REMEMBER, the wire should be running free, the only place where the wire is tied is at the end posts and spacers, never to other posts or droppers.

The number of wires to use is dependant upon the environment and also the species of animal; in wet environments only one wire may be needed for cattle but in in dry regions two would be required. Wire spacing is pretty much determined by the species of animal to be controlled. For large animals, wire spacing less than 6 inches apart is not recommended. The closer the wires are together the greater the inductance (transformer effect) and the higher the power loss. Secondly the closer the wires are together the greater the chance that they will twist, either from animals going under or even the wind. Closer spacing is necessary for small animals such as mice and rats.

FIGURE 9.5. NUTRIENT TRANSFER.

All wires should be properly insulated from the post, even the earth returns, this gives the ability to easily change the charge on any wire, should the need arise. With sheep and goat fences the lower wires should all have a switch or be easily disconnected.

This will allow the animals to graze under the fence and keep the vegetation down (you may want to use a different herd, species, or class of animal to do this). If animals are allowed to graze under the fence, nutrients will be transferred away from the fence line.[13] This will result (with time) in lower forage production under and near the fence than a few feet away. In Hawaii, with cattle on kikuyu grass, a noticeable difference is seen under the fence line within two years as Figure 9.5 shows.

LIVESTOCK FENCES:

Typically the wire(s) should be placed where the animal will try to go through, rather than under. The spacing between wires will depend on the species and class of livestock to be run; it also depends on if they are to be moved frequently or not. Frequent moves allow more flexibility in the fence design; the animal pressure to go through is not as great. In wet environments, with frequent moves, a single hot wire about two thirds the height of the animal is adequate. In a two wire system (earth wire) the top wire (hot) should be about 35 inches high and the ground wire 12 or so inches below this. A three wire cow/calf system has the first wire hot (+), 11 inches off the ground, the second wire is grounded (-) 12 inches above the first and the third hot (+) 12 inches above the ground wire.

For easy to control sheep on flat or level ground, a 4 wire fence is suggested. The bottom wire (+) is 6 inches from the ground, the second (-) 7 inches above, the third (+) 10 inches above and the fourth (+) 12 inches above the third at a height of 35 inches. For hard to control sheep or goats and dogs, on uneven or rough terrain, a 5 wire fence is suggested. The bottom wire (+) is 6 inches from the ground, the second (-) 5 inches above, the third (+) 7 inches above, the fourth (-) 8 inches above and the top(+) 9 inches above the fourth; total height of 35 inches. The actual distances may vary some, depending on terrain and size of animals. Goats may require an additional bottom ground return wire strained to a tight tension, 2 inches off the ground. Figure 9.6 show some multiwire fence designs.

GAME FENCES:

Electric fences can be an effective non-lethal method of controlling wildlife. If improperly designed and constructed, they can be a total disaster. The fence must be designed for the animal that is to be controlled. Wild animals are a bit above domestic livestock when it comes to ingenuity and finding new and novel ways to test the fence. Control is the key word, total exclusion is seldom justified. The fence will be tested continuously, if not by the same animal, then by new ones moving into the area. Consequently, only the best materials and construction techniques should be used; anything less spells disaster and even possible liability suits.

While there are some notable exceptions (elephant fences), most game fences require more than four wires, usually at least six but some may have as many as 11 wires. The wires are usually alternating in charge (+,-,+,-, etc.), also the bottom wire is usually a ground wire, one or two inches off the surface. The low wire causes those animals that like to go under the fence as their first, second or third choice, to go over the ground wire, so when it touches the hot wire, good electrical contact is made. Low level offsets are another way to discourage going under fences.

Jumpers are a slightly different problem. There are two approaches, the first is using a visual barrier fence in front of the main one. This in effect widens the fence so

[13] An animal will only defecate or urinate on a hot wire, once.

that the jumper must not only clear the height, but also has to go for distance. The second approach is to cant (tip) all or part of the fence towards the direction of animal approach.

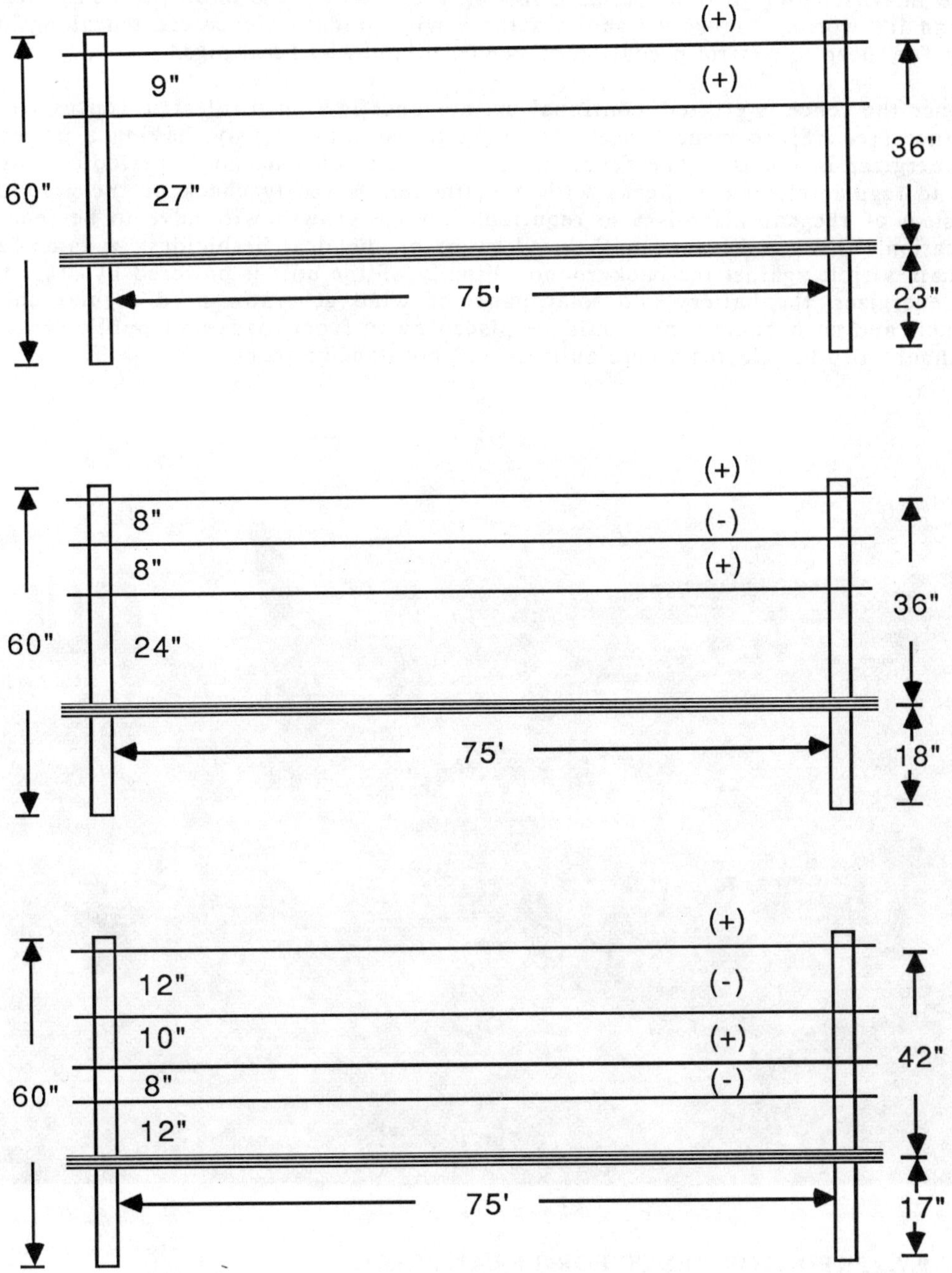

FIGURE 9.6. MULTI-WIRE DESIGNS.

Quite often in game fences use is made of net wire as well as electric, either as a way to get a good ground as in the case of bears, or to provide the visual aspect. Net wire sheep fences with electrical offsets provide a reasonable predator fence especially if the bottom of the fence is flush with the ground. The non-electrified part of the fence should be in reasonably good shape, as a rotting fence will cause more problems through shorts than it's worth. Figure 9.7 shows a three wire predator fence erected alongside a net fence for sheep to control predation of newborn lambs by feral pigs.

Once the fence is erected, continual maintenance is a must. Electric fences are not maintenance free. Since game fences are being tested continuously, having a powerful, quality energizer is a must. The fence must be visually checked on a periodic basis, in addition to regular electrical checks with a voltmeter. A yearly check of the ground or earth system of the energizer is also required. Forage growth will have to be removed upon occasion, this can be done with weed eaters or chemical herbicides; a game fence must remain visible against the background. Finally, if the unit is powered by other than a mains energizer, the battery and solar panel or wind generator need regular checks. Solar panels and such equipment should be placed away from roads and public areas, locations should also be selected where animals will not damage them.

FIGURE 9.7. PREDATOR FENCE FOR FERAL PIGS.

MISCELLANEOUS FENCE DESIGNS:

VISUAL BARRIER FENCES: These are normally used for preventing wild animals from penetrating a field or enclosure. Essentially they are a regular fence, but about 4 to 6 feet in front (on the wild side) one or more hot wires are placed, 2 to 3 feet high. The idea is that the animal will approach slowly, as the combined fence and stand off fence will be too high and/or far to jump, get shocked and retreat. In time, the wild population learns to respect the fence and gives it a wide berth. This does not work as well if the animal is being chased by hunters, but even here, the animals tend to avoid the fence if they have been exposed to it long enough. Appendix 6 illustrates several visual barrier fences.

GRASS FENCES: A grass fence is a recent introduction and is somewhat similar to a visual barrier fence. This fence has twin wires, about 12 inches high and 2 feet apart. A variation may have a third wire in the middle at about 30 inches in height. Wooden pegs (stakes) are used for the low wires and a regular post for the higher wire. They have not been properly evaluated as stock fences, and seem to be used for hard to control animals. A major problem is the forage growth between wires.

SPECIAL PURPOSE DESIGNS: There are rabbit, opossum, bear, deer, dog, monkey, coyote, moose, tiger, rat, etc. and even man fences. Designs can be obtained from two fencing seminars (suggested reading section) or from your local fence dealer.

REJUVENATING NON-ELECTRIC FENCES:

There are two methods used for adding life to a terminally ill non-electric fence, or even a new one for that matter. The first is the use of off-sets, already discussed. The other method is to electrify the top wire. If the fence has a smooth top wire, only insulators will be needed. If the top wire is barb wire, it will have to be replaced with smooth wire. If the second wire is closer than 6 inches to the top, it should also be removed (induction again). The rest of the fence is all connected together and then hooked to the ground rod. If a net fence is used, the top should be folded down and a smooth wire run along the top, at least 6 inches from the net wire, the net portion is connected to the ground rod. If the wire is particularly rotten, a new fence is in order, as the problem with shorts increases as the age of the fence increases.

GATES:

The main thing to remember about gates is that while they may not be used very often, when they are, the entire herd may pass through in a fairly short period of time. So make them wide enough to allow this to happen without having to re-build the gate every time you shift your stock. The actual size will depend on the number of stock, but should be no less than 10 feet; which will allow you to drive a pickup through. If you are shifting hundreds of animals then make the gates 16 feet or even wider. Gate post should be well braced; if a lot of traffic is anticipated, use a H or N brace. Horses, being what they are, often get quite nervous around electric gates. You may want a short stretch of net, barb wire or wooden section on either side of the gate. This is especially important if you are running a dude trail ride enterprise.[14]

Two spring gates connected together (two on each wire) will make a suitable gate and can be placed anywhere along the fence line. Offset lift gates have value on long stretches of fence, but are rather expensive to make. Knock down gates are easily made,

[14] Most dudes just want to look like a cowboy, they don't want to have to ride like one.

since the fence, if constructed properly, should be able to be laid on the ground anywhere between two post. All that is required is something to hold the middle spacer down against the ground while you drive over it; this can be nothing more than a stake driven into the ground with some sort of hook on it to hold the spacer. Figure 9.8 shows a gate made out of discarded elevator cable and home made hooks. Looks a little crude but it works, is cheap, easy to make and replace. One thing about gates in intensive grazing, they may only be open 12 or 13 times a year unless your into half daily shifts, in which case they will be open twice as often.

FIGURE 9.8. LOW COST GATE MADE FROM DISCARDED ELEVATOR CABLE.

Of course, gates can be more of the conventional type. In this instance the problem becomes how to get the electricity from one side of the gate to the other when the gate is open. This is best done by passing the wires (both ground and hot) overhead, about 10 feet above the ground. Overhead transmission is better, if there are any problems they are easily seen and corrected. However, some folks don't like to see wires up in the air. If this is the case the wires can be buried below the gateway. Insulated wire must be used and should be in some sort of pipe, electrical conduit, old hydraulic hose, or black plastic water pipe (don't use PVC as it is too brittle). The ends of the pipe or tubing should be bent down (opening pointing to the ground) and plugged with putty; this is to avoid water collecting in the bottom of the conduit or pipe. Needless to say, if a problem develops it will be hard to detect and harder to repair. Switches on both sides of the gateway will make detection easier.

If the fence has more than one hot wire, they should be connected to a single 10 gage insulated wire before going across the gateway; regular fence wire may be use on an overhead transmission. The ground return wire, if used, should also be brought across with the hot wire; the ground return wire should be continuous the entire length of the fence.

FLOODGATES:

Floodgates or watergates (not the Nixon kind), are used to fence across a creek bottom, to prevent stock from moving up or down the stream channel during periods of low water. They are particularly important where there is intermittent water flow. Figure 9.9 shows one such flood gate on an intermittent stream in northern Texas. The selection of the site to construct a floodgate is just as important as the actual construction. In many cases a lot of time, effort and money can be saved by just moving the crossing site up or down stream 20 or so yards. It should be mentioned that floodgates must be checked after each major rain or flood; remember Murphy's Law, if anything can go wrong, it will.

FIGURE 9.9. A WELL CONSTRUCTED FLOODGATE.

The first step in construction is to place two good brace posts, one on each side of the channel, very, very firmly in the ground. This may well mean doing some cement work, setting up H or N braces, guy wires, dead men, whatever it takes to make sure those post don't move when that area gets wet and ten million gallons of water comes down that channel and starts leaning on things.

The second step is to pass both a hot and a ground wire from one brace post to the other; use insulators on both wires so that either can be made hot. This is to make sure that regardless of what's going on in the channel, electricity gets across and is able to power the fence on the other side. The third step is to hook up a flood control switch so that the electricity going to the floodgate will be turned off when high water is flowing. When the hot wires of the floodgate are placed in water, a direct, or dead short is the result. This direct short causes all of the fence below the floodgate to go dead; in addition, it puts a heavy strain on the energizer and may cause it to blow. The fourth step (usually) is to decide what type of floodgate you're going to build. There are four general types of floodgates, the push aside, the knock up, the knock down and the disposable.

THE PUSH ASIDE: In this type of floodgate the two wires (hot and ground wire) crossing the channel are mounted so that they are at least 6 inches apart and are both parallel to the surface of the ground (both horizontal). Light galvanized chain is attached at one foot intervals along each wire and allowed to hang down to six inches or a foot above the grounds surface. The "hang-down" chains are staggered so that if one looks up the streams channel the chains would appear to be six inches apart, each with a different charge (+,-,+,-, etc.). If the "hang-down" part is much longer than 6 to 7 feet, or if in a high wind area, a stiffener should be threaded through all the chains of the same charge. This stiffener could be another chain, barb wire fence stay, or a straight piece of fence wire; the purpose is to keep the "hang-down" chains from wrapping around each other and shorting out. In high wind areas a piece of insulating material may need to be tied between stiffeners on the hot and the ground side. Another trick is to tied a piece of fiberglass fence post to the end of the chain so that it slightly drags on the ground, this will help keep the chain from blowing around; rocks, if not tied with wire, can also be used. This design is best in steep or narrow gullies, the amount of material (chains) required makes it un-economical for large spans.

THE KNOCK UP: This design requires a cable strung between the two brace post on each side of the channel, in addition two post are set in the stream channel, as close to the bank as possible (if the channel is very wide a third post may need to be set midway). The electric fence spanning the channel is a separate fence. Its end post are made from two by fours or four by fours, the height of which is determined by how high the water is expected to reach. Each end post is attached to the stream post with a single pin so that the post will swing out and up. The cable should come as close to the pivot points as possible. From the cable is suspended 6 to 10 inch planks, about a foot to a foot and a half apart. Wire is now strung from the end post on the down stream side of the planks; insulators are used to attach the wire to the planks. Four or five wires are used with alternating charges, the spacing is similar to the regular fence. The space between the brace post on the bank and the stream post is handled with a push through type of gate, or in some cases made semi-solid with planks. The basic idea of the knock up is that as the water rises, pressure will be placed on the planks and they will float the fence up on top of the water. A variation of this is to use a standard non-electric floodgate design, wooden panels, 6 to 10 feet in length, hung from a cable stretched between the brace posts (stream posts are not used). The electric wires are simply hooked to the panels either with insulators or outriggers (stand-offs) on the down stream side. Because each panel tries to work independently the wires should not be strung tight, but loose enough to allow a high degree of movement.

THE KNOCK DOWN: The knock down flood gates are usually placed in intermittent shallow, but wide creek or river bottoms. The idea behind this design is to construct the fence so that when water flows, pressure, due to the weight of the on-rushing water and debris, will build up to some point, at which time the fence will simply lay down. With the fence flat on the creek bottom, theory has it that the water will rush over the top and will not damage the fence. After things have dried up the fence is cleaned off and stood back up. It doesn't always happen that way, but that's the way it's suppose to happen. Construction is similar to the "knock up" in that two "stream posts" are placed in the stream bed and a steel cable is hung between the two brace posts on the bank and to the stream post for bracing. The knock down part is a separate section of fence; its end post are made out of either two by fours or four by fours. The end post (two by four) is pinned at the bottom to the stream post so that it can pivot when pressure is placed on the top portion. Make sure that you build it so that it falls downstream.[15] The break away unit (the part that holds it until enough pressure has built up) can be a simple jam or wedging action, or it can be a spring loaded clamp; the important thing is for it to let go when it's supposed to. A push through type watergate will often be erected between the brace post and the stream post on each side of the bank.

DISPOSABLE: This is what most of the above designs usually turn out to be, right after a snaggy old log comes floating down that hasn't read the rule book. A disposable fence is one that is built to wash away after a major rain or flood. It should be made out of poly wire or tape with either temporary plastic or fiberglass posts; trees can also be used. After each major rain it is simply re-built as soon as the water level drops sufficiently. A temporary fence might be a better name for it.

TESTING:

You don't build an electric fence and then forget about it; they are not maintenance free. Each fence around the paddock should be tested before stock are turned in. A good voltmeter is strongly suggested for testing; remember, hook up the ground wire first, then the hot one. It requires at least 1500 volts to turn a cow and 2000 volts for sheep. If the design is well thought out, testing can be done from one or two locations. In a wheel design the logical place is the hub. In block designs, the alley way or where several paddock come together. If the energizer is centrally located, this aids in testing, as does the ample use of switches. A good procedure is to turn off all switches and test the energizer, throw on one switch, test that line, turn that one off and turn on another, etc. If you get a voltage drop on one line that's where the problem is, now all you have to do is to find it.

When checking a line watch for the obvious, fallen branches, old wire, trash on the fence, etc. Sometimes shorts develop that are hard to see; try walking the line at night, arcing will be more noticeable then. Keep your ears open and check out any strange clicks. A small transistor radio (walk-man) off tuned to a station will pick up arcing that may otherwise go un-noticed. A dead short is often the hardest to detect as there may not be any arcing. Often voltage will drop during a rain but pick up afterwards, this may mean a cracked insulator; about the only way to find this is to go out with your radio while it's raining, or right after it stops.

If you have to work on the line and the energizer is on there are at least two ways to handle it. The first is to turn the energizer off; one can walk back to the barn for

[15] You may think that's a pretty dumb statement, but we actually saw one made that would only fall up stream; it had been replaced with a more conventional type when I saw it again the following year.

this or if one has a special attachment one can simply send a pulse up the wire to turn the unit off, another pulse turns it back on. The second method, if you don't have the special attachment and don't like to walk, is to short out the hot wire to the ground wire. Shorting out can be done by twisting the hot and ground wires together with a stick, or by using a short piece of copper wire with clamps on each end (always clamp to the ground wire first, then to the hot). This works reasonably well, if the connections are good and the short wire is of sufficient diameter to carry all the current; if not you'll find out soon enough.

SUGGESTED READING:

Ag-Links. Numerous information sheets on fencing: For electric fence construction (FPP-183, 184, 185, 186, 816, 817, 818, 819, 820, 821, 822); Ministry of Agriculture and Fisheries, Private Bag, Wellington, N.Z.

Australian Power Fencing Manual. Gallagher R.S.M. Pty Ltd., 9 Mary Parade, P.O. Box 122, Rydalmere, Sydney, N.S.W. 2116.

Farming of Deer. 1982. D. Yerex. Agric. Promotion Assoc. Ltd.

Gallagher/Snell Power Fence Catalog, 4th ed. 1984. Snell Systems, Inc. P.O. Box 17769 San Antonio, TX 78217.

Goat Farming in New Zealand. 1985. C. Rumble, Ed. Pub. Wrightson NMA Ltd. P.O. Box 1895, Wellington, N.Z.

How to Build Fence with Max-Ten 200. 1980. United States Steel. Cat. #111575; Pittsburg, Penn. 15230

PEL Fence Controller: Instruction Manual. 1976. PEL Products Ltd. Hamilton, New Zealand.

Power Fencing Manual, 7th ed. #6J-5790, Gallagher Electronics Ltd., Private Bag. Hamilton, New Zealand.

Proceedings: Gallagher, First Elephant and Wildlife Fencing Seminar. 1983. Hamilton, New Zealand.

Proceedings: Gallagher 2nd World Wildlife Power Fencing Seminar. Dubbo, N.S.W., Australia, Nov. 1985,

NOTE: Several companies that manufacture energizers and other components of electric fencing have available pamphlets on various aspects of fence construction that are quite good.

Used with permission of NEA.

Seeing is in the eye of the beholder.

10 GRAZING UNIT DESIGN

There are as many grazing unit designs as there are units, each is unique. Since each is different, the design of a grazing unit needs to be carefully thought out, as it is a major undertaking; once in place, you're going to have to live with it, so better do it right. The initial planning should be done without any consideration given to existing fences, buildings or water supplies, or other capital items. The unit should be designed to do the job at hand, the way you want it. Existing fences and facilities were (hopefully) place where they are now for some reason; THOSE REASONS HAD ABSOLUTELY NOTHING TO DO WITH IGM. Later, ranch inventories should be carefully gone over (Chapter 2) as no doubt successive plans will be able to incorporate some of the existing facilities. However, there needs to be a definite, practical reason for keeping an old capital item rather than nostalgia. While each grazing unit is unique there are some basic considerations that apply to all.

ENVIRONMENT: This more than any other factor will determine the overall design criteria. In low rainfall, less than 25 to 30 inches (extensive environments), designs will tend towards the wheel while in areas that receive more than 25 to 30 inches of rain block units will prevail. There are a number of reasons for this change in design criteria, but the major one is that the wheel lacks the degree of flexibility required in intensive (high rainfall) areas. Other factors to consider are the length and severity of seasonal changes, soil fertility, availability of water and ease of access.

FIGURE 10.1. SOME HUB DESIGNS BECOME QUITE ELABORATE. This particular unit has two center pivoted gates that allow cattle access only a portion of the hub and water trough at a time, it also allows two herds to be run simultaneously.

TOPOGRAPHY: The environment has the greatest effect on the overall design; topography has it's major effect on the size and configuration of the design. Steep or level terrain, shade or sun exposure, canyons, bodies of water, rocks or soil, outcrops, etc. all are considered.

VEGETATION: The type of vegetation that is on a piece of land obviously has a great deal to do with what can be done with that land without massive investments in money and labor. While different species and mixes of species can greatly affect animal performance, initially the main concerns are what percent of the vegetation is grazable and what percent is not. Vegetation and topography (V-T zones) are a major concern when laying out any grazing unit.

CARRYING CAPACITY: The environment, topography, present vegetation and the level of management that will be used, determine the present carrying capacity of a piece of land. With management, vegetation can be changed and greater utilization can be made out of what is grown, but initially you'll have to live with what you have and your inventories will tell you what that is. It has often been said that going intensive will double your carrying capacity, in many cases this is so; but it is not automatic. It all depends on what level of management the ranch has been operating at in the past and most importantly, what level the ranch will be operating at in the future. Doubling numbers is a goal or objective, it is not an inherent right of intensive grazing. For initial planning purposes use the present carrying capacity of the land, but keep the door open.

HERD STRUCTURE: What kinds of animals (species), their size, class and how many in each class are to be run? Along with this is the personal preference of the manager; how many animals of each class will be run as a herd? How many head is the operator comfortable in working in a single day? Your projected stock flows should be determined, at least in a rough form (Chapter 3). NOTE: If you are a large operation you had better have your unit managers in on the initial planning.

PLANNING THE WHEEL DESIGN UNIT AND HUB: by Jack Sprague.

Before any planning is done there are several questions the operator should consider:

1. How many paddocks will I need to achieve the desired amount of rest for the vegetation in this unit?

2. Is the available water adequate to carry the planned number of head and can it be delivered fast enough to meet the needs of a larger herd?

3. Are there any physical features on the site which will limit where the hub (center), or radial fences can be built?

4. Can the hub be located on an easily accessible area which is well drained so it is not a mud hole after every rain?

5. Will the design be a conventional "wagon" wheel, fan, satellite or one that incorporates a block design?

6. Will the center be a water hub, a dry hub, a split hub, and/or one with working facilities?

After the above has been determined the next stage is planning the layout. Accurate field boundaries, hub locations and physical features that may be a problem are placed on a map to some desired scale. The total acres of the field should be determined reasonably accurately. The fences will radiate out from the hub so that all paddocks are near equal size (assuming a similar V-T zone). Fences may have to be planned to avoid certain physical features such as, sink holes, rocks, canyons, or other obstructions. In actual practice the amount of forage in any one paddock will never equal that in another, however it does simply things if they are close. The location of the hub must first be selected as this will effect the overall shape of the wheel. In the traditional "wagon" wheel, the hub is located at, or near, the center of the area; the fences radiate out from there like the spokes on a wheel. In a fan the hub is located on, or near, one of the perimeter boundaries and the fences radiate out. A comparison between a fan and wagon wheel, with the same number of paddocks on the same area of land would have the fan using twice as much fencing as the wagon wheel. A satellite unit is several wheel designs connected to each other by one or more corridors; Figure 10.2 shows several of these designs.

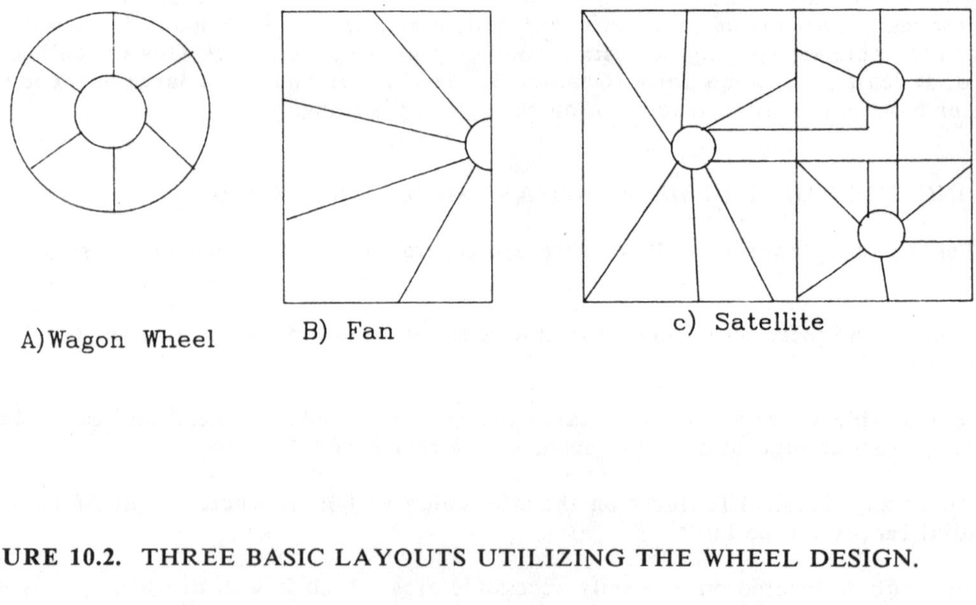

FIGURE 10.2. THREE BASIC LAYOUTS UTILIZING THE WHEEL DESIGN.

Once the hub has been located the next item of consideration is the planning of radial fences. First select a point that is visible from the hub and can be accurately located on the map. This can be the location of the first fence or your base line. Place a line where you think the next fence should be located. Measure the distance along the longest line or base of the triangle formed by the 2 lines. Measure the distance from the base line to the corner, at right angles to the base line. This is the altitude. The formula for the area of the triangle is: area = [(base)/2] X altitude. If your scale was in feet, the answer will be in square feet. Now divide the square feet by the number of feet in an acre (43,560) to get the acres. Adjust the second line until the acres are acceptable and darken the line. If the fences do not form a triangle, the area can be divided into triangles, each area figured separately, then all the triangles added together to find the total acres.

Repeat the process until all paddocks are drawn on the map. The layout can be done by using the first line as 0 degrees and figuring the number of degrees between each fence. If the boundary fence can be seen from the hub, measurements can be made along this fence to show where the fence will intersect the boundary fence. A line of sight can then be used between the hub and then marked to lay out the actual fence lines.

In most cases there will be more paddocks on one side of the hub than on the other, due to hub location and the shape of the field. This may cause some problem in getting the gate for each paddock to enter the hub fairly straight. To overcome this problem stop the radial fence 200 to 300 feet from the hub and make a slight corner in the fence (dog leg) from there to the gate post (Figure 10.3).

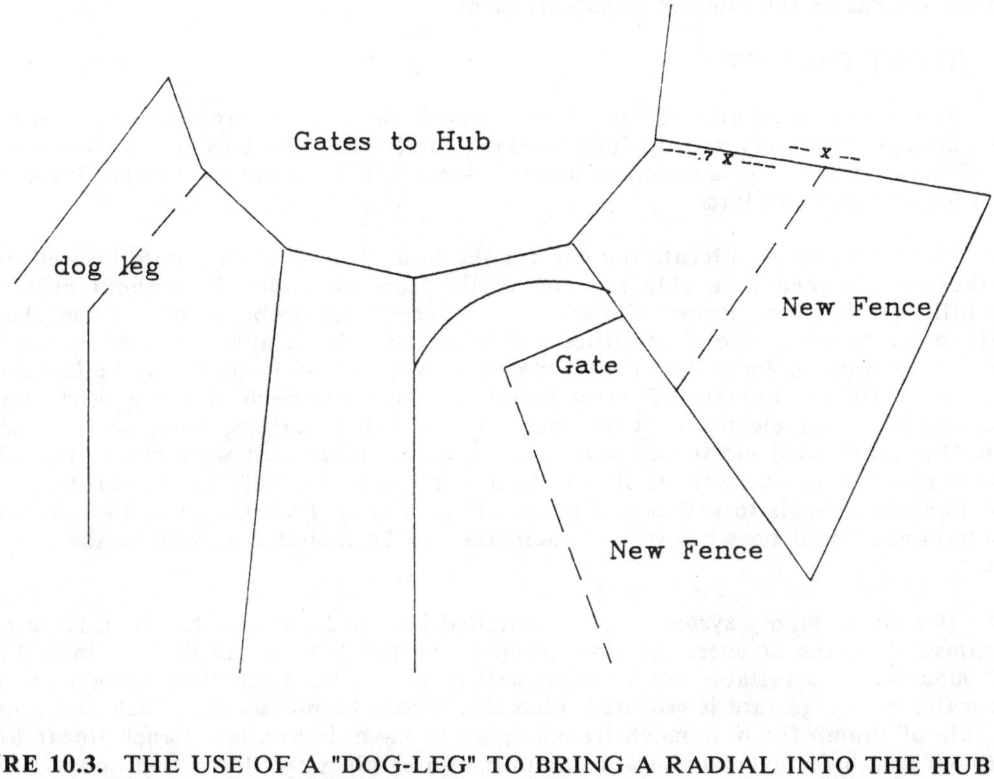

FIGURE 10.3. THE USE OF A "DOG-LEG" TO BRING A RADIAL INTO THE HUB AND TWO METHODS OF FURTHER SUB-DIVIDING A RADIAL PADDOCK.

Usually only 2 or 3 post need to be set for each line. They should be set where they may be sighted on by the operator, so as to set the rest of the post in a straight line. Minor bends in the fence may be necessary due to physical features, rocks etc. These can be done by the fence builder at the time of construction.

A good job of planning will make the job easier to lay out and construct. It is better to spend a little extra time planning than to be forced to make changes in the field during layout. However, don't let the plan get in the way of common sense.

After the system has been in operation you may decide that you wish to divide the paddock again, but you only allowed for enough room in the hub for the number of gates needed in the first design. Don't despair. There are two ways this can be done. First, the existing paddock can be divided by constructing a fence (with a gate) across the paddock, connecting the two radial fences. In a triangular shaped paddock this distance is a little over 70% of the distance from the hub to the boundary fence, measured along a radial. This method requires less fence but a gate must be installed. To use this method, the half closest to the hub is grazed (1) and then the gate is opened to the other half (2) to be grazed. This design has been found to increase the amount on manure in the portion closest to the hub and is only suggested where the other design is not appropriate.

The second method would be to build a new radial fence dividing the existing paddocks in half. Stop the new fence where the distance across the old paddock is twice the size of the desired gate. Install a gate on the end post (a wire gate will suffice), which can then block either of the new paddocks. Figure 10.3 shows both of these variations for increasing the number of sub-divisions.

LAYING OUT THE HUB:

The most critical area of the wheel design is the hub. It provides livestock control, water, minerals and may also include working facilities. The hub is the most expensive part of the unit and if it is found to be inadequate will be costly to change. Planning can save a lot of headaches later.

The first step is to locate the site for the hub. The location should be well drained and the animals should be able to enter easily from all paddocks without problems of steep hills, gulches, etc. Proper site selection reduces later problems of erosion. The next step is to decide what type of facilities will be in the hub. A hub can be a dry hub where the drinking water is located in the paddocks, in which case it must only be large enough to accommodate the number of gates required. This approach is being used more and more, especially for electrified gates made out of cable, springs, wire, or in some cases chain. One ranch used electrified gates for its water center that were made from old cattle neck chains (Figure 10.4). If it is to be a water hub then sufficient area must be allocated to allow animals to gather and drink. If it is to be a working hub then its size will be determined based on what sort of facilities will be included as well as the number of gates.

If a water supply system is to be installed it must be able to handle a large number of animals drinking at once. Shallow troughs, the top 1/3 cut out of a 12 inch diameter PVC, mounted in a suitable cradle work well (Figure 6.3). If shallow troughs are used a high water discharge rate is required, pipe size should be no less than inch and a quarter. One rule of thumb for how much trough space to have, is to allow 1 inch linear distance of trough length for every animal in the unit. Water pipes should be laid under ground to keep water cool, if possible. If working facilities are to be added to the hub their design should first be worked up then incorporated into the overall hub design (working facilities are discussed later).

FIGURE 10.4. A WATER CENTER FOR 16 PADDOCKS. The outside boundary of the hub is composed of 16 electrified gates. Each 10 foot gate is made out of old cattle neck chains, tied together. One switch controls two gates.

There is no special shape for a hub, it can be round, square, rectangular, or any other product of the imagination. The important thing is that it should fit the site and the gate pattern desired, plus any other included facilities. If the hub is to include working facilities, a corridor or race way around the outside perimeter is recommended. If the hub is simply a water hub or used only to move animals from one paddock to another (dry hub), no corridor is needed and the gates will determine the minimum size.

As an example the minimum size for a 16 paddock unit with 12 foot gates will be determined. Since a gate has got to hang on something, allow one addition foot for each gate post, a total of 13 feet (it is possible to use 12 feet, if you hang the gates a certain way, however for this example 13 feet is used). The circumference will be: 16 x 13 = 208 feet. The diameter of the circle is the circumference divided by Pi (3.14): 192/3.14 = 66.2 (use 67 feet). To figure the working area take the radius (one half the diameter) multiply it by itself, then multiply it by Pi: 67/2=33.5; 33.5 x 33.5 x 3.14 = 3524 square feed inside the hub. Divide the hub area by the number of animals that you expect to be in there at any one time. If you think it is too small then increase the diameter some until you have the area per animal that suits you; one rule of thumb is allow 18 to 20 square feet per animal. Table 10.1 will help in determining the size circle needed for different size and number of gates.

TABLE 10.1. CELL HUB PLANNING GUIDE.

	gate opening							
	8 ft.		12 ft.		16 ft.		20 ft.	
# of paddocks	dia. (ft.)	sq. ft.	dia. (ft.)	sq. ft.	dia. (ft.)	sq. ft.	dia. (ft.)	sq. ft.
4	11.5	104	16.6	216	21.6	366	26.7	560
5	14.3	161	20.7	337	27.6	577	33.4	876
6	17.2	232	24.8	483	32.5	830	40.1	1263
7	20.1	317	29.0	661	37.9	1128	46.8	1720
8	22.9	412	33.1	861	43.3	1473	53.5	2248
9	25.8	523	37.2	1087	48.7	1863	60.2	2846
10	28.6	642	41.4	1346	54.1	2299	66.8	3505
11	31.5	779	45.5	1626	59.	52781	73.5	4243
12	34.4	929	49.7	1940	64.9	3308	80.2	5052
13	37.2	1087	53.8	2273	70.3	3882	86.8	5931
14	40.1	1263	57.9	2633	75.8	4513	93.6	6881
15	43.0	1452	62.1	3029	81.2	5179	100.3	7901
16	45.8	1648	66.2	3442	86.6	5890	106.9	8976
17	48.7	1863	70.3	3882	92.0	6648	113.6	10136
18	51.6	2091	74.5	4359	97.4	7451	120.3	11367

A hub without a raceway lacks flexibility as all paddocks open directly into it. A hub with a raceway not only has all paddocks opening into it, but in turn opens to the center (working facilities or just a holding pen) or may have parts of it closed off, making temporary holding pens (Figure 10.5). Without a raceway it becomes very difficult to graze more than one herd at a time, unless other water facilities are built. The width of the raceway should be a minimum of 12 feet, more if it is a small unit and vehicles are to drive through it.[1] Many water centers are split in half (Figure 10.6), this increases the flexibility but still makes it difficult to run more than a single herd in a follow through.

Many hubs are now being built with the outside perimeter of electric and the inside of a net wire, poles or other construction; figure 10.7 shows this type of construction. Gates are electric and are either commercial or home made. Home made gates are best made from light chain or cable, elevator or marine cable work well. An eye loop is made in each end of the cable and a hook attached; switches are usually attached to each gate, or pair of gates, for opening and closing. Some like to hang short pieces of polywire from a cable gate, but this is not necessary. It is recommended that the electric wires be run around the outside perimeter, on poles; feed lines can then go to the radials and gates via switches. Make sure that the poles are high enough to allow all necessary equipment to pass underneath. Underground cables and wire in the hub area is not generally recommended, but if used they must be well insulated and water proof.

[1] A good idea is to stake it out at the site then try to drive your pickup around it without hitting any stakes.

FIGURE 10.5. A WELL DESIGNED AND SOLIDLY CONSTRUCTED CIRCULAR RACEWAY AROUND A WATER HUB. However, the modern trend is to make the outer fence of the raceway electric, with three to five wires, rather than relying on a physical barrier.

FIGURE 10.6. A WATER CENTER THAT HAS BEEN DIVIDED INTO TWO SECTIONS. Compare this design with that in Figure 11.1; the results are essentially the same.

FIGURE 10.7. CENTER HUB WITH OUTSIDE RACEWAY FENCE CONSTRUCTED WITH ELECTRIFIED WIRE AND GATES. Working facilities are in the center.

BLOCK DESIGN:

Much of what has already been discussed applies equally well to block designs. A block design, for the same number of paddocks, with the same acres, requires about 29% less fencing than a wheel design; if a corridor is connected to all paddocks there is only a 12.5% advantage (Table 10.2). However, from just a fencing cost point of view (no water development) the saving of a block over a wheel is closer to 10%. The reason is all the little parts, clamps, switches, ties, insulators, etc; more are used in a block design than in a wheel. The block does require more water troughs than a wheel but the erosion potential is less. The major advantages in a block design is that of flexibility. The ability to make additional subdivisions quickly allows one to take advantage of all the modern tools of grazing management, such as, multiple herds, different levels of forage removal (residual dry matter), forage allocation and conservation.

TABLE 10.2. COMPARISON OF BLOCK VS WAGON WHEEL DESIGN.
Both covering one square mile of area, both with 16 paddocks:

	BLOCK	WHEEL
Total fencing (in feet)	52,800	68,310
(fencing required to put in corridor to all paddocks)	7,920	
Total with corridor (ft)	60,720	
Minimum number of water points (one water feeds 4 paddocks in block)	4	1
Maximum distance to water (ft)	1,444	3,218

Block designs fall into three general categories, based on a combination of the the overall design and the environment: Texas Block (extensive), New Zealand Block (moderately intensive), Break (very intensive). The basic designs for each are shown in Figure 10.8.

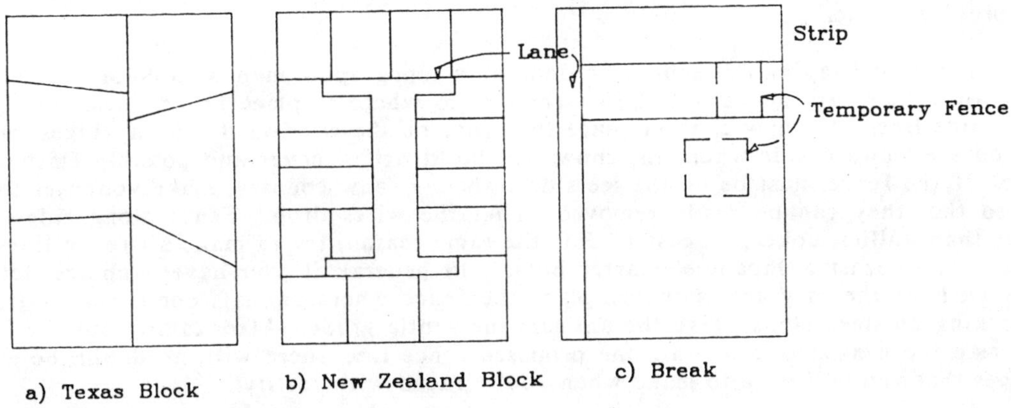

FIGURE 10.8. THE THREE MAJOR BLOCK DESIGNS FOR GRAZING UNITS.

TEXAS BLOCK:

As the name suggest, the area that the herd will be grazing is cut up into large various shaped paddocks, which generally will not have the same acreage. The Texas block lends itself to large acreages (200 to 400 acre paddocks in extensive environments) and is being used with considerable success in some very arid areas (6 to 10 inches of rainfall) in the central highlands of the South Island of New Zealand (very similar to the U.S. Intermountain West).

In setting up a Texas block, attention is directed towards making each block as nearly the same with respect to vegetation and topography as possible (the V-T zone). This will require a knowledge of the area and a good map. Thus, the area along a creek might constitute one block, the hill area adjacent to the creek another and the ridge tops a third. After the V-T zones have been defined, consideration is given as to the type of livestock and how the herd(s) are to be run. Further subdivision (sub-blocks) would then be made according to the number of permanent paddocks desired for each zone, or to multiple use of the different zones.

One aspect in setting up V-T zones that often goes un-noticed is that of southern or northern exposure. In the northern hemisphere the south facing slopes will green up sooner in the spring and stay greener longer in the fall (if there is sufficient soil moisture) than those facing north; however, in the summer the reverse is often true. In the southern hemisphere it's just the other way around. It has been demonstrated in New Zealand that shifting young stock based on how the paddocks lie with respect to the summer and winter sun, resulted in increased performance. As one gets closer to the equator this effect loses its importance. In planning a Texas block don't pay any attention to existing fences or other facilities. Most ranches and farms already have, at least on a map, a block design; but as mentioned before, the reason that those fences are there, have nothing to do with IGM and usually nothing to do with V-T zones. After the first plan has been drawn up, then is the time to see if any existing facilities fit. It is important to make that first plan fit the environment and your plans, not some unknown plan of a previous owner.

The actual laying out should be done on a topographic map with colored pencil or pen. While there are no hard and fast rules as to where to place fence lines; there are some guide lines. In snow country place the fence on the windward side of ridges, rather than on the leeward side where the snow will build drifts, cover and possibly flatten the fence. If the fence must be on the lee side in heavy snow country, make your ties to the post so that they can be easily removed to let the wires drop. Fence along ridge tops rather than gullies, upkeep is easier. For the same reason, try to make all fence lines accessible by at least a Japanese quarter horse. In general, if you have a choice, let the livestock have the easy access routes, place the fence where animals don't usually travel. If working on steep slopes, give the animals the gentle grade. After laying out the fence on a map the next step is to walk the proposed fence line, there will, no doubt, be a few changes that you will wish to make when you're faced with reality.

The use of a bulldozer to smooth out the fence line prior to construction is becoming standard practice. There is no question that some areas require a little smoothing, just as there is no question that an awful lot of fences got put in this country, long before every ranch had its own inventory of "heavy metal". Dozing out fence lines makes for easy fencing and provides a nice access road. These access roads will not only be used by you for maintenance and inspection but also your livestock will use them, which is usually good, but not always as there might be some erosion problems develop due to trailing and camping.

Fence line dozing can also take a lot of land out of production and more than double the fencing cost. It is not uncommon for units in the one to three hundred acre size with many paddocks, to lose between 5-15% of their total productive base by dozing cuts for permanent fences. A couple of outfits got into serious forage problems the first year of operation because they did not take into account this loss of production. It may take several years to get this land back into production. Consequently if you are going to doze, re-seed either prior to, just after erecting the fence, or before the wet season. Not only does this cut down on erosion it also returns a significant portion of the unit to production much sooner than if left to nature. The economics of the use of "heavy metal" on fencing projects should be well thought out. The ability of the modern electric fence to zig or zag around obstacles, removes much of the need to scrape clean an area.

If you have to cross a gully, try to pick a place where building a watergate won't turn into a Federal project. Some gullies and canyons don't need a gate, nothing is going to be traveling up or down them anyway, however most do. Remember that even the worse canyon has some place to cross that's better than other places. Try to get some idea of the type of flooding that is likely to occur. If 20 foot walls of water come down the canyon every wet season, then the type of watergate to be built would be different than an existing creek that may just double its size.

A Texas block does not require the use of alleyways or corridors, animals can shift from sub-block to sub-block by using interconnecting gates. This may cause some problems in shifting animals, as they must go into adjacent paddocks, either that, or trailed to where you want them. Alleyways add flexibility and where feasible should be used.

NEW ZEALAND BLOCK:

The New Zealand block design has it best application in moderately intensive environments, 25 to 40 inch rainfall with moderate seasonal variations. The main features are small size paddocks (less than 40 acres) that can be easily sub-divided with temporary fencing. Each paddock is normally, but not necessarily, connected by a corridor or lane for ease of moving stock.

Because the permanently fenced paddocks are smaller it is easier to lay out each one so that it only contain one V-T zone (vegetational and topographic) per paddock. The greater the level of management on the land the more important it is that the area in each grazing block be as near the same as possible. For those areas that lie above 23 degrees north or south latitude, this also means paying attention to the north or south facing slopes.

The shape of the permanently fenced paddocks should be rectangular if feasible, however this in not critical. Rectangular paddocks allows temporary fencing to be set up quickly and economically; it also makes estimations of the area to allocate for grazing easier. The use of temporary fences during a portion of the year allows the operator to tailor his management to the different seasons. He may use set stocking during lambing, then strip grazing after weaning and into a slow rotation during the winter or dry period. By the same token, if hay or silage is made, the fields are large enough for machinery to be used effectively.

Because the land base is smaller, the chances of an "ideal" site for a fence line, corral or some other structure is less. Because of this, smaller acreages actually require more planning than do large acreages, on a per acre basis. The permanent fences need to be laid out with the same care as described for the Texas Block and the working facilities should be convenient and easy to get to from any portion of the ranch.

Water should be developed so that each permanent paddock has its own supply. When the paddock is further sub-divided with temporary fence, small water tanks can be connected to the water supply and moved when the animals are, such a tank is shown in Chapter 6, Figure 6.4. A portable mineral feeder is also suggested.

The New Zealand Block allows considerable flexibility of use, especially in grazing multiple herds. The main reason for use of multiple herds is to better tailor the nutrient demands of each herd to the available forage. Normally each herd will consist of a different class of animal. The group requiring the highest nutrition will go first, followed by the group that doesn't require as high a plane and so on. The classic three herd example is stocker animals followed by cows and calves followed by dry cows; in a dairy context it would be number one milkers, followed by number two's, followed by dry cows. This type of utilization allows maximal individual performance as well as high per acre gains. Recently a Hawaiian ranch reported on two herds of stocker animals, steers followed by heifers, grazing in a lightly fertilized New Zealand design unit. The steers gained 2.46 lbs per day, the heifers averaged 1.24 lbs per day, the stocking rate was such that 8.6 lbs (liveweight) per acre per day was recorded.

STRIP AND BREAK GRAZING:

The best use of break grazing is in intense environments, 25 or so inches of rain per year, during a rapid growth period in a drier environment, or under irrigation. It has its greatest potential in continuous or long growing season environments but, can be used anywhere the growing season is long enough to put on sufficient weight to pay for the effort. Typically used for only one or two classes of animal per farm or ranch, such as dairy stock, stocker operations, feeder bulls, etc. In both designs the amount of area to allocate, each day, for each herd depends on the number of animals, the quality and quantity of forage and the performance targets selected. There are two sub-headings under this design, both are similar, but different; strip grazing and break grazing.[2]

In strip grazing designs, the permanent paddocks are sized into rectangular shapes of 8 to 20 acres in size, preferably rectangular. The width of the paddock is such that a temporary fence can be easily erected. Stock water is usually provide with a portable trough.

Generally in shifting, the front fence is moved forward the required distance and the back fence then brought up behind the herd. The amount of area to allow the herd to rest and move around in, depends on how much area is being given the herd in each shift. Some operators only allow the animals to have the area allocated for each shift; others will allow the animals to have a larger area, up to about three days grazing. The best approach is to allow the animals to have some old grazing area to lounge around in, this way they are not as likely to trample and soil the fresh area. However, do not give them so much area that you lose the density effect or allow them to re-graze new growth.

Shifting may take place once, twice or even three or more times a day. The reason for multiple shifts during a day is simply that animals gain more with twice a day shift than with once a day shift; one study with dairy animals found that twice a day shifts resulted in 10% more milk production than in daily shifts. Shifting stimulates appetite, the more they eat, the more they gain. There used to be a dairy outside of Mexico City that hired a bunch of children to stand in a line out in their main pasture with switches

[2] Did you hear about the fellow that had two clocks hanging on his wall, each with a different time? When asked why he had two clocks, each set for a different time, his reply was: "If they were the same time, I wouldn't need two clocks."

in their hands. About every hour a signal was given and the children would step back one or two paces, the cows would then graze up to the new line.

Because animals will not normally dung or manure closer than two and a half to three feet from an electric fence, stripping offers a method of obtaining forage utilizations in the mid to high 90% bracket. If the fence is only moved forward two and a half feet at a shift, soiling and trampling is reduced to a minimum. Figure 10.9 shows this approach used on Marshal rye grass in Texas. This operator, near San Antonio, shifted the animals twice a day; 13 mature cows were carried over the growing season on two acres without irrigation.

FIGURE 10.9. STRIP GRAZING IN TEXAS. Photo courtesy of Snell Systems.

Multiple herds can also be used in strip grazing, just as they can in any design, the primary consideration is available water. The lead herd is shifted first by moving the wire forward onto new forage, the wire separating the herds is then brought forward, the second herd moves into the area vacated by the first herd and the back wire brought up. If you feel a little nervous about having two herds separated by only a single strand of polywire, then maintain a space (one shift) between them, this will require one extra fence. With three herds and daily shifts, you will have to get over your nervousness and have each herd separated by only one wire. Much more than a total of three days grazing increases the chances of the animals grazing the new re-growth, especially during the fast growth season and this is a no-no. If it will make you feel any better, in New Zealand cattle are commonly grazed along the highway right of way with only a single strand of electric wire between them and the hazards of modern transportation (figure 10.10).

FIGURE 10.10. CATTLE GRAZING THE LONG ACRE IN NEW ZEALAND

In true break grazing there is only one perimeter fence, usually enclosing 20 to 40 acres or so. Each day the operator must decide how much area to give the herd that day. This decision is based on the desired animal performance and the existing quantity and quality of the forage. After the area has been determined the operator "breaks" out the new area with temporary fencing. The animals then shift themselves to the new forage and the fence line brought up behind them. This continues around and around the allocated area. Multiple herds can also be used as discussed above. The operator must continually juggle the present growth rate, the potential growth rate, the amount of forage on hand and its quality, the amount of residual to leave to obtain the desired growth rate and quality, the number of animals and class, the necessary intake for a particular class of animal grazing a particular forage to give the desired performance, going to church and paying taxes. It is the ultimate form of grass/livestock management.

IN ACTUAL FACT:

In actual fact, the design that you are likely to end up with will probably consist of different parts of several of the above. There is nothing wrong with this, as long as you know what you are doing and why. I suspect that after you have been in IGM for a few years, the urge to try break grazing will start to take hold. Go for it, even if it means using part of your hay field. While break grazing is the ultimate form of grass/livestock management, it is also the ultimate teacher in forage and livestock management.

CORRIDORS AND ALLEYWAYS:

Alleyways or corridors can be added to any design, even a wheel or break, to increase the ease of livestock movement from paddock to paddock. Corridors makes it possible to move to a non-adjacent paddock without having to trail through a couple of paddocks in the process. Alleyways can also be used to supply water, mineral, supplements, or simply a lane to get the animals from a paddock to the working facilities. The New Zealand Block design often has water in the lane and may use the lane as a water hub for a portion of the year. With a couple of spring gates placed at appropriate places in the lane, paddock gates may simply be open and animals allowed to shift themselves into the lane and then to the desired paddock, rather than the fuss and bother of driving the animals out of the paddock with a bike or horseback.

The minimum width for an corridor depends to some extent on the number of animals that will be moved down it at one time, usually 20 feet is considered a minimum. A corridor can also be made large so that it contains one or two days grazing, in effect a small paddock. A lane as a small paddock has particular application in the Texas Block, but also is used quite frequently in the New Zealand block. Construction of the corridor is usually of the permanent type on the Texas Block but may be either permanent or temporary with other designs. Right angles are to be avoided in corridor construction, all corners should be round. Before construction drive your pickup or truck down the corridor to make sure it fits.

If the alleyway is to be used a lot, it may pay to surface it with gravel or grade it up so that it drains. There are certain areas on a ranch and farm that are termed "sacrifice areas". These areas receive so much traffic that no attempt should be made to keep them in grass. The problem then is to apply some sort of treatment so that the area doesn't turn into one big bog hole, or blow away; gravel, grading up, concrete or asphalt works wonders on these areas.

A raceway is simply a small corridor, usually placed in, or around a working facility. The primary use has been in sheep corrals, either as a circular race or as a long chute leading up to a dodge gate. However, their use in cattle corrals has been on the increase, particular for dodge gates, diamond sorting pens and their use around the hub in wheel designs. In general they are a maximum of 20 feet wide and may be as small as a animal width, in which case they are generally called chutes in the U.S. Properly designed they greatly ease the work of separating and sorting animals.

WORKING FACILITIES:

There are any number of designs for working facilities for different species of animals. While there are features common to each species, animals work best in facilities designed for that particular animal. A common feature is the use of circular raceways and crowding pens. An outside holding pen is suggested for deer, however deer work best in low ceiling facilities that are only dimly lighted. Figure 10.11 shows one type of outdoor sorting pen for deer. If working goats in sheep corrals the raceways should be shorten to approximately 10 feet, a minimum height of 4 feet and a width of 2.5 feet. Drafting races should be shorter and narrower than for sheep. shorten to Only sheep and cattle corrals will be briefly discussed here. The suggested reading sections list several sources for both exotic and domestic animals. Assistance and various plans are also available from County Extension Offices and/or the Soil Conservation Service.

FIGURE 10.11 OUTDOOR SORTING PEN FOR RED DEER. Notice the use of a strip of canvas as a very effective temporary fence.

SHEEP:

The most common shape is the rectangular yard, but circular yards are easier to design, erect, to work stock in, and cheaper to build. Figure 10.12 illustrates the main components of a standard rectangular yard and a circular yard. The components listed below could serve as the basis for any set of yards, but will need to be tailored for each individual situation. Yards vary in size, but one rule of thumb is to allow 5.5 square feet per sheep, however for overnight stays allow twice this much (11 square feet) per animal.

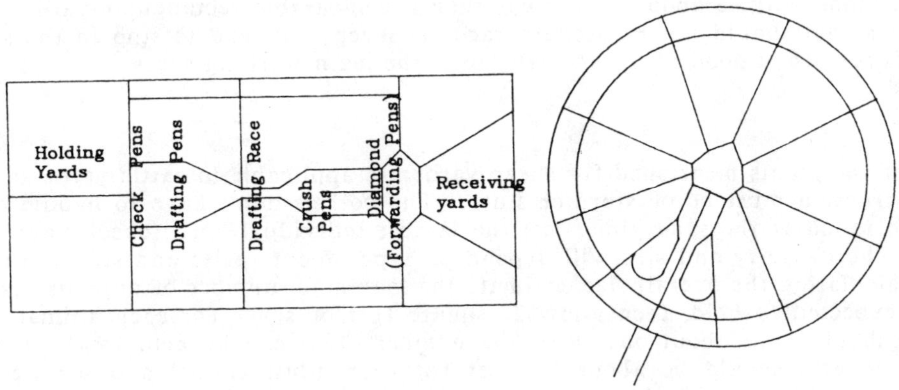

FIGURE 10.12. DIAGRAM OF A RECTANGULAR AND A CIRCULAR SHEEP CORRAL.

The following descriptions of the various yard components has been adapted from the 1983 Farm Budget Manual.

COMPONENTS:

RECEIVING YARDS: They have a large capacity and are the first yard entered. They communicate with the -

FORWARDING PENS: These pens, often diamond-shaped, are used to sort and lead either into holding pens or the -

CRUSH OR FORCING PENS: One or two long, small pens, tapering at one end. They hold about 50 sheep and flow, by means of a two-way gate, into the -

DRAFTING RACE (sorting): Long and narrow, wide enough for one sheep. It is used to divide a mob into several smaller groups, this being accomplish by one or more drafting gates at the exit which guide the sheep into the -

DRAFTING PENS: Small pens to accommodate the drafted groups of sheep, may open into Check Pens (to correct for mistakes or treatment of individuals) or open directly into -

HOLDING YARDS: Corresponding in size to the receiving yards.

Small yards do not need to have all of the components, nor do they have to be in any particular arrangement as long as the sequence is followed. Pens and fences should be designed to serve a dual purpose whenever possible. In circular yards the central diamond pen is an essential part of the design; the outside raceway, while not essential, does serve as a connecting link to the majority of pens, as such it makes the unit much more versatile. Circular yards are better in just about all respects to rectangular yards, construction cost alone will be about 13% lower than a comparable rectangular yard. Trees are a good idea, but should not be along a race, as sheep will tend to stop in the shade. Covered yards are also a good idea, especially over the main working areas.

CATTLE:

Most of the points mentioned for sheep yards are applicable to cattle yards as well, however construction must be of sturdier stuff. The corrals don't have to handle every animal on the ranch at the same time, just the largest individual herd (check your stock flow chart). The capacity and size will depend on type of enterprise and size of the operation. In calculating the overall size, estimate the maximum number of animals that the unit will be expected to hold, then allow 27 square feet of space for each animal. The forcing yard should hold about one third the number that can be held in the holding pens. Chute lengths should be about 6.5 feet for each adult animal and sorting pens should allow about 22 square feet per animal.

Cattle, like sheep have no discernible color vision, but they do relate to contrast in brightness. Sharp contrast in dark and light surfaces will cause cattle to balk. To help eliminate shadows in covered corrals use fluorescence lights rather than incandescent bulbs. Loading ramps should not face into the early morning or late afternoon sun. Any lead-in to a chute or squeeze should have a curved approach, though if the angle of the curve is greater than 15 degrees, a straight portion, the length of the cow should be built immediately before the squeeze. The social order within a group of cattle tends to work against a free movement of animals at points of constriction in yards and races. The dominant animal often gets to a narrow point and stops (her flight zone becomes constricted); less dominant animals are afraid to pass and the whole thing comes to a screeching halt. If the lead up to the squeeze is made without any abrupt angles this will help eliminate the problem. If the problem persists, sell the offending cow.

A good chute limits the side vision and directs it to the area ahead. This is of little use however, if the cow is required to walk into a blind alley; cattle will stop (up to 6-8 yards) from a blind end. The gate ahead should be of the grill type so stock can see ahead. Single file chutes should not be longer than 24 feet without a curve or bend. Overhead walkways frighten cattle; catwalks should be on each side of the chute at waist-height. Rubber bumpers on the side or bottoms of gates allow animals to push by without bruising and also reduces noise, which frightens cattle.

New concrete floors should have grooves 0.3 to 0.4 inches deep and three fourths of an inch wide; these should form a grid pattern, 4 inches between lines. If cattle have to unload downhill, steps are preferred to ramps; ramps soon become slippery with dung and urine. Stair steps with an 3 to 3.25 inch rise and a 13 to 14 inch width work well for cattle. Make sure any concrete is made to sufficient strength, 3000 is preferred.

Hot shots have little use in a well designed corral. If a hot shot must be used, apply it to one end or the other (the part of the head in front of the ears, or the tail head), any other place only confuses the animal. A hot shot hurts, the animal (like us), instinctively tries to move away from the source of the pain. If you stick it to him in the ribs while he is in a chute, he'll try to go sidewards, but can't; he quickly becomes confused and terrified and is liable to go in any direction. Once you have the animal terrified, further application of the hot shot may stroke your ego, but has little application to the job at hand, which is to maintain a continuous flow of animals down the chute.

A great deal has been written on rectangular corrals, there seems little reason to write more; numerous plans are available from Extension and Soil Conservation offices, as well as many commercial sources. The circular cattle corral, like the circular sheep corral is again, the best way to go, all things considered, especially cost, which as for sheep yards will be as much as 13% lower than a comparable set of rectangular corrals. Circular corrals have the additional advantage in that they reduce stress (on both man and beast) and speed up the working of stock. Circular corrals, when used in conjunction with the wheel design, allow one or two men, to gather and work a considerable number of animals, a point to consider in arid environments. Figure 10.13 shows two designs for circular corrals.

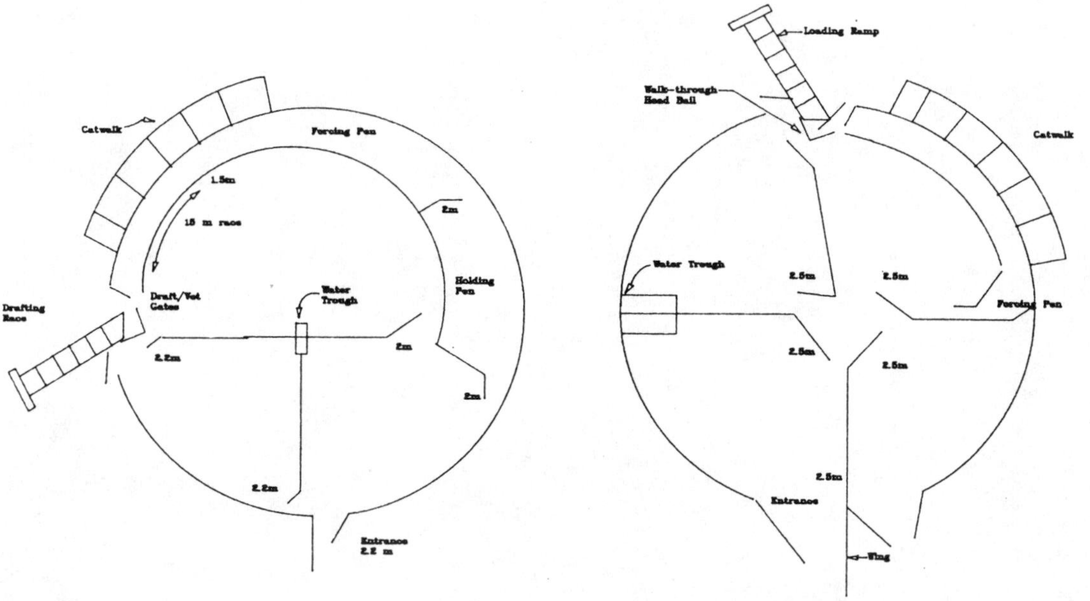

FIGURE 10.13. DIAGRAM OF TWO CIRCULAR CATTLE CORRALS.

SUGGESTED READING:

Beef Housing and Equipment Handbook, 3rd ed. 1975. Midwest Plan Service, Iowa State Univ., Ames Iowa 50011

Farm Budget Manual: Technical, vol 2. 1983. Ed. M.B. Clark & G.P. Rennie. Lincoln College, Canterbury, New Zealand

Farming of Deer. 1982. D. Yerex. Agric. Promotion Assoc. Ltd.

Goat Farming in New Zealand. 1985. C. Rumble, Ed. Wrightson NMA Ltd. P.O. Box 1895, Wellington, N.Z.

Handbook of Livestock Management Techniques. 1981. R.A. Battaglia and V.B. Mayrose. Burgess Pub. Co.

Range Development and Improvements. 1971. J.F. Vallentine, Brigham Young Univ. Press.

Structural Range Improvement Handbook. 1972. FSH 2209.22 R3, U.S. Forest Service.

Used with permission of NEA.

Sometimes even when you know where you're going, it's still difficult to get there.

11 BASIC GRAZING

Grazing comes natural to animals, man has a few problems with it. In the wild, the objective of grazing was to keep the animals healthy so they could breed and keep their species from going extinct. It was not to keep all the lions and tigers fat. Man has taken the grazing animal and changed its major objective to suit his own needs, which is to keep man fat and happy. Because we look at grazing differently than the grazing animal does, often what we want, or do or do not do, runs counter to what the animal wants, or in many cases needs. The trick is to allow the animal to fulfill its obligations but at the same time to get whatever it is that we want. To do that requires an understanding of the basic grazing techniques.

ESSENTIAL REQUIREMENTS:

What is required for ANY grazing program to work? CONTROL!

FIGURE 11.1. The degree of animal control does not have to be this tight, but you do need to be able to put the animals where your want them, when you want, for how long as you want.

Without control, there can be no management; without management you are at best a hunter. Regardless of what sort of grazing program you have, or are thinking of changing to, you must have some measure of control over the animals. There are four basic areas of control, or they might be thought of as dimensions to a successful grazing program. They are: 1. Control of space. 2. Control of numbers. 3. Control of time. 4. Most importantly, control of the animal.

1. Control of space; or how much area is to be grazed. This is done with fences, either permanent or temporary. It can be done by turning water on or off. In some societies it is done with herders, or it can be done with a length of rope, tether grazing.
2. Control of numbers; or how many animals to be placed in the area to be grazed. The ability to control the stocking density; the number of animals in any given paddock at a stated point in time.
3. Control of time. The length of time the animals are to remain in a given area (paddock or pasture). The length of time, area and the numbers of animals determine the grazing pressure that is being applied.
4. Control of the animal. The operator must be able to place the animal where he wants them, when he wants them there, for as long as he wants to keep them there. The animal should have precious little to say about it.

The above four control functions are basic to any grazing program, whether continuous or a multi-paddock unit with half day moves. The only difference between different grazing programs lies in the degree each of the above are applied.

TETHER GRAZING:

Strange as it might seem, tether grazing can be the ultimate form of grazing control. The reasons are simply that the animal is under strict control at all times. The grazier decides where the animal is to graze, how much area to allocate to the animal (length of the rope) and how long to leave it there. The fact that it is not practiced on a commercial scale has something to do with the number of animals an individual can stake out and change water and mineral for in one day. We're not sure that anyone has ever done a study but changing an animal every 10 minutes would seem to be hopping right along.

Tether grazing is one of the oldest forms of controlled grazing known to man. It got its start around the beginning of animal domestication and reached the peak of its popularity in the late 1800's. It is still widely used today. It has a lot going for it; the main thing is since you have total control of the animal you know where to find the critter. This may not sound like such a big deal; but after you have gotten up at 5 AM and spent a hour or so tramping around out there in the cold and wet with a milk pail clutched in your cold little hand, the idea of knowing exactly where to find the milk cow has a certain appeal.

The animal is attached by means of a rope or chain to a stake or tree. The length of the rope determines the amount of area that the animal is allowed to graze. If the animal is allowed to graze in a complete circle the area is determined by multiplying the length of the rope by itself, then by 3.14. For example, if the rope is 10 feet long the area that the animal can graze is: 10 X 10 X 3.14 = 314 square feet. The rope should be attached by a swivel to the stake, this keeps the rope from twisting or knotting up. NEVER use a chain around the animals neck, even if passed through a hose. The weight of the chain, or heavy rope, presses on the muscles, ligaments and nerves in the neck and can result in severe pain for the animals as well as permanent injury; horses are particularly prone to this type of injury.

In tether grazing you have complete control of the animal. You determine how much feed the animal gets and when, what minerals are given, how much water and where the animal does its grazing. Properly done, tether grazing can result in excellent production either in beef or milk; improperly done and you can end up with a real "psycho" on your hands. The animal is at the mercy of the weather, dogs, cars and kids. He can run only as far as the end of the rope and that's not far enough to get away from a pack of dogs or rock throwing kids.

An animal on pasture selects its own feed and water, in tether grazing you select it for him. Most people's idea of what is good animal feed bears little comparison to what the animal thinks is good feed. Animals like fresh feed on offer each day, they like young tender forage and they like a variety of different plants. Tether grazing can provide this only if the animal is moved at least once a day. At least 10 gallons of fresh clean water should be provided each day, more if it is hot. If a small tub is used it is best to stake it down so the animal won't knock it over. The animal should be wormed, vaccinated for diseases and ear tags put in for fly control. Tether grazing works but to make it work well you have to care.

NATURAL, CONTINUOUS GRAZING AND SET STOCKING:

In natural grazing the animal is free to pretty much go where it will. There are some constraints, availability of feed and water, physical features such as deserts, oceans, cliffs, canyons and such that may act to keep the animal within certain boundaries. But if the species is successful, population pressure will build up and the species will spread to adjoining areas. The grazing patterns that a particular species of grazer may develop depends on the kinds of plants and the part of the plant that it eats, other grazing species, as well as the kinds of predators it is faced with. Large herd animals are typically migratory, often moving several thousands of miles during a year. In natural grazing the animals are under control, but Mother Nature is doing the controlling; man has little to say about it.

A definition of continuous grazing is "the grazing of a specific unit by livestock throughout a year or a part of the year that grazing is feasible". It is argued that continuous grazing is a natural method of grazing. It's not. The moment fences are built that restrict the flow of animals, or the moment exotic grazing animals (livestock) are introduced to an environment where they did not exist before, or the moment predators are controlled or eliminated, or the moment man injects a new energy source or removes an old one, the natural order of things has ceased to be. A new order, for better or worse, is now present. The failure to recognize this simple truth, is what has caused all the problems of range deterioration and the resulting arguments over what to do about it.

Continuous grazing, as it has been "traditionally" practiced, is not management, it is the lack of management. It arose in response to a set of environmental and social conditions, that for all effective purposes ceased to exist with the stringing of barb wire across the ranges. It has hung on this long because it was easy, and because it was so easy, everybody did it, and because everybody did it, there had to be a rational reason behind it. Consequently, a lot of time, effort and money has been spent, rationalizing about continuous grazing.

A great deal has been written on how to continuous graze and it all boils down to adjusting the stocking rate so that you can ensure having enough feed to last the herd during the course of the year. In theory this means first determining the average stocking rate of the ranch over many years and then stocking at 75% of this number. Forage quality is ignored or made up for by supplements. In actual practice attempting to run at this level results in either being over or under stocked. This is due to the variability of the rainfall and thus forage production, the fact that the calf crop will lag the season by one year and will also be different from one year to the next, changing markets conditions, the length of time required to get replacement cows on line to replace those culled earlier. It might all be worth it if it resulted in range improvement, but it doesn't. The fact that the animal selects its diet is the basic reason that continuous grazing does not result in range improvement, unless the stocking rate is extremely light.

Continuous grazing will usually, but not always, put on more weight per animal than intensive grazing, more so in temperate regions than in tropical. However, to do so, the stocking rate must be such that the animal can select the high quality forage necessary to obtain those high gains. This means a light stocking rate. Today, the price of land rules out light stocking for most commercial operation in developed countries as being un-economical.

Set stocking arose as a way to deal with the increasingly apparent problems of continuous grazing. Set stocking is a form of continuous grazing in which an effort has been made to inject some management into the program. In set stocking, as in continuous grazing, animals are placed in an area for the entire grazing period. However, the stocking rate is varied in set stocking; it's called "put and take". As the growth rate increases, animals are added (put) to the herd and when the growth rate slows, are removed (take). In set stocking a predetermined level or height of the forage is usually the deciding criteria for putting or taking.

Set stocking will put on more weight per individual animal than most rotational programs and because one is continuously adjusting stocking rate to the amount of available forage, it can obtain excellent gains on a per acre basis. As a small plot or laboratory tool for training or research purposes it's an excellent technique; as a year round commercial ranch program it's a disaster.

Set stocking is a PRECISION MANAGEMENT program. To be effective as a year long program it should only be used on a mono-culture (only one species of forage), if more than one species is present the animals will select the most palatable species until it is grazed out (this takes a little longer with temperate grasses than with tropical). Since growing conditions are continually changing the stocking rate must be changed to match (put and take). This continual adjustment, while bothersome is not the major problem. But where, on a commercial operation, do you put them and from where do you take them? One can go to the sale barn to buy or sell extra animals, lease pasture, borrow the neighbors animals, hold out extra land on the ranch somewhere to keep the extra animals (but what sort of program will this piece have?), or just run considerably below the optimum stocking rate.

There is nothing wrong with continuous grazing or set stocking as tools. What is wrong, is to expect either one of these tools to do all jobs equally well. A pair of pliers is a wonderful tool for the jobs it was designed for; but it makes a lousy wrench and even a worse screwdriver. In continuous grazing there was no control of time, little in the way of control of numbers or space and precious little control of the animals. If you doubt that, have you ever tried to clear a 4000 acre paddock by yourself? The set stock program does not have control of time, it does have control of animal numbers (by definition), paddocks are usually small, one to five acres and because most experiment stations have a lot of help to push animals around, better control of the animals than continuous grazing. That the scientific community was a bit uncomfortable with the one tool theory can be noted by the admonition found in virtually every published article on grazing: "The grazing program must be tailored to each individual operation." The how, other than de-stocking, was seldom mentioned.

SUB-DIVISION:

The most economical method to obtain the necessary controls is through sub-division of the grazing unit.[1] Table 11.1 illustrates the effects of sub-division on a grazing unit; the amount of time that each paddock will be rested (no animals present) and the grazing pressure (animals per acre) as the number of paddocks increases. The paddock numbers double, starting at 1 and ending at 64; there is no magic in doubling, it is simply a convenient way to illustrate the effects of an increased in paddock numbers.

TABLE 11.1. EFFECTS OF THE NUMBER OF PADDOCKS ON A GRAZING UNIT.

Days rest (R), per paddock (P) = 30 days. Total acreage in grazing unit (TA) = 100 acres. Stocking rate for 1 year = 1 animal per acre, or a total of 100 animals. Comparison is for a period of 1 year.

(P) number of paddocks	(G) days grazed each paddock (days)	(CL) length of 1 grazing cycle (days)	(C) cycles per year (#/yr)	(%R) % total rest each paddock (per yr)	(PS) average paddock size (acres)	(D) density, # of animals per acre per paddock (#/Ac)
1	365.0	365.0	1	0	100	1
2	30.0	60.0	6.08	50.0	50	2
4	10.0	40.0	9.13	75.0	25	4
8	4.92	34.3	10.64	87.49	12.5	8
16	2.0	32.0	11.41	93.75	6.25	16
32	0.97	31.0	11.77	96.87	3.13	32
64	0.48	30.5	11.97	98.43	1.56	64

Equations Used: Days grazed; G=R(P-1). Cycle length; CL=R+G.
Number of cycles per year; C=365/(CL). Percent rest each paddock;
%R={(CL-G)/CL}{100}. Average paddock size; PS=100/P. Density of animals in grazed paddock; D=100/PS.

Some interesting things take place as the number of paddocks increase. The greatest effect occurs when the paddock is simply divided in two and run as a flip-flop; from zero rest, each paddock is now rested 50% of the time. As the paddock numbers

[1] Sub-division in grazing management terms does not mean a housing development, even though that might return the most money.

further increase, the percent of rest does not increase at the same rate. Thus a paddock level of 4 gives each paddock a total rest of 75%; with 8 paddocks, each will be rested 87.5% of the time and so on. If the total time that each paddock is to be rested was the only reason for sub-division, then it would make little economic sense to go much beyond the 16 paddock level, as the cost of each unit of additional fencing remains the same but the returns for each increment get less and less.

There are other reasons than just the percent of total rest for each paddock that make sub-division beyond the 16 paddock level desirable. But in general the economics are such that higher paddocks levels only have significance in intensive environments or in extensive environments during drought. The techniques used to increase paddock numbers during drought are essentially to combine herds from two or more grazing units and shift the animals from paddock to paddock throughout all of the combined units; drought management is discussed in Chapter 14.

The major benefits from paddock numbers in excess of around the 16 level is due primarily to nutrition and the stimulation of forage intake that is caused by frequent shifts to fresh feed. At high performance targets two things are important; feed quality must be high and it must be consistent. Well managed feed yards do not change feed from one day to the next; any changes are made gradually. If a paddock has sufficient forage for a graze period of 7 days, the first day the animals will eat more than one seventh of the available forage. Not only will they eat more, the quality of what they eat will be higher than on the last day. This means that on the last day, the quality of forage consumed will be lower than on the first day and also the amount consumed will be less; often less than is needed to give the animal it's fill. The result is that the animal will spend more time walking and less time grazing, which means more energy expended and less forage consumed than on day one.

Figure 11.2 depicts the decline in quality of a seven day graze unit compared to daily shifts. Bear in mind that the "average" quality is the same in both units but because of animal selection, the animal will take the best first and leave the rest till later if given the opportunity. The result is that the rumen micro-flora has to continually adapt to a changing diet. This means a continual shuffling of numbers of cellulose digesters to carbohydrate digesters, with the result that efficiency of conversion is not as good as if a constant diet had been provided. The daily shift in quality from day 1 to day 7 is relatively small and the micro-flora are not under too severe a stress. However, the shift of quality on day 7 to day 1 of the new paddock can be quite large (several percentage points of digestible protein and energy) consequently, the populations of the different rumen organisms never stabilize. Not only is the animal on a low quality forage the last few days, it is also not getting its fill; this is especially true where the animals are left in a pasture until they "gruboff" the available forage. Upon changing pastures the animal is suddenly faced with a higher quality forage, plus all he can eat. As a result the animal gorges itself, the rumen flora can't handle the sudden change in quality, the animal gets the diarrhea and we say it was "washy" feed. The longer the graze period (except in light continuous or set stocking) the worse the effect.

Other advantages of high paddock numbers have to do with the social behavior of the animals in a herd and the increased animal density. With high density and daily shifts the majority of animals enter the paddock with their heads down, grazing; sort of "I'm going to get mine before the others get it" attitude. Shifting stimulates increased forage intake; even if the shift occurs after a graze period. Social facilitation it's called; this can increase intake by as much as 20%.

High animal density coupled with short graze periods lays large quantities of manure and urine down in a short period of time; similar to a fertilizer application. Observations indicate that the trampling and spreading action of the animals hooves spread the majority of cow pies. The action of dung beetles is more evident, apparently due to the large amounts of available manure in a small area. Trampling, compaction and plant damage is reduced as the time element is less, even though numbers are higher.[2]

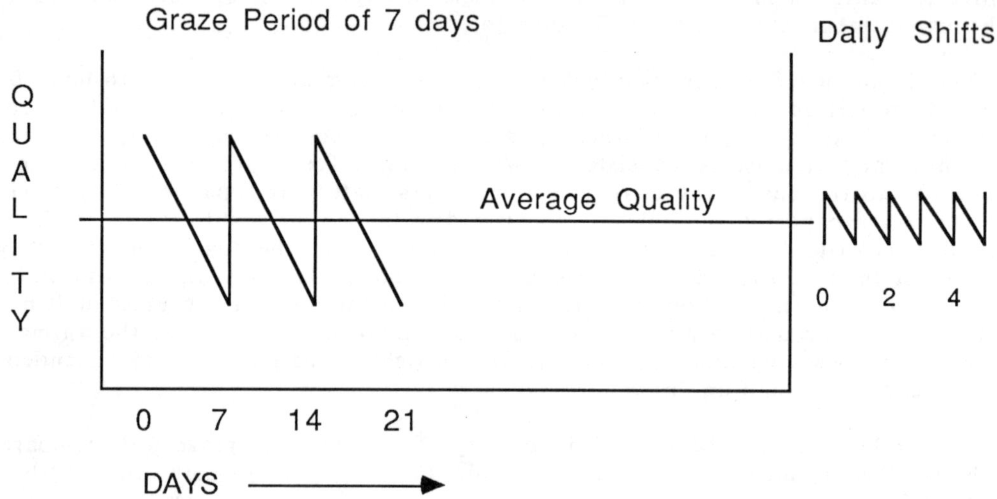

FIGURE 11.2. EFFECT OF GRAZE PERIOD ON THE QUALITY OF DAILY FORAGE INTAKE.

ROTATIONAL GRAZING:

Before discussing the principals of rotational grazing, a bit of history is in order.

A LITTLE HISTORY:

Rotational grazing probably got its start in Biblical times and was practice to some extent by the Romans. However, the basic principals of intensive livestock grazing involving rotational grazing were first set down in 1760 in an anonymous French

[2] If you walk across a lawn at the same place every day for a year, a path will develop; if you walk that same path 365 times in one day, a path will develop, but in a couple of weeks it will be grassed over.

dictionary titled "The Agronomist, pocket dictionary of the farmer"; these were later published in 1768 in the "Maison Rustique". In 1778 Abbot Rozier published articles on rotational grazing, this was followed by articles written by James Anderson (inventor of the Scottish plough) in 1797. However little attention was paid to these concepts until they re-emerged in a translation by Johnstone-Wallace in 1944. In 1959 an English translation appeared of the works of Andre Voisin, titled "Grass Productivity". Voisin was a French farmer who had applied the above concepts and worked out the practical details during the 1930's and 40's; he called his program "rational grazing". The basic principals of rotation were known in New Zealand as early as the 1930's but didn't really get moving until the 1950's.

Prior to and just after World War II, there was a surprising amount of rotational grazing actually taking place in the U.S., mostly on dairies, and mostly in the eastern part of the country. The work of Andre Voisin was known, he had visited parts of the eastern U.S. and some had read his book in the original French. However, the subsequent cheap fossil fuels soon put an end to "Rational Grazing". Beginning about 1982 the concepts of Voisin were revived in Vermont and other parts of the northeastern U.S., where there is now considerable interest in the techniques of "rational grazing".

After World War II, for reasons discussed in Chapter 8 on Ecology, there was little work done in the U.S. on different methods of grazing, other than by a few hard-headed individuals that just knew there had to be a better way. Mr. Gus Hormay, first of the U.S. Forest Service and later of the Bureau of Land Management, developed a basic three or four paddock program that saw success in various forest allotments. In Texas Mr. Leo Merrill worked on a three herd, four paddock system that demonstrated improved performance as well as range improvements. But by in large the siren of cheap fossil fuel energy had everybodies attention.

Adding to the problem in the U.S. was the fact that shortly after World War II, a group of individuals interested in the management of the more arid sections of the West, split from the American Agronomy Society and formed the Society for Range Management. The new group striving for self identity seemingly rejected much of the work on paddock and stock manipulation in favor of changing the environment to fit the animal through inputs of fossil fuel, range fertilization, brush clearing and herbicides. Continuous grazing was adopted as the "natural" system and the only one that could work on the vast areas of the west. It is of interest to note that they adopted as normal, a grazing program that has resulted in either a deteriorating environment or an uneconomical range livestock operation. However, some work was continued on manipulations of paddocks and animals by agronomist in the south and eastern portions of the country.

Beginning in the 50's, building on the work of Voisin and others, rotational programs were introduced in Kenya on tribal lands; the multi-kamp designs, which were essentially a wheel design with animals being rotated through the different paddocks. The successes reported found their way to South Africa and what was then Rhodesia and were picked up by workers and ranchers there, the most notable of which was Mr. Allan Savory. In the mid 70's Mr. Savory came to the U.S. and in a series of appearances, managed to get the livestock industry off of dead center (at least they were talking about grazing management and a few started putting in grazing units, the first being the Spurlock Ranch in Arizona) and at the same time antagonized many of the "scientific professionals" in the Range Society.

The late 70's saw a spurt in activity to prove that rotational programs were not better than continuous grazing, not only in the U.S. but in Australia as well. The scientific community developed and tested numerous rotational programs involving different lengths of stay, numbers of animals and rest periods. About the only things they all had in common were they all seem to go by different initials and they were all very rigid (for statistical reasons) in the different treatments they imposed (the move every Monday syndrome). Many, not all, found that rotational grazing produced more pounds per acre but continuous grazing, with light to moderate stocking rates, produced more per animal. About the only real insight that was gained from all this, was that any grazing program, if it's rigid enough, will fail.[3]

PRINCIPLES OF ROTATIONAL GRAZING, OR THE "HANG LOOSE" APPROACH TO GRAZING:

The term rotation or rotational grazing implies that one is going in a circle, or that one has assigned regular turns or a sequence to a series of acts. This implied rigidity is what led Andre Voisin to coin the term "rational grazing", the New Zealanders "controlled grazing" and Allan Savory the "Savory or Holistic Grazing Method". There is a very real danger of letting a word define ones actions with the result that soon one is into a rigid grazing program (move every Monday syndrome); this is the first sign of eventual failure. As far we're concerned, "a rose by any other name would smell as sweet", it doesn't matter what you call it as long as you remain flexible. However, for those that like to identify with a word or words, we offer the "Hang Loose" approach to grazing. We can think of few other phrases that imply a greater degree of flexibility. From now on whenever you see the word rotation - think "Hang Loose".

There is really no one reason that is more important in rotation than another, but if one was forced to pick a reason for rotating, it would have to be the rest given the forages when the animals are shifted out. Table 11.1 illustrated the effect that paddock numbers have on the amount of rest that each paddock enjoys after its guests leave. Giving the plant time to replenish it energy reserves after grazing, can not be over emphasized. Some of the other important factors are the ability to accurately allocate forage to the grazing animal, favorable changes in animal behavior, higher utilization of available forage, the ability to time entry into a paddock so that it matches the growth rate of the forage, as well as herd effects, better allocation of nutrition and many other smaller effects, but none the less important since they all make up the total picture.

Rotation allows the operator to determine when to introduce animals to a particular paddock based upon the amount of rest the forage needs. Chapter 5 discusses the reasons why plants need rest and how this varies depending upon seasons. On rye grass in rapid growing conditions this might be as soon as 10 to 12 days after the last graze period, whereas the large bunch grasses, such as elephant, might require 35 to 40 days. However in the winter or dormant season both grasses may require a rest period of 90 to 110 days. The only time a grass doesn't need a rest is when it's dead and even then one may wish to allocate. The point is that the rest period for any particular grazing unit will change from season to season and in some cases from rainfall to rainfall. The required rest is not something that you can look at the calendar and determine, you have to look at the grass, check you production targets and make a decision.

In general when growth rate slows down, the grazing cycle (the time it takes to make a turn through all paddocks in the grazing unit) slows down, when the growth rate speeds up, the grazing cycle speeds up. For example in winter the growth rate of the rye grass has slowed down until it takes three weeks to grow one inch in height (you know

[3] Like staying up all night to find out where the sun goes; pretty soon it will dawn on you.

this because you have been measuring it). You have decided that you will not graze a paddock until the height has reached 4 inches, which if everything stays the same will require 12 weeks or 84 days of rest. You have 10 paddocks. How long will you have to stay in each paddock so that each will have the required 84 days rest?

Use the grazing equation for one herd: G = R/(P-1), where G = the days rest, R = the days rest and P is the number of paddocks. Plugging in the numbers gives G = 84/(10-1) = 84/9 = 9.33 days. Each paddock should be grazed a total of 9 and a third days, this will give the other paddocks the necessary rest of 84 days. In actual practice you would move the herd into a paddock, leave it there for either 9 or 10 days, then move to another paddock and graze it for 9 or 10 days then to the third, etc.

So you keep chugging along, shifting the herd every 9 or 10 days until you have grazed 7 paddocks in all. Suddenly you get an early spring warm up and the growth rate jumps to one inch in 10 days. Now it only takes 40 days to grow 4 inches. Plug the new numbers in: G = 40/(10-1) = 40/9 = 4.44 days. Now instead of waiting 9 or 10 days to move you are shifting every 4 and a half days. Notice how fractions are handled. Always round off high. Don't attempt to divide the grazing day up into thirds, or smaller (unless you've a milk herd and are milking three times a day), half day moves are as small as you need to go. Say the first move was on Sunday at 6:00 AM the next move would be four and a half days later on Thursday at 6:00 PM.

Now it's late spring and the growth rate is really hot, one inch in three days, you still will be grazing at 4 inches in height. However the contractor has started working on your road and they have tore down the fence on one of your paddocks so they can park their equipment; the paddock will be out of commission for at least a month. The new days grazing are: G = 12/(9-1) = 12/8 = 1.5 days. So every day and a half you shift your herd. This is an introduction to the "Hang Loose" concept; the growth rate changes, you change, lose a paddock, make the necessary adjustments, but keep moving.

The length of time it takes to come back to the original paddock is a cycle. If the days grazing were 1.5 days and the days rest were 12 days, one cycle would be the sum of the two: G + R = cycle or 1.5 + 12 = 13.5 days to complete one cycle. To find the number of cycles in a year divide 365 days by the number of days in one cycle. In this case the number of cycles in one year are: 365 days/13.5 days = 27. If you were to stay at 1.5 days graze and 12 days rest, that's how many times you would have to open the gate.

There is a wee bit more to it of course. All that has been considered so far is simply adapting to the growth rate or a change in paddock numbers, nothing has been said about numbers of animals, or how much they will eat.

FORAGE ALLOCATION:

It is the dead of winter, snow is up to your neck, there is no more grazing. You have 200 tons of hay and 400 dry cows, 100 non-pregnant replacement heifers, 20 bulls, and 3 horses. How long is that 200 tons of hay going to last? Everyone in snow country, or in wet-dry country goes through a exercise similar to this every year. Do I have enough hay (silage) to make it until the grass starts to grow again?

Before you can figure out how many days that hay is going to last you need to know the weights of the animals. Let the cows average 1000 lbs, the replacements 700 lbs, the bulls 1600 lbs and the horses 1500 lbs each. You now need to decide one more thing and that is: Are you going to feed the replacements to gain or just hold them at maintenance? Lets say you are going to try to let them gain a half a pound a day.

There are several ways to figure this, the animal unit concept is one way (Chapter 6). Using your scientific calculator enter 1000 in the power to an exponent key, and raise this to 0.75 this gives 177.8, this is your divisor. Now enter the weight of the replacements, 700 and raise this to 0.75, this is 136; divide this by 177.8, this equals 0.76 (the AU equivalent); however since you want the heifers to gain add on a 0.1 fudge factor so they would become 0.86 AU's. The bulls would be done the same and would be 1.4 AU's. The horses are arbitrarily given an AU of 1.5 each. The total AU's are: 400 cow at 1 AU = 400 AU's; 100 replacements to gain, 100 X 0.86 = 86 AU's; 20 bulls at 1.4 AU's each = 28 AU's; 3 horses at 1.5 AU's each = 4.5 (better figure 5). Adding them all up gives a daily demand of 400 + 86 + 28 + 5 = 519 AU's.

Depending on the quality of hay, the animal will need to consume between 1.6 to 2.0% of their liveweight in dry matter every day just for maintenance. Assume that the hay is good quality grass hay, so we will use 1.8 %. However, this 1.8% times the live weight of the animal, is the amount of DRY MATTER that the animal should eat; hay has between a 15 to 20% moisture content. So a 1000 lb cow times 1.8% (1000 X 0.018 = 18 lbs) gives 18 lbs of dry matter, if you feed her that much hay she will be short. Figure the hay has a 15% moisture content,[4] this means that it contains 85% dry matter (100% - 15% = 85%). Divide the amount of dry matter in the hay (in decimal form) into the amount of dry matter you need to feed: 18 lbs/0.85 = 21.1 lbs. This is the amount of hay you will need to feed each AU on the place.

You now have the total AU daily demand, 519 AU's and how much hay is required to feed each AU, 21.1 lbs. The two are multiplied together to give 10,951 lbs (5.48 tons) of hay required each day. To find out how many days that 200 tons will last, divide by 5.48; you have enough hay to last 36.5 days.

Well that was really a lot of fun, but we're talking forage, not hay. Quite true, but the idea is the same, the only differences are the quality of the forage will often be less than good hay and the animal will do its own harvesting, you won't have to pitch it out, but you will have to see that it's rationed out (allocated) correctly.

There are two types of forage allocation programs. The first is when you have so much green feed and so many animals to eat it. The second and usually the most critical is, when you have so much feed (usually brown or getting that way), so many animals to eat it and there is going to be so many days before any more feed becomes available (winter, dry season or drought). The first, allocating green feed will be cover later in this chapter under feed budgets, the second approach follows.

Chapter 5 discussed several methods used to determine how much forage is available for animal use. Whichever method is used (clipping, pasture probe or eyeball) the available forage is determined for each grazing unit or pasture on the ranch. The results are then added up, this is the total available forage on the ranch. The next step is to determine the percent utilization. This is the amount of available forage that actually gets into the animal. Once you get to estimating forage, either with your eye or a probe, you will be able to determine percent utilization for your grazing unit with a fair bit of

[4] To determine the amount of dry matter in the hay, weigh a sample, then dry it in a oven over night, low heat, or use a microwave, then weight it again and figure the percent moisture present in the hay. Divide the second weight [dry weight] by the first weight [wet weight], times 100.

accuracy; the amount of forage left when the animals leave, divided by the amount that was there before they were moved in, times 100 gives the percent utilization. Until you know your own values use the following: 50% for continuous or set stocking, 55% for long term rotation (one month or longer graze periods), 60% for short term rotation (week), 70% for daily shifts and 75% for intensive strip grazing. Multiply the percent utilization number (in decimal form) by the amount of available forage, this gives the amount of usable forage under the particular type grazing program you're using. Divide this number by the amount of forage that each head will eat and the result will be total cow-days of available grazing. Divide this number by the total number of animals on the ranch, this gives you the number of days you have left until you run out of feed.

For example: You measure the grazing unit with a pasture probe and determine that you have 950 lbs of available forage per acre. You will be grazing each paddock 7 days before shifting to a fresh paddock, your percent utilization is 60%. Multiply 950 times the percent utilization; 950 lbs/acre X 0.6 = 570 lbs/acre of usable forage. You are going to carry dry cows (based on their weight and the fact that the forage is dead) you figure that they will eat 20 lbs per head per day. Divide this number into the usable forage per acre; (570 lbs/acre)/(20 lbs/head/day) = 28.5 head/day. You have sufficient forage to carry 28.5 cows for one day on one acre. You have 200 head and 700 acres. Multiply the 28.5 cow days/acre by the 700 acres; 28.5 cow days/acre X 700 acres = 19,950 cow days. Now divide this number by your 200 cows; 19,950 cow days/200 cows = 99.8 days. You have sufficient forage to feed your herd for 100 days. If the dry season last longer than 100 days and you do not get any forage growth during this time, you are in trouble.

Another method of determining the amount of forage on hand is to estimate the area of pasture that would feed one cow for one day. This usually requires two people although one can do it if stakes are used. Estimate the area that you think would contain sufficient feed for one cow for one day, place the stakes at the corners. Either measure it with a tape, or pace it off (if you know the length of your pace) and determine the number of square feet in this area (multiply the length of two sides together). Divide this number into the square feet in one acre (43,560 sq ft). This result is the number of cow days that one acre will support; multiply this by the number of acres in the pasture and you have the total cow days contained in the pasture. Do this same procedure for each pasture in the entire ranch, add up the results and divide by the total number of animals on the ranch; this gives you the number of grazing days left.

For example: You estimate the area to feed one cow for one day to be 38 feet on one side and 29 feet on the other. The total area is: 38 X 29 = 1,102 square feet required to feed one cow for one day. Divide this number into the square feet in one acre: (43560 sq. ft./acre)/(1,102 sq. ft./cow) = 39.5 cow days per acre. You have 700 acres and 200 cows. Multiply the number of acres by the cow days per acre: 700 acres X 39.5 cow days/acre = 27,650 cow days of grazing. Divide this by the number of cows: 27,650 cow days/ 200 cows = 138 days of grazing left.

Both of the above methods have problems: Clipping is time consuming but if reasonable care is taken, fairly accurate. The pasture probe is fast and reasonably accurate, if correctly calibrated for your area and type of forage. Visual observation, both as to amount of forage available or the area to feed one cow for one day, are also fast but unless you have a good eye for estimating, can be inaccurate. The saving grace of the estimation method is that it is usually conservative; you tend to allow more forage than the animal actually requires. The best approach is to do the estimation but also do a few clips as a spot check.

These methods give you the number of days that you can carry your herd into the future, based on the PRESENT forage condition. They measure standing crop only and are somewhat conservative. They are a worst case scenario. Neither approach takes into account future forage growth, due to rain or present soil moisture levels (Chapter 14 discusses how to project into the future).

FEED BUDGETS:

A feed budget is simply the amount of forage that is required to feed a given number of animals, at a predetermined level of performance, for some set period of time. Feed budgets are tied very closely to stock flows. Once the stock flow has been determined, a feed budget is easy to calculate. In the simplest case, say one bull, kept in a pasture for one year, it would equal 365 bull-days.

There are three aspects to a feed budget:

1. The amount of forage (quantity) required per animal or herd.

2. The quality of the forage required.

3. At what time of year that quality and quantity is required.

The quality required depends on the performance levels you had set for the bull during the year. The actual pounds of forage required would be determined based on the weight of the bull; Chapter 6 discussed methods to determine quantity. For purposes here, figure 2.5% of his body weight, in pounds of dry matter per day, will be required to gain a little weight and 2% of his body weight for maintenance. For most feed budgets animal days will suffice. There is a tendency to get carried away with accuracy, it makes little sense to worry about the difference of one or two pounds a day of intake when the error on the amount of forage present may be hundreds of pounds. The major thing is to always be conservative, that is, figure they will always eat a little more than the tables say and that you always have a little less than what your measurements show. To do this always round off on the high side when working with consumption and round down on the low side when figuring amount of forage.

DETERMINATIONS OF FORAGE QUALITY:

Regardless of where the ranch is located, all forage on it can be broken down into 4 different categories of quality. These groupings will be relative to a particular ranch; the highest quality on one ranch may or may not be the same on another. The major groups are alive or dead. Dead forage (brown) has survival value only and that is what it is called; survival forage. The alive group (green) is further sub-divided into three groups: 1. Stocker. 2. Cow/calf. 3. Bull.

SURVIVAL FORAGE: This is used only during dry periods, or drought. Its main value is that of filler. It is characterized by low crude protein and digestibility levels. The allocation method is the usual way this material is utilized. Dry cows or bulls could be grazed on it with supplements. Weaners and lactating cows, even with supplementation, will generally lose weight, the quality is simply not there.

STOCKER FORAGE: This is the best the ranch has to offer. It is young vigorous growth with lots of leaf and little stem, the "cream" of a pasture or paddock. All classes of animals gain weight on Stocker quality forage. Crude protein content of 12% or higher.

COW/CALF FORAGE: This is older material, with more stem to leaf ratio, it is the second best the ranch has to offer. Weaners placed on cow/calf forage will generally experience "weaner slump", actually lose weight for several months. Yearlings will make modest gains. Crude protein content of 10% or higher.

BULL FORAGE: This is what is left of the green material. This material is generally suitable for maintenance rations for mature animals. This is the type of forage that animals on pasture maintenance duty generally get to eat. Crude protein content usually 8% or higher, although may be lower depending on the amount of dead material mixed in.

To get a better picture of the different quality classes, divide the height of a green pasture sward into three equal parts; stocker feed would be the top third, cow/calf quality the second third and bull forage the bottom third. Survival forage would be the dead leaves and stems in the bottom third.

These quality groupings are relative. The actual nutrient content of each will vary depending on the season, the species or mixture of the sward and from ranch to ranch. As a result the actual performance of the animals grazing them will also vary, depending on the amount of feed on offer of each type and the season of the year. They are what the ranch has to offer at any given time during the course of the year.

SAMPLE FEED BUDGET FOR SMALL COW/CALF OUTFIT:

For purposes of demonstration it will be assumed that this ranch has 1 bull, 25 mature cows, 4 H1's and 8 H0's (replacement heifers). Forage production can be considered as coming during a specific growing season and the surplus conserved (hay or silage being put up as two or three different qualities), or the ranch is in a continuous growth situation with what surplus forage that is not grazed being conserved until it is needed (hay or silage).

For example, consider a breeding bull, with a 2 month breeding season, beginning 1 May. From the end of the season, 1 July you decide to put him on cow/calf forage for two months, to gain back a little of the weight he lost. From 1 September to 1 March the bull will be on bull forage (maintenance), from 1 March to 1 May he will be on stocker forage to get in condition, from 1 May till the end of the season on 1 July he will be with the cows on cow/calf forage (where he may or may not eat, depending on the interest he takes in his work). Figure conservative and allow 2.5% of his body weight for his daily dry matter requirement when he is gaining weight and 2% when he is not; assume he weighs 1600 lbs. A feed budget for him would look something like that in Table 11.2.

The cow herd to go with this bull might be run as follows: Calves to start dropping 8 February, end 9 April; breeding season starts 1 May to 1 July (it's important to make it the same time as the bulls). Weaning will take place when calves are 6 months old, on or about the 10 of September. Cows on maintenance until 8 January, from 8 January until 9 April on cow/calf feed, place on stocker feed from 9 April to 1 May (flushing). Assume the cows will eat 2% of body weight at maintenance (bull feed), 2.5% on cow/calf feed and 2.8% when on stocker feed; the cows weigh 1000 lbs. The feed budget for the cow herd might look like that in Table 11.3.

TABLE 11.2. FEED BUDGET FOR BREEDING BULLS:

	date			
	3/1-5/1	5/1-7/1	7/1-9/1	9/1-3/1
Forage Quality	s	c/c	c/c	b
Number of Days	6161	62	181	
Forage Quantity (lbs/day)	40	32	40	32
Number of Animals	1	1	1	1
Total Required (lbs)	2,440	1,952	2,840	5,792

TABLE 11.3. FEED BUDGET FOR MATURE COWS:

	date			
	1/8-4/9	4/9-5/1	5/1-9/10	9/10-1/8
Forage Quality	c/c	s	c/c	b
Number of Days	91	22	101	151
Forage Quantity (lbs/day)	25	28*	25	20
Number of Animals	25	22*	25	25
Total Required (lbs)	56,875	13,552	63,125	75,500

* Three cows culled for non-performance.

The final feed budget for a cow/calf operation consist of the replacement heifers. In this particular operation the heifers will be bred (AI) to calve as 2 year olds; two weeks before the cow herd. At weaning (H0's) they will go into stocker feed, bred as yearlings (H1's) and remain there until ready for breeding as 2 year olds (H2's), at which time they will be transferred to the cow herd. The cow herd requires 3 head of replacements per year, at 80% weaned calf for the H2's this means that there will have to be: 3/.8 = 4 H1's to be exposed to the bull the second season. Since AI is going to be used, figure 50% weaned, this means that 8 heifers must be bred AI to get the 4 H1's to get the 3 H0's. The heifers weigh 450 lbs at weaning (H0's), 650 lbs at first breeding (H1's), 850 lbs at their second breeding (H2's). Since these animals will be on good forage allow 3% of body weight for amount consumed. Table 11.4 shows the calculated feed budget.

Putting all three feed budgets together, this operation would have to produce 302,280 lbs of dry matter each year; 96,200 lbs of stocker feed, 124,790 lbs of cow/calf feed and 81,290 lbs of bull feed during the course of the year. Another way of looking at it is that 31.8% of all forage produced would have to be of stocker quality, 41.3% would have to be of cow/calf quality, and the balance, 26.9%, bull feed.

Obviously this is not the real world with it's changes in seasons, differing rainfall patterns and all the rest, but it does show the procedure. In actual practice you would break the year into seasons, 12 months or even weeks, depending upon the level of intensity. The quality and quantity of feed required would then be set down for each season, month or week based upon your stock flow projections. After the feed requirements, both quality and quantity, have been totaled for the time interval you're using, you then decide whether or not you can realistically meet those requirements. If you can, well and good, if you can not, then you at least know that something (fertilization, forage conservation, irrigation, buying or selling, etc.) had best be done before you get too far down the road.[5]

TABLE 11.4. FEED BUDGET FOR REPLACEMENT HEIFERS:

	date	
	9/10-5/1	5/1-9/10
Feed Quality	s	s
Class and number		
H0	8	-
H1	4	8
Number of days	233	132
Average weight (lbs)	550	750
Quantity (lbs)		
H0	17	-
H1	26	23
Feed needed (lbs)		
H0	31,688	-
H1	24,232	24,288
Total forage (lbs)	55,920	24,288

THE ALL PURPOSE CHART - APC:

A rather tiresome amount of charts have been proposed as aids for various aspects of grazing management. Here's another one (Figure 11.3). It's really neat, nothing on it but lines. In fact it's so simple that anyone can make one of their own; all you need is a ruler, pen or pencil and a piece of blank paper. The beauty of it is that it's cheap and you can make it fit your own operation, simply by changing the width of the columns or length of the rows. Whatever headings you wish to put in the rows and columns is your business. We have found that it works very well for stock flows, feed budgets, work scheduling, even as a calendar, if you like to fill in the blanks.

[5] You may be still be able to make ends meet, but you may not be able to make them like each other.

FIGURE 11.3. ALL PURPOSE CHART (APC).

SHIFTING:

One of the first things noticed when a switch is made to IGM is that the behavior of the animals changes; they no longer behave in quite the same way. For one thing they don't run from you; they're gentler. A major difference is that they are more herd oriented. The trend is not as noticeable in sheep but cattle tend to group themselves into mobs of 40 to 80 head. Each mob acts as a distinct unit, even though the individual members change from mob to mob.

However, in high density small paddock grazing (50 or more head per acre), it has been noticed that if there's a lot of forage in the paddock, relative to the number of animals entering, the herding effect is stronger than if the forage is at maintenance levels. When the animals enter a paddock with ample forage, the whole herd tends to move together, starting at the gate, moving to the far end, back to the gate, etc. Each pass the height of the forage drops, the animals eating the top portion first, then the next and so on. But if the forage levels are low when the herd enters, they tend to split up, with each individual moving rather slowly and eating almost to ground level before moving on; the effect, after the animals have been in the paddock for a hour or so, is similar to "patchy grazing". In this case the animal still graze the forage off by layers, but stays put until most all of it is gone. Of course not all animals do this, but the majority will.

If frequent shifts are made the animals begin to anticipate the shift, this is especially true if the shift is made at the same time each day. The result is that when you arrive to open gates, you find the majority of the herd waiting for you, bawling their heads off. One study in Hawaii found that 75% of the herd anticipated the move. A different study noted that 99% of the herd had moved themselves into the fresh paddock within 30 minutes after the gate had been open.

It is strongly advised that a special signal be given the animals prior to actually opening the gate; such as a bell, whistle, horn or just a particular yell. If not, the animals will move to the gate every time a person or vehicle approaches. One Hawaiian ranch shifts animals with a motor bike, after a while every time a helicopter flew over (the sound is somewhat similar to a motor bike), the animals would move to gate, ready to shift. There are a lot of sight seeing helicopters hauling tourist around the islands and the animals would often trot to the gate 2 or 3 times in a single day.[6]

Shifting on small paddocks is not much of a problem, animals will either see you or hear your signal, or they will see others beginning to move and will follow. Because the paddocks are small, they are easy to clean out, all stock can be checked as they pass through the gate. One problem that does arise especially with young animals and particularly on the wheel designs, but also occurs in block designs, is that the lead animals, as they go through the gate, do a 180 and go down along the fence line in the new paddock, opposite to the direction of the other animals moving towards the gate. The "tail enders", being follower type animals, see the lead animals going in the other direction, stop, turn around and follow them; the only problem being that they are still in the old paddock. This is no "big deal" if you are behind them pushing them out, but if you're letting them shift themselves, it can get you hopping. One solution is to place a gate about half way down the radial and shift through that. You can also try standing part way down the fence line, on the fresh paddock side, to discourage the lead animals from doing their 180 thing. If either of these don't work, you're just going to have go and push them out the best you can. Make a game out of it, maybe they'll see you having fun and quit it.

Opinions vary as to the best approach for shifting cattle on large sized paddocks, some ranchers like to force move, that is, ride behind the herd and clean out the paddock at the time of move. The feeling here is that all animals will be seen and any problems can be taken care of at that time. Others prefer to open the gate and leave, then come back later and clean out the paddock. A few will have two paddocks open at all times, when they shift a new paddock will be open and the back paddock cleaned out and closed. While the back paddock is being cleaned the majority of the animals will shift into the fresh paddock; since there is only a few head in the back paddock, the time involved is not great. This last method is suggested where feasible.

Many worry about shifting young animals or while calving or lambing are taking place. With sheep you are advised to set stock during the lambing period; stay out of the area, unless you are shed lambing. If you are shuffling animals around, or disturbing them you'll end up with a lot of "bum" lambs. Remember sheep are a follower species, those lambs will follow almost anything for the first 3 to 4 days.

Cattle are a laying out species, the new born will typically be "hidden" for the first week or so. On large paddocks this can present a problem when trying to clear a paddock. There are several things you can do. Leave several paddocks open behind the main herd. On large paddocks, generally the herd stays in a paddock for 2 to 4 days, thus there may be 2 to 3 paddocks open at once. The majority of the herd (if they have been in the unit for some time) will be in the fresh paddock, those that have calves back are free to go to them when they want to. This does not present any great problem as far as allowing the second bite on the forage, unless you are in a very rapid growth phase. You may still

[6] While we are concerned about animal health, jogging is not recommended.

have to ride to make sure, as there is always that few that never get the word. If you're in a rapid growth situation, or even if you're not, place the bottom wire of the electric fence 24 inches off the surface of the ground and make it the earth wire (negative charge), the hot wire should be 8 to 10 inches above this. This allows the calves to pass underneath without getting shocked. If you don't want the calves creeping ahead of the cows, you will need to put a switch on the bottom wire so you can either make it ground or hot. Don't use this technique with sheep, there are enough problems with sheep going through electric fences without training them to do it; if you want to creep sheep, make a special gate.

In arid country with large paddocks, the use of a water, or one-way gate can be used to shift stock. A one-way gate is placed so that when the animals come to water they must pass through it first; it is important that the animals pass through each and every time. At first the barriers are hooked open so the animals can go either way. These barriers, or bayonets, are hung off plumb so that they will swing close, but are easily pushed aside as the animal enters. After the animals get use to the new structure and when you want to shift, just unhook the bayonets. Figure 11.4 shows such a one-way gate in Queensland Australia.

FIGURE 11.4. A one-way gate, such as this one shown in Queensland, is often used to shift cattle from one large paddock to another in arid environments.

It is possible in some arid areas, or where the natural water is fenced off, to shift cattle by turning off the water in one paddock and turing it on in another. The paddock needs to be ridden for a couple of days afterwards to make sure all the animals have been moved. In some range areas that are unfenced, where water is a long ways apart, animals can be shifted by simply turning off the water in one area and turning it on in the new one.[7] It is important that the animals know the country, either that, or you should ride the old water hole a few time and push them to the new area. Don't leave the old water on, just make sure you ride it every day for three or four days. Probably a good idea to check the old water anyway, even if the critters were born and raised in the country; if something can go wrong, it will.

EXTENSIVE VS INTENSIVE:

Intensive means highly concentrated, while extensive is just the opposite. These words are used in two contexts, one dealing with the environment, the other with the degree of management one brings to that environment. Both will be discussed.

MANAGEMENT:

In the economics that the rancher finds himself in today, there are just two management strategies that are economical (that's assuming you're in the business to make money and not for other reasons). One must either manage extensively, or intensively. There is no middle ground; unless you are prepared to pay to raise cattle, or have some other good thing going for you. Either way, a lot of work. But there's a certain beauty to a positive cash flow, you can afford to take a vacations now and again without feeling so guilty about the cost and chances are, you'll have something to leave the kids.

EXTENSIVE MANAGEMENT:

"How many cows do you have? Wal, I don't rightly know, but I put out 37 bulls last spring." Most of us at one time or other, have met someone like that. They usually drove around in a "rattlely" old pickup, or sometimes a beat up sedan; however, there was almost always a big fancy car parked in a run down shed out back. Corrals were always falling down and their equipment consisted of a little old tractor, usually with a front end loader as the only attachment. Their homes used to be a cross between a clap-board shack and a oil rigger's tar paper cabin (now days they're more modern, not fancy, but nice). Their cows were either fat or thin, depending on the season; usually had an old pug horse they rode and were often in the saddle. They always seem to have a little hay or pasture for sale. Eccentrics we called them, and smiled. In the bad years they would complain as much as anyone else, but they were the only ones smiling when they walked out of the bank.

Their secret, they managed extensively. They were, for the most part, exceptionally good money managers; they wouldn't spend a nickel unless they knew a dime was coming back. The animals pretty much shifted for themselves, those that couldn't were sold. Roundups were infrequent, but when they occurred, seemed like half the town was in on it (free labor, well maybe not free, they usually provided some beer). Any big job that needed doing, haying, fence building, dozer work, etc., was contracted out. Little jobs were often done on week ends by friends or relatives looking for a chance to get away from town, or in exchange for hunting or fishing rights. They didn't like to pay in cash, not if a trade could be made, but if cash was the only way, they were

[7] These are the kinds of places where jack rabbits have to pack a lunch to get across.

usually fair. The extensive manager is a tight money manager, some might even say cheap. He doesn't spend much, he doesn't make much, but he always seems to make more than he spends. They own very little in the way of capital items that depreciate.[8] Some could be accused of raping the land, most however return just enough to maintain it.

Extensive management has been normally associated with extensive environments, where land has traditionally been relatively cheap in relation to the value of the animal. It can of course work in intensive regions but today, only where the land is owned free and clear. Extensive management is geared towards exceptionally low cost per unit of production; production levels are also low. Not much is given to the animals, and not a great deal is expected in return. It is a 1 to 3% return on investment operation, profits are steady, but not large. An extensive managed operation is as close to simple hunting as it's possible for a livestock operation to get.

INTENSIVE MANAGEMENT:

The main difference between extensive and intensive management is that the intensive manager views livestock production as an agri-business, he fully expects, in fact demands, to have a return on investment greater than what can be made on the money markets. The basics are the same regardless of the environment, only the approach is different. There is not much more we can say about intensive management, except that's what this whole book is about.

ENVIRONMENTS:

Management on an extensive outfit is every bit as intense and necessary as on the most intensive strip-grazing farm, even more so. The difference between one animal per 100 acres and one animal per one acre is one of climate, the level of management should be exactly the same, if in fact not greater on the lower stocking rate. However the reverse seems to be true.

EXTENSIVE ENVIRONMENTS:

Managing an extensive operation means setting overall goals, specific objectives and production targets, then doing it. It means cash flows, stock flows, balance sheets, margin analysis; in short, running an agri-business. The difference is in the primary emphasis, or priorities if you will. On extensive environments the primary emphasis must, of necessity, be on the unit of production, the animal, with secondary emphasis on the land.

The animal must be selected to perform within the environment, not the other way around. Rigid selection is a necessity; having a calf every other year just won't do. In central Nevada, with a average precipitation of 6 to 8 inches per year, the average percent calf crop is around 60%. It was said that those old desert cows just couldn't produce a calf every year. Well, if the calf crop was less than 50%, that statement is believable. But since it was over 50%, that meant that some of those sookies, were having a calf every year. Which ones, nobody knew, as nobody identified individual cattle; but they were there. It would take some time and effort, but if those cows are identified, progress can start.

---

[8] Paul, a very wise old Idaho trader, once told the senior author: "Son, don't ever borrow money to buy anything that depreciates." He, of course, forget that advice every time he buys a car.

Tom Lasater, founder of the Beefmaster breed, has some excellent observations about selection of range cattle. Sell all cows that fail to wean a merchantable calf for any reason. "You may get rid of a few good ones, but you get all the lemons". Keep at least 80% of all heifer calves and breed all for at least one breeding season. After you get your reproductive efficiency up, start selecting based on: 1. Disposition. 2. Performance testing in a constant environment. 3. Short breeding season (60 days maximum), but never let up on reproductive efficiency.

Most extensive operations have a calving season that usually extends more than 3 months, many are year long. The one characteristic that extensive operations have in common is that the forage production is highly erratic; some years only a little, the next year "gobs". In short, the operation is highly changeable from year to year, yet with a year-round or a long breeding season you are locked into a rigid program and the only option you have is to "ride it out".

With a short breeding season, 45 to 60 days, you have a great deal of flexibility in dealing with the variable environment that you are operating in. If you are getting below normal rains you can wean your calves early and either sell them, or if you have a well watered pasture some place, put them there. This takes the load off your cows; the nutritional requirements of a dry cow is just about half that of a lactating one. Also, it is much more efficient to have the calf obtain it's nutrition directly from forage, rather than first pass it through the cow and then to the calf (through milk); as a result your available forage goes farther. A six month weaning age becomes feasible. Then, since you kept the cow in better condition this year you will have a better calf crop next year. In addition you will have a uniform bunch of calves and the average weight is higher, simply because their age is not spread out over 4 or 5 months. In a close calving interval, calves born one estrous cycle back (18 to 20 days) will weigh 35 to 45 lbs less, those two cycles back 60 lbs less.

Sixty day breeding season sound too short? Not really. There's an outfit in the central part of the south island in New Zealand that gets between 90 to 95% weaned calf crop with a 45 day breeding season. They do this on an extensive environment, 6 to 10 inches of rain a year, very similar to the high desert country in the intermountain west of the U.S. In addition to sheep, they run 1500 herford cows and cull rigidly, they also test their bulls for libido (this is discussed in Chapter 6).

An extensive operation is geared to a program of low labor inputs per animal. One can not afford to spend much time or money caring for one animal. If one gets sick and if you find her, treat her, but then think real hard about selling her. The animal must be able to care for itself, "survival of the fittest". This does not mean that they are "wild". Cattle are not born wild, humans make them that way with stupid, careless handling. Since you are selecting on disposition the animals will be gentle and easy to work, especially if you "pattern" them. That is, run them through your various facilities several times a year without hurting them.

If you go for intensive management that means getting rid of a lot of "Heavy Metal"; those extra pickups, vehicles, tractors, machinery, fancy sheds and buildings, in fact anything that is not used for 9 or so months each year. If that sounds a trifle Spartan, consider how many 35 cent, 450 lb calves it takes just to pay the yearly interest on a $14,000 pickup at 13%; 12 head.[9] You will be doing a lot of riding, either horseback or with a motorcycle. The average cow to cowboy ratio had better be somewhere around

[9] Based on 1985-86 Hawaiian statistics; using mainland U.S. values about 6 calves required.

a 1000 to one, or better (extra labor, is hired when needed). Initially worming and fly control must be practiced, later select only those animals that show a measure of resistance; pesticides are used on an as needed basis. Handling is done as infrequent as possible and then in correctly design facilities, not fancy, but well designed with the cow in mind and not so much the comfort of the cowboy.

Supplementation is practiced but not forage substitution; more than 5 lbs of hay or silage per day and you're in a forage substitution program. Protein, energy and minerals are provided only when they are known to be in short supply for the class and physiological state of the animal. Basic forage requirements (bulk) are met through feed banking, living, or standing hay, not by feeding them a total ration of hay or silage. There are exceptions to this in the higher latitudes or elevation with deep snow cover, but even here standing hay can be used for a portion of the winter. When the cost of the yearly supplement for a cow is much more than 15 to 20% the cost of the animal, something is seriously wrong.

Subdivision of the ranch into paddocks should be done but only at a modest level. The cost of subdivision rapidly exceeds the expected benefits in extensive environments. The actual number of paddocks per herd will vary with the environment; 6 to 8 paddocks is probably the minimum. The drier the environment the fewer the paddocks and the slower development must be, to be economical. If hay is fed, each winter feed it out in a different paddock.

INTENSIVE ENVIRONMENTS:

Going intensive is no bunch of roses either. It means a radical re-adjustment of attitude, no more preparing for the local rodeo with your yearlings; your neighbors maybe, but not yours. It means being there or having a responsible person there, every day checking animals and paddocks. It means setting goals, cash flows, stock flows, balance sheets, margin analysis; in short, the whole bit, just like the extensive. The difference is on the emphasis.

Intensive operations place the emphasis on the unit of production, the land, with secondary emphasis on the animal. The animal becomes a means to an end, that end being the conversion of forage into a comfortable living for you. The modern dairy cow has been called a genetic monster, an udder with legs. Genetic selection has been towards higher production. The many traits that would allow her to survive in the range have either been bred out or simply allowed to disappear. The energy that she would normally use to survive has been channeled into production instead. In the U.S. the dairy cow has been turned into a converter of grain to milk, in New Zealand she converts quality forage into milk.[10] The ideal animal for an intensive operation should be similar. The operator will provide for its daily needs, all the animal has to do is to convert quality forage into meat, milk and/or fiber.

Consequently, nutrition becomes an important item. Forage conservation must be used to maintain production levels, rather than as a way to "keep 'em standing til the drought breaks". Intensive grazing is a must as are multiple herds, pasture maintenance, break grazing, forage allocation, residual dry matter allotments, etc. The manager must provide as near optimum an environment as possible, the animal simply converts forage into human food as near its genetic potential as possible.

[10] By the way the genetics that make for good grain converters, do not necessarily mean that they will be good forage converters, and vice versa.

This is not to imply that there is no selection program, there is, a most rigorous one. Performance standards are naturally higher than in extensive programs. The basic selection program just has a different basis than the one in an extensive environment. Since the environment is modified for the animal, those traits that the extensive animal needs to survive and produce are not as important in the intensive animal. The intensive animal does not need as strong a mothering ability as the range animal, she shouldn't have to fight off coyotes or dogs. An intensive animal shouldn't have to walk 4 miles to water, or search a thousand acres to find enough to eat, or raise young without assistance, or any of those things that a good range animal has to do to survive. The emphasis is simply on fertility and how well the offspring converts quality forage.

But above all else - HANG LOOSE.

SUGGESTED READING:

Dairy Production from Pasture. 1982. Ed. K.L. Macmillian and V.K. Taufe. New Zealand Society of Animal Production.

Economics of Intensive Grazing: A case in Hawaii. 1984. P.S. Leung and B. Smith. Research Extension Series 045. Univ. of Hawaii.

Grazing Animals. 1981. Ed. F.H.W. Morley. Publisher, Elsevier Scientific Co.

Grass Productivity. 1959. Andre Voisin. Crosby Lockwood & Son Ltd.

Lasater Philosophy of Cattle Raising. 1972. L.M. Lasater. Texas Western Press.

Savory Grazing Method or Holistic Resource Management. Numerous articles by Mr. Allan Savory and Dr. Stan Parsons may be found in Beef Cattle Science Handbook, volumes 15, 16 and 17; 1978, '79 and '80. Ed. M.E. Ensminger. Agriservices Foundation, 3699 E. Sierra Ave. Clovis, CA 93612.

The XXX Ranch by John Merrill. USDA Yearbook of Agriculture, 1983. Superintendent of Documents, Washington, D.C. 20402.

Wortham Lectures in Animal Science. 1965. J.C. Bonsma. Texas A&M University Press.

Used with permission of NEA.

Sometimes the obvious move is to go back and re-define your goals.

12 FORAGE CONSERVATION

When one thinks of forage conservation, if one thinks of it at all, one is tempted to think only of hay and/or silage; putting it up all summer and feeding it out all winter. Well that's the way it's been done in the past and there are certain environments where if livestock are going to be run, it will likely always have to be done that way. Putting up hay and silage is expensive and will likely remain that way. If the total cost of grazing grass is a penny a pound, then it will cost around 4.5 cents a pound for grass silage (corn silage will run between 8 to 9 cents), and about 9 cents a pound for hay (all on dry matter basis). At those kinds of cost one can not feed a great deal of hay or silage to animals and still stay in the black. Feeding 15 pounds of hay a day would cost $1.35 per animal; a 60 day feeding period would cost $81.00. So how many times are you going to buy that $200 cow during the course of her life?

Hay and silage are high value forages, their main purpose should be feed for high value animals. As discussed in chapter 14 under drought, there are other ways to keep 'em standing and there are alternative methods to get the most out of your conserved forages. Also, hay and silage are not the only methods of conserving forage, there are others; feed banking, crops and living hay to name a few. What follows is a brief description of each method, the reader is invited to check the suggested reading section for additional information.

HAY AND SILAGE:

The difference between hay and silage, if both are put up properly, are small from an animals point of view. The efficiency of utilization of nitrogen by the animal for growth or milk production may be lower with silage than with hay. This effect is greater if silage is the sole feed, rather than if a another supplement is fed. Cattle fed silage tend to put on a higher proportion of fatty tissue than those fed hay. However, there seems to be little difference between hay and silage in the efficiency of energy utilization. The obvious advantage of silage is that it can be put up under weather conditions that make haying all but impossible.

Regardless of which is made, it is important to have the proper equipment in good working order. Running a bunch of junk yard equipment only works if you are a good mechanic and are only using it on 20 to 30 acres. Any more than that and your losses in quality and quantity due to breakdowns and down time will more than eat up any possible savings. You can't afford to have that machinery sitting in the field watching the leaves fall off of the alfalfa, while you or your wife run to town for a two dollar part.

In putting up either hay or silage, quality is the name of the game. Both must be harvested when the quality is high, alfalfa at one-tenth bloom, grass hay before seed heads form (in the boot). The reason is that quality deteriorates rapidly beyond these points. For example Coastal Bermuda grass may have a crude protein content of 19% at three weeks of re-growth, this will drop to around 16.5% at four weeks, also the amount of fiber will increase. The major reason is that at three weeks the percent of leaf to stem is 86% but at four weeks this has dropped to 62%; stems contain less protein and more fiber (undigestible material) than leaves. Once the plant begins to form seeds, quality literally falls out of bed.

So what's quality worth? Well if the protein in 41% cottonseed meal is worth $0.45 a pound, the protein in hay or silage is worth the same. At those numbers for every 1% increase in digestible protein in hay or silage the value of the material would be increased

$7.07 per ton dry weight. If the cost of the cottonseed was twice that much delivered to your place then your hay would be worth twice as much also; and each additional percent of protein would be worth $14.14 a ton more.

SILAGE: By R. Guyton and B. Smith.

Silage is green forage which is stored in such a manner so as to remove all or most of the air (oxygen) around it and then to keep air from entering. Also, the green herbage must be of high enough quality to bring about the production of sufficient acids to preserve the forage.

There are a number of types of silos. Some of them are: Upright or tower (expensive); bunker or clamp; plastic bags (baggies); trench or pit (cheap); piled on the ground (very cheap). Upright silos are made of concrete or metal. Trench or pit silos are dug into the ground (Figure 12.1)and may or may not have have bottoms and sides made of concrete or wood. Bunker or clamp silos are constructed of concrete or wood and sit on top of the ground. Baggies are long sausage looking things that are made of plastic and lay on top of the ground; they have the advantage in that the machines are usually portable and any size quantity can be made right in the field. Stacking on top of the ground, with or without a plastic covering is also used on occasion.

FIGURE 12.1. Good packing is essential in making good silage. Wheeled tractors are best. Make the pit large enough so that the entire surface can be driven over by one or more wheels.

The type of equipment required for making silage are; tractor (only one is required, however a second one will speed up the operation considerably as it can be used for hauling and packing), chopper (flail is best), forage wagon and a front end loader with tines. A pitch fork is also recommended as is some type of silo. The cost of the necessary machinery can be 30 to 40% less for silage than for hay.

The losses associated with the type of storage vary, as might be expected, with the type of silo. One study found that based on a 150 day storage period the losses were 4.5% for a tower, 9% for a trench and 35% for a bunker. The losses for a pile on the ground often reach over 50%. The major reason for losses are poor sealing, loose packing, and overheating in the silo, this is usually due to high ambient temperatures especially in tropical regions. When internal temperatures reach 120 degrees F, much of the nitrogen will be tied up in insoluble compounds and not available to the animal.

The actual steps involved in silage making in a pit, surface pile or bunker are: The material may be direct chopped and hauled directly to the silo or it may be mowed, wilted in the field, chopped and then hauled in. Once the material is at the silo, it is dumped in layers and packed to removed as much air as possible (packing is usually with a wheeled tractor). A silo should be filled in as short as time as possible, preferably in one day. However, if this is not possible the silo must be covered and sealed every night to keep as much air as possible from entering the packed material. When the silo is filled it is covered with plastic or other material to seal out air and water. The silage should be fermented and ready for use in about 30 days.

The ensiling process is as follows: As the green herbage is placed in the silo or pit, it continues to respire and use up oxygen left in the stack. After the oxygen is used up, bacterial (lactobacilli) ferment the plant sugars into acids, mainly lactic acid. When the concentration of the acid reaches a pH of approximately 4.0 the silage will be preserved as long as air is excluded. If the pH does not reach a low enough point, other bacteria will decompose the lactic acid and sugars into other compounds allowing the silage to spoil. If air enters the silage will turn into compost.

Silage can be made from virtually any plant or crop, common ones are: Pasture grass and/or legumes which are surplus to grazing needs; grass and/or legume areas which are planted mainly for cutting; feed crops, such as corn or sorghum; and food crop by-products such as cannery waste, banana, pineapple, citrus, sugar cane and even papaya. The type of material used will depend on the individual operation, its assets and needs, as well as its location to suitable supplies.

Feeding value of silage is basically influenced by three things: First, the quality of the pasture or crop to be ensiled. Second, the moisture content of the ensiling material. Third, the type of fermentation or the ensiling process itself and if secondary fermentation takes place. To get the highest quality material from a pasture or crop it must be harvested at the proper stage of growth. Young, succulent forage is high in protein and low in fiber, while mature forages are low in protein and high in fiber. Grasses should be harvested before they flower or head out; but legumes should be at bud stage or early bloom when harvested.

The moisture content of the ensiling material is important for several reasons. First, if material is very wet there can be a large amount of effluent, or juice, which will drain out of the silo or pit. Effluent is quite smelly and can be a serious pollutant if allowed to contaminate water supplies.[1] The effluent also contains valuable nutrients which are lost as a feed and by reducing the acid content, may cause the silage to spoil. If wet material must be used it should not be packed as tightly in the silo as heavy packing will squeeze out more of the crop juice. Material which is 12% dry matter (88% moisture), will produce from 60 to 120 gallons of effluent per ton of ensilage. If the material has 25% or more dry matter, very little or no effluent will be produced. Second, if material is too dry (much over 40% dry matter), it will be hard to get sufficient packing in the silo. Loose packing will leave too much oxygen in the material and will result in lower quality. Third, the higher the moisture content the more pounds of water one must handle, both into and out of the pit to get the same amount of feed value as found in drier silages.

Fermentation is influenced by, moisture content, packing, sugar content, additives and excluding air. As stated previously, if the material is too wet, problems develop with the effluent; if too dry, with packing. The length of the material also affects packing; shorter cut material packs better than longer cuts. Heavy stalked material must be cut in shorter (1/2 to 3/4 inch) pieces to get the same amount of packing as longer (1 to 2 inches) of fine stem and leaf forages.

The fermentation process is dependent on the amount of water soluble carbohydrates found in the forage. These are changed into different short-chain organic acids, the most desired is lactic acid. If the dry matter and the water soluble carbohydrates are low, secondary fermentation takes place. This process converts lactic acid to butyric acid and at the same time clostridial bacteria are breaking down protein and amino acids. The end result is lower voluntary intake and utilization by the animal. The ideal content of water soluble carbohydrates is 3% of the fresh weight; if it drops below 2%, the chance of secondary fermentation greatly increases. Generally tropical forages have less water soluble carbohydrates than temperate forages; a couple of exceptions are elephant and pangola grasses.

In addition to having lower levels of water soluble carbohydrate, tropical grasses typically contain starches rather than sugars as in the temperate grasses. Sugar and starch are both carbohydrates, however sugars are a simpler form. What happens is that there is one more additional step, the break down of the starch into sugars and from there into acids. Carbohydrate content increases slightly in plants later in the day, especially on bright sunny days, also water content in the plant tends to decreases during the day; as a result it is better to cut in the afternoon, if possible.

In tropical grasses and legumes, especially if cut at late maturity, the carbohydrate content may be too low to get good fermentation. A source of sugar often needs to be added. This can be done by simply adding molasses as the material is placed in the silo. Suggested rates are 1% molasses to temperate grasses and 2% molasses for tropical grasses and legumes (this is 1% or 2% of the wet weight of the material going into storage, eg. 40 lbs of molasses to be added to each ton of fresh tropical grass). The molasses should be cut with water to insure an even distribution.

One way to raise the percentage of carbohydrates in the crop is to field wilt. Essentially this means cut it and let it lay for several hours to bring the moisture level down between 60 and 70%. This of course increases the cost of harvesting as a second pass must be made over the field. The question to wilt or not to wilt is a difficult one to answer. Animal trials with both wilted and non-wilted silage, both properly preserved,

[1] Like the fella that thought it was just his new girls perfume, but later found out she was just spoiled rotten.

show no difference. However, there is much greater chance of making poor silage with non-wilted forages, temperate grasses in particular and especially if cut at a lush stage of growth. Weather conditions also get involved, if it is damp the losses that occur while waiting for the crop to wilt may be greater than if it was direct cut. Additional losses also occur in the process of handling the material again, which typically are 3 to 4% of the total material. There is no perfect solution, but if you direct chop, do it during the driest part of the day and consider using molasses.

Several types of additives can be used to increase fermentation. Organic acids, such as formic or a mixture of organic acids and formaldehyde are often used to improve fermentation and also animal production. It has been found that animals tend to eat more of the treated silage than the untreated stuff; there is some evidence that there is an improved digestibility (9%), with the use of formic acid. These additives are normally given to unwilted crops or grasses, although on wilted temperate crops an increase in animal production has also been shown.

Alkalis have recently been used in an attempt to improve the digestibility of mature grasses; usually sodium hydroxide or anhydrous ammonia. However, the responses for legumes and tropical grasses were less than for temperate grasses. High levels of ammonia (3% of fresh crop weight) raised the digestibility by 10 percentage points, however the economics are questionable. There appears to be little value in this approach for tropical regions.

Innoculants made up of different types of bacteria may also be added as the material is being chopped. These bacteria, being the right sort (Lactobacillus plantarum), help to increase the fermentation after the material is stored. However, research conducted as South Dakota State University indicated that no significant differences in feedlot performance, nutrient digestibility, or nitrogen retention were found between untreated and microbial inoculated corn silage when fed to feedlot steers. All in all the decision to use an additive or not, depends to a great extent on your ability to make good silage, as well as the facilities you have available. Additives take some of the risk out of making silage, but they will not make quality stuff out of junk, nor correct for poor harvesting techniques.

The advantages of silage are: Harvesting losses due to the weather will be less than with hay; protein and energy levels are usually higher when harvested as silage instead of hay; where silage making is part of a grazing operation, it is a more dependable method of cleaning up pastures and ensuring aftermath grazing; mechanization from harvesting to feeding is easier with silage; silage is a safe and palatable feed that is readily eaten by livestock; the cost of the required equipment is less than for a comparable hay operation.

Disadvantages of silage are: Losses can be high if crops are not harvested at the proper stage of maturity; if facilities are inadequate; if the crop is not chopped fine enough, packed well and properly sealed; wet silage can present problems with effluent; silage must be fed soon after removal from storage to prevent spoilage; silage is costly to transport (high water content).

Silage does not have to be made in large silo's. Perfectly good silage has been made in 5 gallon buckets. If you only have a limited need for silage, consider making it in 55 gallon drums. It is best to line the drum with a plastic bag to help keep out the air, an industrial trash bag works well. Pack the material by standing in the barrel and jumping up and down as you fill it up (this is even better than jogging). After it has been packed use a vacuum cleaner to suck out the air, then tie off the bag. It is best to seal the barrel with duck tape as well. Depending on how well you packed and the fineness of chop, 300 to 350 lbs of material can be ensiled per barrel.

In summary, properly made silage will be a high quality, very palatable feed that will be readily eaten by livestock. Silage fits in well with various livestock operations and will give a source of high quality feed for use in maintaining pre-set production levels or in periods of stress.

HAY:

There has been so much written on how to make high quality hay that it's a wonder anybody ever makes poor hay. Making quality hay is easy, "all's" you have to do is cut it at a high stage of quality, field dry to around 13% moisture, bale it and move it to a storage area. The major problem seems to be that when the crop is at a high stage of quality it's usually raining every other day or the relative humidity is so high that it is baled with 20 or more percent moisture and the stuff heats up to temperatures over 100 degrees F, which causes losses in digestibility of both protein and energy, or it molds, or both. Seems about the only place good hay is made on a routine basis are on irrigated fields in the desert. Quality hay can be made in humid environments but to do so requires attention to a number of items that a desert operator can ignore.

The first step is cut the forage at the "right" stage of quality, early bud for alfalfa and just as the seed head emerges from the "boot" in grasses. There is no substitute for this first step. Farmers often delay cutting, waiting for the "right weather". A study in New York showed that cows fed early cut, weathered hay, produced 4 more pounds of milk than did cows fed late cut, non-weathered hay. The quality of the crop decreases fast with maturity.

In Wisconsin, 1 inch of rain in the first 24 hours after cutting reduced carbohydrates 20% and dropped the relative feed value of alfalfa 14%. The same amount of rain on the same alfalfa hay but nearly dried, reduced the carbohydrates 56% and lowered the relative feed value 30%. Energy was leached out which left a higher percent of fiber and reduced digestibility and intake. If your crop is going to get rained on, it's better to get the rain shortly after cutting.

It is difficult to justify haying if you lose much more than one cutting out of five. Having the right equipment goes a long way towards reducing losses. In most environments, a mechanical conditioner, either mounted on, or behind the mower speeds up drying. It takes about 30 hours of sunshine in the Midwest U.S. to field dry non-conditioned hay; about 4 days in spring and 3 in summer. Conditioning may reduce drying time up to 2 days. In arid climates it may be possible to bale conditioned hay the same day.

Leaving the hay in a wide swath rather than in a narrow windrow also helps. A "fluffer" or "tedder" will turn the swath over and can speed drying without excessive leaf loss. Tedders were designed for grass hay, however they work well on alfalfa if the moisture levels are between 50% to 60%. If moisture drops much below this level the loss of leaves will be large. In the northeast farmers will cut alfalfa and windrow it in the late afternoon or evening. The next morning, after the dew dries, but before the leaves begin to wilt, they spread the windrow onto dry stubble with rotary tedders; this cuts drying time by half a day. Tedding is a tool, like all tools it is not necessary to use it each and every time and under all conditions. Making hay requires good management; a tedder can be part of it.

Large round dense bales help the operator get more hay up faster, also large dense round bales shed water reasonably well. In Indiana it was found that large round bales stored outside lost an average of 10.5% compared to almost 17% for small round bales; in Florida dense large round bales (1300 lbs) lost about half as much as looser large round bales (950 lbs). As great as the large round balers are, they do cause considerably more leaf loss in alfalfa than do the small rectangular bales.

The use of chemicals can now cut down on losses. When moisture is over 20% chemical preservatives, usually organic acids (normally propionic or acetic acids) are sprayed on the windrow in front of the baler at rates up to 20 lbs per ton of windrowed hay. In this manner hay can be put up at moisture levels between 25 to 30%.

Another trick is to use a drying agent applied to the entire stem and leaves at the time of mowing and mechanical conditioning. These chemicals work by removing the waxy layer on the surface of stems and leaves. In general this will reduce the moisture content an additional 3 to 10 percentage points during a 24 hour period. However, there appears little benefit when applied on the first cutting, or when the humidity is high. These chemicals work very well on alfalfa, not as well on trefoils and red clover and hardly at all on grasses. The chemicals used are different salts; potassium carbonate, sodium carbonate or sodium silicate. There are other products containing methyl esters of fats, vegetable oils or animal fats, but research has shown that they work no better than the above chemicals. The application rate should be at least 30 gallons per acre; the mix is 1/4 lb of the salt to 1 gallon of water.

If you're in a really wet area consider mow or barn drying, either solar or electrical. The dryer should deliver 500 cubic feet of air per minute per ton of stored hay. Costs can vary, in Wisconsin the costs figured out to be between $3.50 to $15.00 per ton of hay depending on the type of dryer and amount of hay put up. The article "Hay in a Day" in the suggested reading section gives additional information on modern hay making techniques. Make hay while the sun shines is still good advice but now it is possible to extend the definition of sunshine.

A word of caution. There are any number of items that get into hay that can cause problems. Noxious weeds and toxic plants, these can not only cause problems to animals they may, through their seeds, spread weeds over the farm. Pieces of baling wire, nails, fence staples and other small pieces of metal are capable of causing hardware disease and other problems. Weeds can be handled by herbicides and metal objects are not the problem with todays balers that they were a few years ago.

Another problem that appears to be increasing is poisoning due to consumption of blister beetles by livestock, particularly horses. The problem appears mostly in alfalfa hay grown west of the Mississippi River. Only hay that is conditioned appears to cause the problem. This is because hay that is crimped crushes the beetles and they remain on the stems and leaves; un-crimped hay allows the beetles to get off before baling.

Suggestions for making beetle clean hay: Keep weeds out of the hay, some beetles like flowering weeds. Harvest the alfalfa before it blooms. Don't condition your hay. If blister beetles are seen in field, spray with a insecticide.[2] Watch for grasshoppers, the larva of blister beetles prey on grasshopper eggs, the adult beetle eats flowers. First crop hay is usually safer than later cuttings, as beetle numbers build up in the summer.

---

[2] There was this scientist that drank a quart of insecticide every day to prove it didn't hurt people. One day he put on his pants and his fly died.

You can make the best hay in the world but if you don't store it properly it can all be for nothing. The average loss on round bales stored outside, based on a value of $65 per bale, will be 8 to 10 dollars a bale. The outside 6 inches in a round bale contains one third of the entire weight of the bale. Average spoilage will penetrate 4 to 6 inches. At these cost it doesn't take many bales or years to pay for a storage barn. If you must store round bales outside just putting them on rocks or poles will reduce the loss by 15%, covering with plastic also helps.

OTHER FORAGE CONSERVATION TECHNIQUES:

Hay and silage are not the only ways to conserve forages. Anything that extends or adds to the amount of feed normally produced, or on hand should be considered as conservation. All of the techniques described below require animal control, the better the control the better the results; as well as the ability to implement an allocation program.

STANDING HAY:

Leaving a paddock or pasture ungrazed for later use is an old technique often used in seasonal environments. If the animals are simply turned into the paddock and left there (continuous graze) it is of questionable economic value. If, on the other hand, the paddock is sub-divided so that only a small amount of the total forage is allocated to the herd at any one grazing, this has a greater value. The reasons is that the animal when first turned in, will eat more on day 1 than it requires for maintenance (or for a pre-selected rate of gain). On subsequent days the amount of forage consumed will decline, until less than a maintenance amount is eaten. While on first glance this might seem to average out, in actual fact it does not. For example, a one acre paddock contains 450 lbs of available forage of suitable quality for dry cows, if that dry cow requires 15 lbs of the available forage per day for maintenance. A simple calculation would indicate that there are 30 cow days of available feed (450 lbs/15 lbs per day); this of course is wrong.

Each day the animal is in the paddock, a certain amount of the forage will be soiled (dung and urine), trampled or otherwise wasted. The major factor are not these, as losses of this sort will occur regardless of the number of paddocks or length of graze, unless very short strips (2.5 feet) are used. However, these types of losses will be much less, the shorter the graze period. The major reason that you will not get 30 days grazing out of that paddock is that on day 1 the animal will eat between 20 to 30 lbs, much more than required for maintenance.[3] For a few days this same consumption will continue, then slowly decrease as the feed becomes harder to find and/or eat; eventually the amount eaten will be less than the 15 lbs required (this will likely happen sometime around day 15). Remember the number of bites, amount per bite, and the pasture mass relationships discussed in Chapter 6. In addition the palatability as well as the nutrient quality is also declining.

It may be argued that the animal will use the surplus feed to lay on fat (buffering ability), however even if she did (and this is open to question given the quality of the feed) the energy cost of making fat is about 90% of the energy actually stored and the energy cost of breaking down that fat is also about the same. The end result is that if 100 calories were eaten and transformed to fat (10 calories stored as fat) and the fat later broken down, only one to five calories would be available for maintenance; the rest

[3] Squirrels and ants may know how to ration their feed over the winter period but not livestock.

would be lost as heat. Any time energy changes form there is a corresponding loss (Chapter 8). In this instance energy changes form twice and the total loss is multiplied.

Standing hay is mature forage, as such it is limited to use for mature, non-lactating animals. Other classes will require an additional protein and/or energy supplement; non-lactating animals may also require a supplement depending on the quality of the standing hay. Standing hay used as continuous grazing is wasteful. It may have a place if there is a lot of feed available but if feed is in short supply an allocation program is called for.

FEED BANKING:

One of the many definitions of the word bank is: "To store or reserve a supply of anything needed for future use." Putting up hay or silage, or leaving a paddock ungrazed (standing hay) are the traditional ways this has been accomplished in the past and have already been discussed.[4]

Feed banking is the building up of a supply of forage for future use, while at the same time utilizing a portion of it (lax grazing). It utilizes the concept of residual dry matter, discussed in Chapter 5. As might be expected an allocation program in conjunction with a feed budget is a must. You must know approximately much feed is going to be required to see you through the expected period of short feed supplies.

The procedure is to leave increasing amounts of residual in each paddock until such time as a sufficient amount is "stored". From then on additional growth is removed during each graze period, until growth rate slows and a allocation program is begun. Feed banking works best on legumes or temperate grasses.

LIVING HAY:

This is not to be confused with standing hay, living hay is an entirely different concept. In fact, it can almost be considered as a supplement rather than forage conservation. The area used should be central to the grazing area, have reasonably easy access and be laid out so that strip grazing is possible. The size will depend on the number of animals to be fed and the rate of production desired. The area is normally planted to a high quality grass or legume or even an annual, such as barley. Small grain forages make good fattening feed. Steers grazing small grain forages can achieve ADG's of 1.8 lbs or better and have obtained conversions rates of 1 lb liveweight gain per 8 lbs of forage.

Some sort of fertilization and/or irrigation program is usually, but not always required. In a re-seeding program, simply take a section out of the normal grazing area, re-seed and after establishment use the area as the living hay until the entire area can be re-seeded. This same approach can be used with fertilization, the area shifted around from year to year. Don't try this with Bahia, bermuda, star or digit grasses. The quality of these forages is not sufficiently high enough to act as a supplement, grasses of this type are better used at putting on high gains per acre rather than high daily gains on animals.

The procedure for living hay is to graze the animals on regular pastures (poorer forages) during the morning and early afternoon. In the late afternoon the animals are shifted to the living hay. The amount of area allocated to each animal will vary,

[4] Some people are like a river, they need two banks to keep them going in the right direction.

depending on desired performance targets. The animals remain in the paddock until it is grazed to the desired level, usually one or two hours, then moved back to the poorer forage. Fencing is by temporary electric, with the next days "break" being built while the animals are grazing.

This technique has its most obvious application in range areas where summer or winter forages are of poor quality. It can be used very well in combination with the wheel or Texas block designs. Initially the animals will have to be physically shifted, but if the break is made at the same time each afternoon, animals will quickly learn the routine. It is also advisable to use some sort of signaling device, such as a horn or siren. This is NOT a forage substitution program, it is a supplement, a desert if you will, to be eaten after the main meal. Consequently the grazing time and amount of forage allocated must be carefully controlled. For this program to work, you need control of your animals; if you don't have control, don't try it.

INTERCROPPING:

Simply put, intercropping is the planting of different crops on the same area of land, either at the same time or at different times of the year. If broadcast seeding is used on pasture the area should be heavily grazed (flogged) or mowed prior to herbicide application. This can be followed up by using animals to tramp in the seed or with a light harrowing or disking. If no follow-up is used, the seeding rate must be higher. No-till planting also results in a better stand with less seed. An insecticide should be applied either at seeding or shortly there-after, especially in tropical regions or where there is a heavy insect build up as in late summer. There are four procedures of interest: 1. The Zebra Technique. 2. Summer Forage Seeding. 3. Winter Annual Seeding. 4. Double Cropping.

ZEBRA TECHNIQUE: This requires a boom-type sprayer and a broadcast seeder. The sprayer is set to spray alternate bands, 4 to 10 inches wide, using Paraquat or low rates of Dalapon (4 to 6 lbs of the 74% commercial product (Dalapon works better on sod grasses than does Paraquat). The idea is to kill the top growth but not the entire plant. This killing of the top growth gives the pasture a striped look, thus the name Zebra Technique. The broadcast seeder is loaded with usually an appropriate legume, although a different cool or warm season grass, is sometimes used. Variations of this technique are to mount the spray unit in front of a no-till seeder and drill the seed in. If a seed drill is used, an annual such as barley or annual rye grass works well. Direct drilling gives a better response and uses less seed than broadcasting.

SUMMER FORAGE SEEDING: A broadcast application of Paraquat is first applied. The seeding is with Sudan grass, sorghum/sudan cross or millet. This method works best on run-down pastures containing cool season (temperate) grasses. Warm season sod grasses such as bermuda, bahia or pangola do not respond as well as the regrowth is too rapid during the summer months and tends to shade out the new seedlings.

WINTER ANNUAL SEEDING: This procedure lends itself to warm season sod pastures. In late summer or early fall a broadcast spraying of Paraquat or Dalapon is applied at a pint of Paraquat or 4 to 6 lbs of Dalapon per acre. This is followed by seeding in annuals like wheat, barley, rye or rye grass. This method provides a high quality feed during the normally slow growth winter season.

A variation used in Hawaii on dense kikuyu grass pastures is to first flog then roto-till and rip the pasture; the tool bar is mounted directly behind the roto-tiller. A second pass is made about two weeks later with a roto-tiller, a seeder and a culti-packer.

Perennial rye grass is sown at the rate of 30 lbs per acre, Haifa clover at 6 lbs/acre (these rates are about 1/3 higher than normal and are used because of the high levels of organic material incorporated in the seed bed). This results in a nice mix of kikuyu and rye grass/clover; the life of which depends on the grazing techniques used for maintenance.

DOUBLE CROPPING: This is another method to lay in additional forage for later use. The procedure is to plant an annual cereal (winter wheat, barley or rye) in the fall, eg. right after corn silage harvest. The crop is harvested in early spring just before spring planting. As with any grass, cereals should be cut just as the seed head emerges from the boot. A annual legume, such as hairy vetch, can be planted along with the cereal, this will increase the protein content above the 16 to 17% possible with the young cereals alone.

Since cereals are cool season crops, many in the tropics dismiss their use as a temperate option only. However, any place where there is winter moisture and the average temperature drops into the 70 degree F range, they should be considered.[5] Obviously double cropping is not possible in all environments, nor should it be attempted without adequate planning, the necessary equipment and some initial testing.

TESTING: Put in some test plots before you blow a bundle on a massive tillage program. Environments change, often over quite short distances, due to different soils, climates or managers. Test plots will let you know the best crops for your particular environment and also point out some potential planting or harvesting problems that will have to be resolved prior to go-ahead. One outfit in Hawaii was in a big hurry to plant alfalfa, so they plowed up 500 acres and flew on alfalfa seed; they did this four times during the course of a year before they got a halfway decent stand. They now know a great deal about planting alfalfa in Hawaii; however, a crop, not an education, was the original intent. In this case testing would have taken 3 to 4 months, some things would have been learned, considerable money would have been saved and a earlier and better stand would have been the final result.

SUPPLEMENTATION:

Supplementation is just that, not forage substitution. A typical problem often arises when the level of supplement increases, the animals simply substitute the supplement for the forage. The goal of a supplementation program is to provide the limiting factor that is restricting performance, not a total ration geared for maximum performance. Too little supplementation and you've missed your goal, too much and you find yourself in a feedyard situation by accident.

There are seven categories of nutrients which every animal needs in varying amounts. They are; carbohydrates, fat, protein, minerals, vitamins, water and air. Any of these can be the limiting factor in a production program. Air is seldom a problem except in dusty areas. Water often is, especially if stale, too hot, too cold, too salty, or polluted (dirt tanks where the animals have access and drink while standing in them). For most purposes only energy (carbohydrates and fats), proteins, minerals and occasionally vitamins are considered supplements.

In order to know what the animals are lacking to achieve the desired performance targets, two things must be known: 1. The quality of the forage the animals are grazing. 2.

[5] Contrary to popular belief, much of the land located in the tropical belt does not have tropical temperatures 365 days a year.

The amount of the respective nutrients that are required for the class of animal to obtain the desired performance level. Laboratory analysis is the most accurate method to determine the quality of the forage; it also takes time and may be expensive but so far there is no better alternative if high performance is desired. It is important in submitting a forage sample to be sure to sample what the animals are actually eating (Chapter 5). This is very difficult to do in continuous or longterm rotational grazing but is much easier when mob grazing is used in conjunction with short graze periods. In this situation "all's" you have to do is to look at the paddock the animals have just left, compare it to the paddock they just entered and clip accordingly.

Once you know what they are eating, you can compare it to what is required and select the appropriate supplement. However, the possibilities for mistakes are almost endless.

In the tropics, if the animals are on young forage the limiting factor is generally energy, although don't discount phosphorus and occasionally calcium. Molasses is a great source of energy and in the tropics is fairly cheap; it is also high in calcium which can throw the Ca:P ratio all out of whack. A good idea in feeding molasses is to make available a mineral mix of a 1:1 ratio of Ca:P or mix in some mono-sodium phosphate in the salt or water.

In Hawaii, molasses has been fed free choice to steers grazing poor quality kikuyu grass with molasses consumption reaching 17 pounds per head per day with out ill effects.[6] In another trial molasses was fed at the rate of 4 lbs/head/day over a period of 10 months to steers stocked at 2 to the acre. Kikuyu grass quality ranged from very poor to very good (crude protein 8.5 to 15.3%). The average daily gain was 1.8 lbs per animal or an acre basis, 3.6 lbs per acre per day (this trial was economical). The ranch produced close to 1000 lbs per acre in a period of nine months at a cost of 29 cents a pound of gain. In general energy and protein supplements should be fed in the afternoon; trials have shown a greater response (up to 10% better), when fed in the PM as opposed to feeding in the AM (substitution again). The amount to feed is generally 0.5%, or less, of the animals body weight. Energy supplements in the form of roughages have been shown to reduce forage intake if more than 5 lbs are fed (this will vary some, depending on body weight). In extensive operations, protein supplements can be fed every other day without significantly reducing effectiveness.

Putting out salt (sodium chloride) is almost an act of faith with many ranchers. No one questions that animals need salt (or more exactly, sodium), what is beginning to be questioned is do they need a supplemental source of salt. Just because your animals gobble up great gobs of salt every year doesn't mean they need it; liking something and needing something are two different things.[7] Most temperate forages are not deficient in sodium, such as perennial ryegrass, orchard grass, Yorkshire fog, trefoils and white clovers. There are a few temperate forages that are low, timothy, tall fescue, alfalfa, red clover; as well as most tropical grasses. Most forages, once they have died, turned brown and leaching has begun will more than likely be short of sodium. If the forage contains less than 0.10% sodium, then salt is called for. The point is that salt cost money, unless you know that your forages are deficient in sodium or unless you are using salt as a carrier for other minerals or for some other reason, you should seriously question your continued use of salt on a routine basis.

---

[6] No, it was not economical.

[7] I like to eat great gobs of ice cream, I definitely don't need it.

In high performance situations energy, protein and/or minerals may often be limiting. Attempting to find which one(s) may be a long and frustrating experience. A soil analysis can help; obviously if it's not in the soil · it won't be in the plants. Analyzing the animals blood works only on a few minerals, those that the body does not store very well. Taking tissue samples from various organs (biopsies) also works for certain minerals, but is hard on the animals, a two to four percent death loss is not uncommon. The basic problem is that most tissue and blood test only tell you if there is a sever shortage of the mineral, or in some cases a toxic amount, which is fine for the extremes but not when you are trying to maximize production. Forage analysis is the best method, but only if you are sampling exactly what the animals are actually eating. This is no problem in feed yards, but can be a big problem on pastures and ranges, especially in set stocking or continuous grazing. The vast majority of forage analysis that have been taken are totally worthless in attempting to define what might be missing in the animals diet, simply because the forage sampled was not the same as what the animals were actually eating.

Shot-gun mineral supplementation is the normal approach taken by many in the U.S. Sometimes this works, sometimes it doesn't. It will often work where performance targets are low and there is not a severe shortage of one particular element. It is seldom the solution where performance targets are high. The reason is that at higher performance levels things like the calcium to phosphorus ratio, the nitrogen to sulfur ratio, the interactions between copper, molybdenum, sulfur and iron, etc., become increasingly important. This requires that not only the right kinds but the right amount of each mineral be present. For high performance, the mineral program will have to be tailored to the particular forage for pasture and the season of the year. Don't be mislead into thinking that one mineral formulation will do equally well over the entire year or over the entire ranch if yours covers a lot of country. The only time this is likely to happen is if you are in a continuous growth situation and control both crop fertilization and irrigation.

Once a deficiency has been determined the problem remains as to how best to correct it. Pasture applications are one way and in some cases may be the only way. Injections of different minerals such as iodine, selenium and iron have been found to give good short term results. Researchers in the U.K. and Australia have been more interested in developing ways to provide single or a limited number of elements to grazing ruminants than have those in the U.S. For example they have developed heavy pellets which remain in the reticulorumen and slowly release the desired element into the rumen; cobalt, selenium and magnesium are the elements typically used. The cobalt pellets have been around for over 25 years old, selenium and magnesium are more recent. It was found many years ago that copper in the form of oxide wire, needles or particles could be used as a source of copper in the ruminant. These materials remain in the folds of the abomasum for perhaps up to a year where they slowly dissolve. Very recently a company in the U.K. has released a soluble glass pellet which contains the micro element as an inherent part of the glass structure. These pellets remain in the reticulorumen about 18 months and appear to have less problems of an insoluble coating forming on them than the regular pellets.

SUGGESTED READING:

Alfalfa Science and Technology. 1972. C.H. Hanson. Agronomy No. 16, American Society of Agronomy, Madison Wis.

Forages, the Science of Grassland Agriculture, 4th ed. 1985. M.E. Heath, D.S. Metcalfe and R.E. Barnes. Iowa State Univ. Press.

Make Hay in a Day. 1985. D. Rohweder, M. Collins, M. Finner and R. Walgenbach. Crops and Soils Magazine, August-September.

New Holland Grassland Manual. 1962. C.B. Bender. New Holland Machine Co. New Holland, Pa.

Silage Additives USA. 1985. K. Bolsen and J.I. Heidker. Chalcombe.

Silage. 1977. A.J. Davies. Bulletin 37. Ministry of Agriculture, Fisheries and Food, London, England.

Silages made from Tropical and Temperate Crops. 1983. J.M. Wilkinson. World Animal Review, No's, 45 and 46.

Silage Production and Preservation. 1966. Extension Bulletin 308. Univ. of Minnesota, Extension Ser. St. Paul Minnesota.

Silage and Silos. Special Circular 223. Pennsylvania State University, Extension Ser., University Park, Pa.

Used with permission of NEA.

There's a lot of new technology entering the livestock business, and most of it can't be ridden.

13

HI-TECH

While it may be argued that some of the items mentioned in this chapter do not represent high technology, they do represent a new way of doing some old jobs and in a few cases, new ones. One of the major ones, electric "Power" fencing has already been covered (Chapter 9) and will not be discussed further. Others are offered only for consideration, whether or not they apply to your operation will be your decision.

COMPUTERS:

Computers are great and have many uses for a ranching operation but DO NOT purchase a computer just because you feel the computer age is leaving you behind. At a meeting the other day a rancher asked us about purchasing a computer. Our first question was: "What are you going to use it for?" Our question was answered with a question: "What does a computer do?" We finally found out that they wanted to keep their financial records on the computer. However, their books were up to date and weren't taking too much time to keep current. Under those conditions our answer was: "DON'T PURCHASE A COMPUTER." That same day another rancher told us that he was a poor organizer and wanted a computer to keep him organized. We're poor organizers and learned very quickly that disorganization can occur on paper, or in a computer. Again our advice was: "DON'T PURCHASE A COMPUTER." Our main reason for being so conservative about computers is that a good business computer and software can cost around $5,000: more than once computers have been purchased with good intentions, but only ended up gathering dust.

Not all ranchers have the same computing needs. Some suggest that any operation larger than 5,000 head needs a computer to stay competitive. But this does not mean that smaller operations can not benefit from using computers. The following steps will help you determine if a computer is needed and if it is, how to make the right purchase.

1. Find out what a computer does.

2. Talk with people using computers, preferably with people in the same business.

3. Learn a few computer terms.

4. Make a list of your needs.

5. Select the software that meets your needs.

6. Purchase a computer that is capable of using the software that you have selected.

7. Decide on who will use the computer.

8. Test the economic feasibility.

WHAT A COMPUTER DOES:

What a computer does and what it is used for, depends on the user and often has little in common with what others may be using their computer for. In a 1984 survey of computer use in agriculture it was found that the typical on-farm/ranch applications of computers, in order of major use were: 1. Analysis of records and planning. 2. Financial record and book keeping. 3. Physical record keeping. 4. Correspondence. 5. Inventory control. 6. Education and entertainment.

COMPUTERS ARE RECORD-KEEPERS: Financial records such as accounts receivable and payable are good examples. The neat thing about these type of records is that one entry can automatically be entered in several different areas of the accounting system. With a pen in hand the double entry system means that you have at least two places where an entry must be made. We know a businessman that handled public funds and was subject to an audit at any time. With the old pen in hand system these audits took at least two weeks and sometimes more. With the computer he can provide an auditor with all the information (printouts) for him to look at in less than two hours. In addition, the longest the auditor has stayed around was one week.

Production and breeding records are another type. With the computer you can search for certain animals that meet different criteria or parameters that you might decide to set up. For example, list all the animals that have an average daily gain above 2.5 pounds and are between 9 and 12 months of age. Out would come a list showing the animals, their location on the ranch, ancestry, etc.

Inventories are the last major type. In reality, production and breeding programs are a type of inventory. These programs can be used to keep equipment, machinery, land, buildings, and other miscellaneous records. The parameters that the user can request is very large.

COMPUTERS ARE DECISION-MAKING AIDS: Decision-making programs are very popular with agricultural producers. These programs allow the user to play "what-if" games. For example, we have prepared a weaned calf decision analysis program. The user supplies the requested information and four different alternatives can be compared. The comparisons will help the rancher determine if he should sell weaners now, send weaners directly to the feed yard, or grow the weaners out on pasture and then send directly to the slaughter house (grass fats).

Typical questions that ranchers face involve: Should the operations be expanded? How best might forage resources be scheduled in their use over the year? What's the best way to put a budget together to analyze the ranching operation? How best might forecasts be made on what will happen to the business over time? The computer can be an excellent assistant in providing information upon which rational decisions can be made on the above questions, as well as many other related ones.

COMPUTERS ARE PROCESS CONTROLLERS: In addition to decision-making aids and record-keepers, computers are also process controllers. They can control such devices as pumps, gates, irrigation systems, and record weigh gains or milk yields.

Cattle can be weighed on an electronic scale which is hard-wired to a computer for automatic recording of weight gain of each individual animal. A special ear tag provides the computer with an easily identifiable number to sink its microchips into. The rancher can then track the individual progress of any animal or group and compile exact head counts by variables such as sex and breed; similar devices can be used in the milk barn.

COMPUTERS ARE A SUPER-FAST COMMUNICATOR: Computers are used to write letters, memo's and reports; these can generally be done faster and neater than with a typewriter. Ranchers and farmers can also use computers to access on-line data-base networks, such as AgNet, Agristar, and Pro-Farmer. These networks provide up-to-date market and in some areas, weather information.

COMPUTERS ARE FUN MACHINES: Besides using the computer for serious business as mentioned above, they can be used by the families for education and entertainment. There are numerous games and educational programs available for almost any computer.

TERMS:

As is typical with any new product, there is a new set of vocabulary with which potential users must be familiar. Computers are no exceptions. This discussion will be limited to the most popular class of computers, namely the microcomputers which are the type most often used by farmers and ranchers, as well as everybody else. Microcomputers are the most common computer in the market place due to their affordable prices. They formed the basis of the recent "computer revolution".

A typical microcomputer system is composed of "hardware" and "software". The term "hardware" refers specifically to the physical equipment that makes up the computer package, of which there are a number of components (see Figure 13.1). First there must be some sort of input device, normally a keyboard that resembles that of a typewriter. The keyboard is used to enter input information into the computer for processing. The second component is some sort of output device. While there could be several types, the most common are a video screen (like a TV) and the printer.

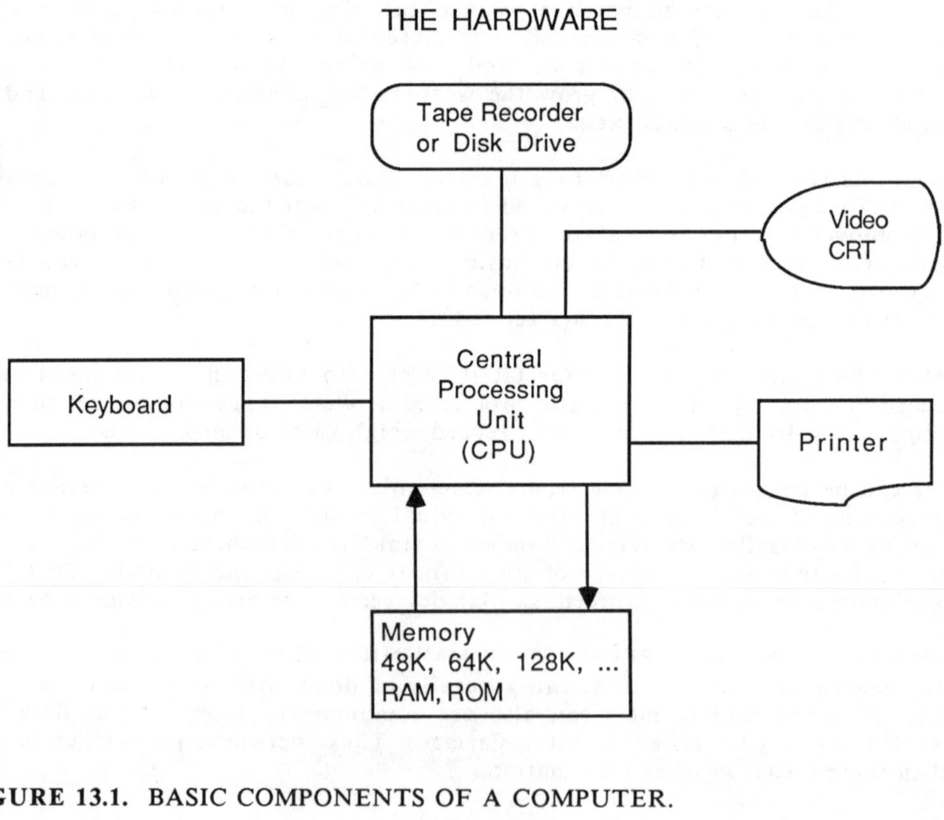

FIGURE 13.1. BASIC COMPONENTS OF A COMPUTER.

A third hardware component is some sort of storage device; used for data or information that will be processed at some later point in time. Typical storage devices are magnetic tapes, such as those used in audio cassette recorders, or disks (diskettes) that resemble phonograph records but are made of the same material as are magnetic tapes. A fourth and final hardware component is the central processing unit or CPU. The CPU is the "heart" of the operation and is the mechanism used to perform the computing operations. There are two essential parts within the CPU; the memory and the microprocessor.

The memory portion is used to store information within the computer itself; not on disk or cassette tapes. Memory can be likened to a set of mailboxes in which specific pieces of information can be stored and retrieved, once the "addresses" of the mailboxes are defined. Operations and manipulations on the stored contents of the memory are performed by the microprocessor.

A unit of computer memory capacity is referred to as a "byte". For example, to store the set of characters and spaces in the phase "intensive grazing management" would require 28 bytes; 26 for the letters and 2 for the spaces. A double spaced typewritten page normally has about 1500 possible spaces and would require a memory capacity of 1500 bytes. Generally the term kilobyte is used to describe the memory capacity. A kilobyte is approximately 1000 bytes (actually 2 raised to the 10th power, or 1024 bytes) and is abbreviated by "K". The capacity of a computer is represented by the size of the memory, commonly referred as "RAM", which stands for Random Access Memory.[1] Most basic systems contain at least 256K of RAM. Auxiliary memory, such as diskettes and tapes are also measured by bytes.

In the 1984 survey of computer uses in agriculture, it was found that a typical microcomputer user in Hawaii owns an IBM or IBM compatible computer with 256K of RAM, two 360K disk drives and a printer.

The term "software" refers to a set of instructions that tells the computer what to do or the specific kinds of operations to be performed. These instructions are written in various computer languages, such as BASIC, FORTRAN and PASCAL. Typical users don't have to worry much about these languages as they will generally be using "application software" (user friendly). There are many application software in the marketplace. The most common software used (from the same 1984 survey) are electronic spreadsheets (such as Lotus 1-2-3), word processors, data-base managers, accounting packages, graphics and telecommunication programs.

You don't need to know how write a program in order to evaluate software programs. But you do need to be familiar with the subject area of the program so that it can be used properly. The different agricultural software programs are no substitute for knowledge about the basic subject content. For example, you don't need to know how to write a 105 day weaning weight program but you do need to know the input into it and how to use the results the program will give you.

This brief description of some of the major terms does not exhaust the new "computerese" vocabulary. To really get an earfull, simply stop by your nearby "friendly" computer store and ask some innocent question, like: "How much RAM does that computer have?" It's a good idea to take a translator with you, one of your teenagers will usually do just fine.

[1] A RAM can mean many things, but in this case it is not a male sheep. For further discussion on this subject, see the never to be released book, "Ewe, Ewe, Ewe".

GETTING STARTED:

Talk to people using computers, there are a lot of them out there. While most will not be in ranching or farming, there is still a lot of "expertise" in the form of clubs, night classes, individual users, extension agents, etc. These sources can be of help in introducing you to the relative merits of different hardware items and even some basic software, like different spreadsheets and word processors. Sales persons can be of great help but remember, they make their money by selling products that they carry, don't expect to get an unbiased opinion of a competitors product. Various consumer product magazines and some computer magazines also carry unbiased reports on different computer items.

Once you know roughly what computers do, you can decide and make a list of what YOU want a computer to do for your operation. Whether it be book keeping, preparing tax returns, keeping detailed breeding records, or simply playing games; they should be clearly listed. In addition you should also list all the anticipated future needs, since this will affect your choice of computer (different capacities).

After you know what it is that you want the computer to do for you, the next step is to select the software that will do it. Again seek outside help, there is a lot of free software produced by the land grant universities, so check with your county agent. There is also a very comprehensive listing of commercial agri-business software published by Moore Data Management Services (see suggested reading section).

Special attention should be paid on documentation and manuals (are they readable), ease of use (is a PhD required) and availability of help. Often the software available on the market will not satisfy your particular needs. In that case you will have to resort to a custom-made software program (expensive), or learn how to program yourself (very time consuming).

Once you have your selected software "all's" you have to do is to find a computer that will run it; be sure to try your software on the machine of your choice. Then estimate the size of the records you will be keeping; a common mistake is to purchase a system which is too small to handle all desired records. Usually you have to purchase equipment on a piece-meal basis; different brands for each component of the system. Make sure that the separate pieces of equipment are compatible. This is usually not a problem if the equipment is purchased from a computer store; it can be a problem when purchased from discount stores, large multi-product outlets or through the mail. It's not the quality of the item or the cost (which is often cheaper); just don't be surprised if their "Ajax" printer fails to plug right into your "Little Gem" computer. Things to look for are the reputation of the vendor, availability of training, warranty, backup and maintenance services. Finally compare prices and select the system that fits your pocketbook.

Deciding on the right hardware and programs are important, equally important is deciding on who will be running the computer. If an additional hire is needed, the extra cost should be noted. However, if you decide to use the machine yourself, make sure you plan for enough time. The use of a computer can often be quite frustrating and time consuming; especially in the initial stages of use. If you are not comfortable with or committed to attention to detail, forget the use of computers. You are probably better off to hire someone to take care of the entire computing needs of the ranch.

The last thing, or perhaps the first thing, is to determine if this investment will pay for itself and if so, when. This is not always easy to do as much of the value in using a computer is personal satisfaction. Often a computer leads to a savings in time in doing a particular job, it also often leads to doing more things which, from a strictly business point of view, may or may not need to be done.

CONTROLLING PASTURE WEEDS WITH HERBICIDES: by Philip S. Motooka.

Herbicide use has grown spectacularly after their introduction in the late 1940's. The new synthetic chemicals were much more economical and provided much better weed control than mechanical methods. Spray equipment were cheaper to buy, operate and maintain than mechanical control equipment. Moreover, chemical control was rapid and large areas could be treated in a short period.

With selective herbicides, it was possible to control broadleaf weeds without injury to grasses. This preserved the ground cover and reduced the likelihood of wind and water erosion. Also, because the grass was uninjured, yield response was immediate.

Finally, herbicides have an outstanding safety record. Contrary to the adverse publicity resulting from unfounded allegations, the scientific evidence is overwhelming that herbicides are safe to the user and the consumer. In fact, it is mechanical implements and powered equipment that makes agriculture one of the most dangerous occupations today. Even the ordinary home lawn mower kills 15 to 25 people a year in the United States and maims tens of thousands more.

FOLIAR APPLICATION:

Despite the tremendous potential of herbicides, their careless use can easily cancel any advantages they may have. Sloppy application will result in excessive costs, poor control and non-target plant injury.

Basically, the applicator's objective is to apply a small amount of herbicide uniformly over a large area. In the past, high volume spraying (e.g. spray-to-wet, spray-to-runoff, drenching) was recommended to facilitate uniform application. High volume spraying, still commonly practiced today, is inefficient primarily because of the large volume of water that must be loaded, transported and sprayed.

Although uniform coverage of the weeds with the herbicide spray is critical for best control, high volume sprays are not necessary to achieve uniformity. In aerial applications for example, volume-rates of 10 gallons per acre (gpa) or less are used. Indeed, there are formulations made to be sprayed without water. Certainly, volumes in the range of 30 gpa would be adequate with wide-swath sprayers. Lower volumes greatly increases labor efficiency over high volume, orchard-gun applications. Not only is a greater area covered per load in a shorter period, there is less frequency of reloading and consequently, less "down time", particularly if the spray area is far from a water source.

The common practice is to prepare a spray mixture using a set concentration, typically 1%. However, if no attention is paid to the volume-rates employed, i.e. gallons of spray per acre, gross inefficiencies will result.

Table 13.1 illustrates what happens to the herbicide rate as the volume-rate increases, assuming a 1% herbicide mixture is sprayed. A powered orchard-gun sprayer can put out around 300 gpa in brushland. If 2,4-D were used, this means at least 36 lbs 2,4-D/Ac is applied. The recommended rate would be between 1 - 4 lbs 2,4-D/Ac. A volume-rate of 100 gpa wold be more than adequate, providing 4 lbs 2,4-D/Ac. Very excessive volume-rates would increase wastage even further.

TABLE 13.1. INCREASING WEED CONTROL COSTS WITH INCREASING VOLUMES OF HERBICIDE SPRAY PER ACRE (volume-rates).

	Volume-rate (gpa)			
	50	100	300	500
2,4-D rate (lbs/Ac)	2	4	12	20
2,4-D cost/Ac at $12/gal ($)	6.00	12.00	36.00	60.00
Picloram rate (lbs/Ac)	1	2	6	10
Picloram cost/Ac at $100/gal	50.00	100	300	500
Surfactant cost/Ac at $7.00/gal*	1.25	3.50	10.50	17.50

* Surfactant concentration 0.5% of spray volume.

NOTE: Herbicide spray dilution 1%, volume basis.
Labor and equipment costs not included.

When a more expensive herbicide such as picloram is used, the cost quickly becomes prohibitive (Table 13.1). Picloram can be effective and economical. For example, catsclaw Caesalpina sepiaria and sensitive plant Mimosa pudica can be controlled with 1/4 lb picloram/Ac (1 pint/Ac). This would cost about $13/Ac. Preserving this economy requires precise application to prevent overdosing with such an expensive product.[2]

The problem of herbicide overdose cannot be solved by diluting the spray mixture while still maintaining a high volume-rate. Too dilute a spray mixture will decrease the potency of the herbicide. Furthermore, the use of excessive volumes means more labor and fuel are used and equipment is more rapidly worn, just to load, transport and spray water not necessary for herbicide efficacy.

CALIBRATION:

In order to apply herbicides precisely, the sprayer and the applicator have to be calibrated. This is easier than it sounds. The payoff is more efficient use of labor and equipment, elimination of herbicide and surfactant wastage, better weed control, and reduced risks of drift.

Following are instructions for calibrating various types of equipment by a simple method. The numbers used are examples only and will vary among different applicators and equipment. The only requirement is that the volumes used should be reasonable.

KNAPSACK SPRAYERS: These should be outfitted with a flat fan nozzle. Nozzles are available that provide flat fan patterns up to six feet wide. Conical spray patterns are inappropriate except for spot treatments, although they can also be calibrated for greater efficiency. To calibrate:

[2] One ranch in Hawaii saved over $20,000 in one year time, just by changing from high to low volume spraying; the results, in terms of control of the weed [guava], was as good as the high volume spraying.

1. Use a constant pressure. If the knapsack sprayer does not have a regulator, a constant pressure can still be achieved by maintaining a constant resistance on the pump handle.

2. Fill the tank with water and mark the level with a dipstick. Spray as you walk a known distance at a constant, comfortable speed, without waving the wand. Do not drench the foliage.

3. Refill the tank to the original level, measuring the amount of water required. Assume 0.42 gal (54 oz) was used.

4. Calculate the area sprayed. Assume the area you sprayed while walking was 100 ft long and 6 ft wide: 100 ft x 6 ft = 600 sq. ft. 5. Convert to acres: (600 sq. ft)/(43,560 sq. ft/Ac) = 0.014 Ac. 6. Calculate the volume-rate: (0.42 gal)/(0.014 Ac) = 30 gpa.

NOTE: If the volume-rate is below 20 gpa, or higher than 40 gpa for low-growing weeds, repeat the operation and try to get the volume-rate in the desirable range. This can be done by walking faster or slower, or by using a nozzle with a smaller or larger orifice, or both.

7. Calculate the herbicide required. Assume 1/2 gal 2,4-D/Ac is recommended. Then, (0.5 gal 2,4-D/Ac)/(30 gal/Ac spray) = 0.017 or 1.7%.

8. Make up spray solution; for example, 2 oz 2,4-D/gal plus 0.6 oz (0.5%) surfactant/gal.

9. Spray in the same manner as in the calibration.

For the knapsack sprayer with a conical spray nozzle for spot spraying taller weeds, the same procedure is used for calibration except:

1. Higher volume-rates are necessary e.g. 40-150 gpa.

2. Calibration should be done on land similar to that to be treated.

3. The wand needs to be waved to get good coverage. Again however, the weeds should not be drenched.

POWER SPRAYER, ORCHARD GUN: To calibrate power sprayers, it is critical to select a pressure that will allow a reasonable output (gal/min) which in turn will allow a moderate volume-rate. Lower pressures also reduce the threat of herbicide drift.

1. Fill tank with water, marking the level on a dipstick if the tank does not have a gage.

2. Spray an area of known size as uniformly as possible without drenching.

3. Refill the tank to the original level, noting how much water was required. Assume this volume was 22.5 gal.

4. Calculate the area sprayed. Assume an area 20 ft x 600 ft was sprayed. Then, (20 ft x 600 ft)/ (43,560 sq. ft/Ac) = 0.28 Ac.

5. Calculate the volume-rate: (22.5 gal)/(0.28 Ac) = 80 gpa.

NOTE: If the volume-rate exceeds 150 gpa or so, recalibrate to get it down to 80 to 150 gpa. In thick brush, it would be difficult to get complete herbicide coverage. However, it would be preferable to concentrate on keeping the volume-rate low than in trying to get complete coverage at excessive volume-rates. In actual spraying, repeat applications will be required at 6-24 month intervals until complete control is achieved. Such repeat treatments will control plants that escape the initial spraying. Therefore, it would be best to calibrate with the objective of keeping the volume-rate reasonable.

6. Calculate the herbicide required. Assume 1/2 gal herbicide per acre is recommended, then: (0.5 gal)/(80 gal) = 0.006 or 0.6%.

7. Make up the herbicide mix, plus 0.5% surfactant and spray as in the calibration procedure.

WIDE-SWATH SPRAYERS: Before calibrating these, their nozzles must be checked for uniformity. Boom sprayers are simple to check. Just collect the output of each nozzle for a set period of time. This can be done by simply lowering the boom and placing a coffee can under each nozzle and running until the cans are about 3/4 full. Each nozzle should deliver a volume within 10% of the others, or be replaced. Collecting the output of jet type nozzles is difficult, so uniformity of the spray pattern may be ensured by replacing the nozzles periodically. To Calibrate the wide swath sprayer:

1. Fill the tank and mark the volume.

2. Spray a measured length of land typical of the area to be sprayed with herbicide. Spray at constant speed and pressure. Assume the area sprayed was 200 ft and the swath width was 50 ft. The area is then: (800 ft x 50 ft) = 4,000 sq. ft., or, (4000 sq, ft)/(43,560 sq ft) = 0.09 Ac.

3. Measure the volume of water required to refill the tank to the original level. (Assume this was 3.5 gal).

4. Calculate the volume-rate: (3.5 gal)/().09 Ac) = 39 gpa.

5. Calculate the herbicide concentration needed. Assume 0.5 gal of herbicide per acre is recommended, then: (0.5 gal/Ac)/(39 gal/Ac) = 0.0128 or 1.3%.

6. Make up the spray solution, including 0.5% surfactant and spray as in calibration.

MANAGING HERBICIDES FOR EFFICACY AND ECONOMY.

Certain principals and strategies applied in chemical weed control will go a long way towards increasing the efficacy (power to produce an effect) and economy of weed control.

1. Select the best registered herbicide for the job on the basis of effectiveness on the target weed species, cost per acre (not necessarily cost per gallon or pound) and if pertinent, on other properties of the herbicide, for example, drift potential, persistence and mobility of the chemical in the environment (Table 13.2.).

TABLE 13.2. EFFICACY AND PROPERTIES OF HERBICIDES FOR PASTURE-RANGE WEED CONTROL.

Herbicide	Efficacy	Properties
2,4-D	Herbaceous (soft) broad-leaves, guava, strawberry guava, sourbush, joee, sensitive plant.	Esters somewhat volatile. Amines essentially non-volatile. Not mobile, not persistent in soils. Loses activity in 1-2 weeks.
Dicamba	Guava, false elephantsfoot, many broadleaves sensitive. Smartweeds very susceptible.	Somewhat mobile in soils. Somewhat persistent, soil active for a month or two, when applied at recommended rates.
Picloram	Most broadleaves sensitive. Hilahila, burbush, and catsclaw very susceptible.	Very mobile and persistent in soils.May remain active in soils 6 months or more, depending on climate.
Dalpon	Grasses sensitive. Must be applied by spot spraying.	Not mobile, not persistent in soils. Loses activity in 2-4 weeks.

2. Employ appropriate application techniques as discussed above.

3. Apply herbicides early in the infestation. Dense, tall brush has already affected yields; it is harder to kill, requires more herbicide, surfactant, fuel and labor. Spraying tall brush increases the danger of drift and furthermore, working in dense brush is difficult and dangerous.[3]

4. Apply herbicides when the weeds are actively growing.Foliage should be full enough to intercept the herbicide spray.

5. Apply when weather conditions are good for plant growth. Foliar, systemic herbicides travel from the leaves into the stem and roots with carbohydrates ("food") produced in the leaves by photosynthesis. A drought stressed plant will not photosynthesize, hence there is no movement of "food" and consequently, of herbicides out of the leaves. A rain free period is needed for maximum absorption of herbicide by the leaves. The rain free period required may be minutes or hours, depending on the species and herbicide and surfactant used. To be on the safe side, several hours should be allowed.

[3] Understory forage production generally decreases when the canopy cover exceeds 30% of the total land area.

6. Plan applications to prolong suppression of the weeds. In areas of seasonal rainfall, apply herbicides early in the rainy season when plants have just flushed a full foliage, but before its food reserves, depleted in fueling regrowth, are replenished. Likewise, brush that has been chopped, mowed, or sprayed previously, should be sprayed when they begin to recover. Do not spray too early, before downward translocation has become strongly established. On the other hand, waiting too long will allow a plant to rebuild its food reserves. At some point in between, downward translocation will be strong while the food reserve of the plant is still depleted and the plant is weak.

SURFACTANTS:

Addition of a surfactant (spreader-sticker) to the spray mixture aids in increasing the rate of herbicide penetration into the leaves. By spreading the spray droplet, surfactants cause greater surface area contact between the droplet and the leaf surface. Surfactants also retard drying of the droplet and washing in light rain, both of which enhance herbicide penetration. However, the most significant effect of surfactants is that they penetrate the leaf surface and thus form a pathway for the herbicide to follow into the leaf.

There are many kinds of surfactants, the spreader-sticker being the most commonly used. The second most common type used in spraying are the drift retardants. These are designed to keep spray from misting.

Common household detergents and soap are also surfactants. However, they are too weak for pesticide application. As with pesticides, use surfactants recommended, where recommended and as recommended.

The amount of surfactant recommended does not depend on the herbicide rate or concentration. It depends only upon the volume-rate (volume of spray per acre). Typically, 0.5% surfactant is recommended. Thus if a lower volume-rate is used, a lower volume of surfactant is required (Table 13.1). This is one of the many advantages of using moderate volume-rates in spraying.

DRIFT AND DRIFT CONTROL:

The most serious hazard in chemical weed control is drift and the unintentional plant injury that may result. There are two types of drift:

1. Physical, in which liquid or solid particles are carried away from the target area by air movement.

2. Vapor, in which the chemical volatilizes from the spray droplet, or from plant and soil surfaces into the air and is thus carried away from the target area as a gas.

The minor consequence of drift is the loss of herbicides from the target area. The major consequence is injury to non-target plants which may be extremely costly.[4]

There are number of things the applicator can do to reduce drift to inconsequential levels:

1. Do not spray on windy days.

[4] Like law suits, not to mention hard feelings.

2. Use lower pressures.
3. Use large orifice nozzles that produce a coarse spray.
4. Keep the nozzle as low as possible. Plan to spray before the weeds get tall.
5. Use drift retardants where spraying under poor conditions is necessary.
6. Use a non-volatile formulation where possible.
7. Use another method of application e.g. cut-surface, basal bark, soil application.

Probably the worst situation is to apply a volatile herbicide with a mist blower aimed into the canopy of tall brush. A mist blower produces fine droplets which remain in the air longer than the larger droplets of other application techniques. Fine droplets also facilitate vaporization of volatile herbicides. Aiming high into the air allows the droplets to travel farther before landing.

SPOT TREATMENTS:

Foliar spraying can be directed at individual plants where there are just a few plants to be controlled. Such spot spraying is also useful where the herbicide employed may be harmful to forage grasses or desirable legumes. For difficult-to-control species, there are several other methods of herbicide application more effective than foliar sprays. However, very large trees are difficult to control by any method.

NOTCHING: Notching allows the applicator to place herbicide into the sapwood (xylem) of woody species. The procedure is simple.

With an ax or machete cut notches around the base of the tree. The notches may be made at intervals, or, for tolerant species or larger trees, may be made end-to-end in a continuous ring. A concentrated herbicide is then squirted or brushed into the wound.

Herbicides effective in foliar applications tend to be more effective by cut-surface treatments. Glyphosate is effective on many species by this method. However, it is not cleared for use in pasture and ranges in the U.S. Still, glyphosate applied to cut-surfaces may be useful around homes and other parts of the ranch not subject to grazing.

CUT-STUMP TREATMENTS: In this method, the tree is cut down and a herbicide concentrate is applied to the cut-surface of the stump. The objective is to let the herbicide be drawn as deep down into the stump as possible. Under dry conditions, the sap in the sapwood will begin to recede immediately after the tree is topped. If herbicide treatment is delayed, air will be sucked into the sapwood as the sap recedes. This will form a "vapor lock" that will impede the downward movement of herbicide in the sapwood. Therefore, the herbicide must be applied immediately after topping.

Under high soil moisture conditions, the sap will not recede significantly, if at all. The herbicide will not be drawn deep enough to kill bud sprouts. This method works best under dry or even drought conditions.

BASAL BARK METHOD: This treatment employs oil soluble ester formulations of 2,4-D. The herbicide is mixed into diesel oil to a concentration of 4%. This solution is brushed or sprayed on the bark of the target tree. The bark from ground level to 18" or 24" is treated completely around the trunk. The herbicide solution should be allowed to run down to the soil line in order to kill the bud sprouts below ground level. Because only 2,4-D is commercially available in the ester formulation in the U.S., this method is limited to those species susceptible to 2,4-D. A newer herbicide, triclopyr (Garlon 4) may be used this way on non-crop areas. It has not yet been cleared for pasture-range use. The oil-herbicide solution may also be used on stumps. Where allowed, 2,4,5-T esters are effective on a broad spectrum of woody plant species.

GRANULAR HERBICIDES: There are herbicides available in granular form designed to be applied to the soil. One of these is picloram (Tordon 10K) which is in the process of being withdrawn from the U.S. market. The other granular herbicide is "spike" (tebuthiuron) which is a broad acting herbicide. Spike will control a wide range of plants, including forage plants. Application rate determines the type of control obtained. This herbicide works best on fibrous rooted plants. Check the label and/or your county agent for different rates of application.

ROPE WICK: Figure 13.2 shows a home made rope wick applicator for mounting on a tractor or pickup. Holes are drilled in a PVC pipe, rubber grommets are fitted over cotton or nylon rope segments, then inserted into the pipe. The ends of the pipe are sealed with one end fitted with a removable cap, so that a concentrated solution of herbicide can be added, or removed. The unit is mounted at a height that will wipe herbicide on the weeds to be controlled but not the lower lying forage. In applying, go slow; also some brushy weeds may require application from more than one direction

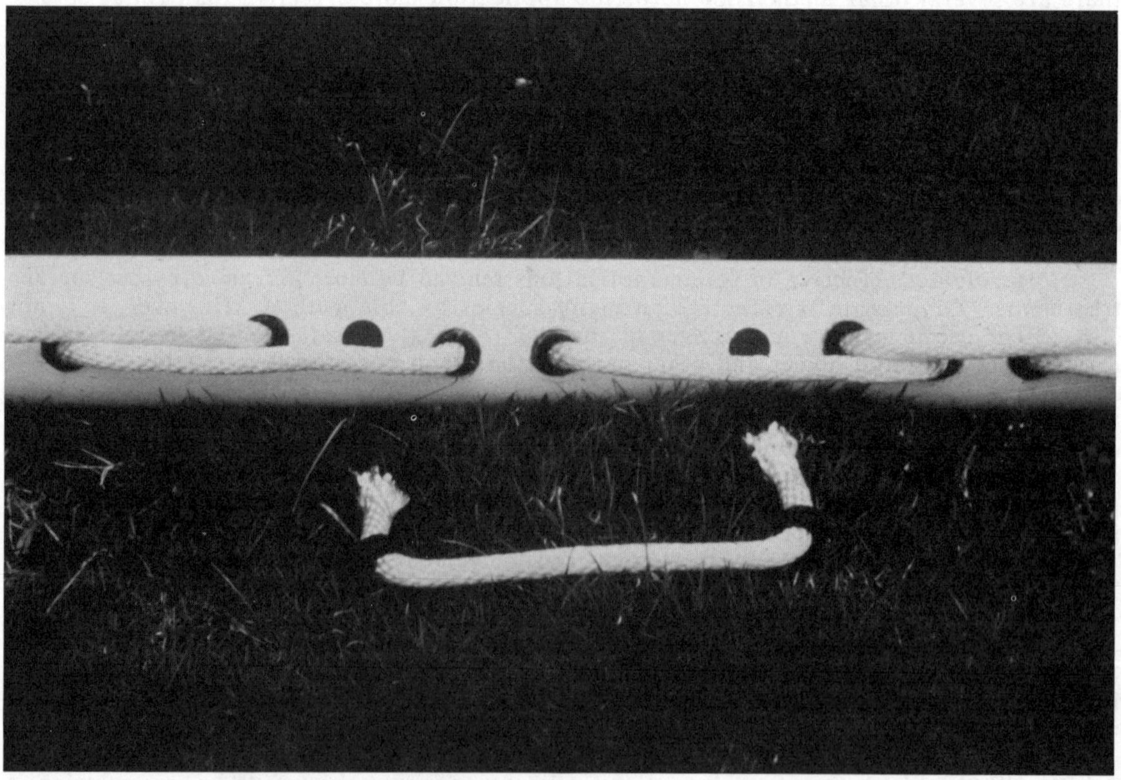

FIGURE 13.2. ROPE WICK APPLICATOR.

PRECAUTION: In working with any pesticide, label directions should be followed exactly, except when special directions from the State or controlling agency is allowed.

JAPANESE QUARTER HORSES:

The Japanese quarter horse or as it is otherwise known, the all terrain vehicle or dirt bike, is gradually replacing that stalwart symbol of the old west, the horse. This is probably to be expected as all you have to do is climb on and go, whereas with a horse you first have to catch it; not always a speedy procedure. Then to the horse had to be saddled, maybe a shoe tacked on, fed, etc. The time element alone makes the dirt bike attractive, especially in IGM. Often the time it takes to climb on a bike, check a unit and return is less than what it would take to catch a horse and get ready to ride. This is not to imply that there is still not occasion where a horse can do a better job than a bike but those occasions are getting fewer and fewer. Time is money.[5]

In choosing your first bike there are several criteria other than price that should be considered. 1. Will it be used on flat, rolling or hill country? 2. Will it be used solely by the owner, workers, kids, or wife? 3. Are you a mechanical idiot or do you know machinery? 4. What sort of loads are likely to be carried, eg. rider plus dogs or trailer? Table 13.3 list some factors to consider when choosing the size of a farm bike.

TABLE 13.3. FARM BIKE SIZE ACCORDING TO VARIOUS FACTORS.

	Engine size		
	up to 100 cc	up to 175 cc	over 175 cc
TERRAIN			
flat	X		
rolling		X	
hilly		X	X
RIDER			
owner		X	X
worker	X	X	
kids	X		
MAINTENANCE			
owner	X	X	X
garage	X	X	
never		X	X
LOAD			
rider only	X	X	
rider - dog	X	X	
trailer/side car		X	X

This table is adapted from Ag-Link #FPP 637, MAF, New Zealand

[5] In spite of the cost of living, it's still popular.

There are all sorts of bikes, two wheelers, three wheelers and a much scaled down version of the "Jeep". Each type has its own characteristics, the regular bike does tend to buck one off if only the front brakes are applied. The three wheeler requires a learning period so that one can steer, seems as though one has to shift weight in the opposite direction, they also have a habit of tipping over if going straight up a hill and a bump is hit. However, the three wheeler is great in wet areas, whereas the bike is better used on dry terrain or where there are scattered rocks.

For the average rancher it is best if a "farm bike" is purchased, rather than a souped up version of a dirt bike or something from a motor-cross. It is a good idea to get one or more reputable dealers to bring their bikes out to your place for actual field test. Additional things to consider are; economy, easily parked, pulls a trailer easily, comfort, reliability, maneuverability and quietness.

MAINTENANCE: Its a good idea to hose off the bike at the end of every ride if you have it covered with mud or manure; otherwise once a week. DAILY: Checks should be made of fuel, oil and tire pressure. WEEKLY: Checks should be made of the chain tension (20-25 mm of free play) and lubricate with oil. Brakes, both front and rear, put a drop of oil into cables and make sure they work freely. Throttle, open wide and let it snap back. It should close with a snap, if it doesn't, fix it, a throttle that doesn't snap back can be real dangerous. Clutch, should be about 3 mm clearance between lever and lever bracket, oil cables. Check the oil level of the gear box in 2 stroke engines. Check the battery, tighten loose bolts, screws, nuts and take a look at the lights. MONTHLY: Go over the cables, chain and sprockets. QUARTERLY: Give it a minor tune up, change oil, check spark plugs, remove and clean oil and carbon build up on 2 stroke engine exhaust baffles, check tappets, clean and replace air and fuel filters, check the front forks (they are filled with a light non-foaming oil) and make sure that they compress smoothly and return at a slower rate.

RIDING: Two major things, think ahead and wear a helmet; if you have a $10 head then buy a $10 helmet, if your head is worth a little more, then spend accordingly. On rough ground take all or some of your weight on the footrests, this lowers the center of gravity, also your legs can absorb sharp bumps better than your back. In mud or slick ground, use low rpm's but stay in as high a gear as possible. When climbing a steep bank ease your weight back as you open the throttle, when your front wheel starts up, lean forward on the handlebars to keep the front wheel on the ground. One major problem is to be climbing a steep hill and run out of power; if you have planned ahead this should not be a problem. However, keep one foot on the brake (usually the right one), put the left on the ground to support the machine. Apply the rear brake, put in the clutch and let the bike slowly roll back and around the support leg until you are facing across the slope. In braking use both front and back if going in a straight line, if going down hill, use more back than front, if turning use more back than front. Many riders use only one or two fingers on the front brake, the others along with the thumb are on the throttle, this tends to let the back brake do most of the work. In straight line braking the front brake can account for 85% of the braking. If the front brake is applied hard at the wrong time, that Japanese quarter horse can turn into a real western bronc in a hurry.

ELECTRONICS:

The electronic revolution continues. Most of the new devices simply allows the same job to be done easier, faster and with greater accuracy. Consequently, procedures or actions that were un-economical just a few years ago can now make a positive economical contribution in today environment. The future holds even greater promise. It is not the intent to go into every possible electronic gadget that might ever have a place on a farm or ranch, only two will be briefly mentioned; pasture meters and scales.

PASTURE METERS:

It's no wonder that the emphasis in grazing has always been on the animal, not only does it move (attracts the eye), its easy to measure, which makes it convenient to work with. Once you're able to accurately measure a thing it becomes possible to do all sorts of "neat" experiments to find out more about it. Which is one of the reasons that a great deal more is known about animals than about pasture-animal interactions. The problem with forages is that they are very difficult and time consuming to measure accurately. The greater the accuracy required the more time is required. If you want accuracy to within 5 to 10 pounds per acre, you have to clip, not mow, the whole acre.

Up until recently the best that could be done was to develop a "grazers eye", which took some time but was surprising accurate at pasture masses under 2000 lbs/acre (within a 150 lbs/acre or so). Of course one could always break out the clipping frames, as scientist do but even then one had to take many clips (15 or so) to get a reasonable estimate. Because of this inability to rapidly and accurately determine how much is there, a lot of things that could have been done, weren't.

It would be nice to be able to say that the wizards of modern electronics have developed a device that will quickly, accurately and cheaply measure the amount of forage in a field under all conditions. Well, they haven't. But they're working on it. What they have developed are different devices, that at least in temperate areas, under intensive grazing, allows one to rapidly measure forages quantity to a fair level of accuracy. This allows one to develop a "grazers eye" quickly. The accuracy is not 100%, in fact it can be off ten or more percent; however considering the average level of forage estimating ability on most ranches and farms, is quite an improvement.

The major stumbling block in measuring forages is water; water in the plant, water on the plant and water in the air. There are two basic directions that have been taken to deal with this; radiation devices and electrical ones. Both have problems. Radiation has a bad image to start with and there are limits as to what can be done. One type of device is the Beta attenuation meter, which uses Beta particles. This device works well on wet (green) forages (within 5%), but not on dry, dead material (70 to 75% error).

The other approach is to use various characteristics of electrical circuits, capacitance, inductance and resistance (the combined effect of these different properties is termed reactance). These meters or "pasture probes" are available commercially and sell for less than a $1,000. The best that have been developed thus far are reasonably accurate with under a 10% error; but are calibrated only for temperate pastures less than 2500 lbs/acre and 12 inches in height. Work is presently under way to calibrate these meters on higher levels and various tropical pastures. The probes are fast and most will average up the number of readings taken. This allows one to take many reading over an entire paddock and thus get a better estimate of what's actually there. For greatest accuracy, 20 separate observations, scattered over each paddock should be taken. Later, after your graziers eye has been developed, you don't need to take so many, just enough to keep your eye honest.

However, even if the meters are not calibrated for your particular type of grass or forage, they can still be put to good use, even though you don't know exactly how much forage is there. This sort of occasion occurs: 1. When you want a group of animals to graze only high quality forage. 2. When you are running two herds and you need to pull the first herd out while there is still sufficient forage for the second herd. 3. When you want to move the herd so that you utilize the most forage (50%), but can still expect fast re-growth. In these situations you don't really need to know how much feed is actually there; you are interested in the percent left or removed, not the actual amount. Since we're dealing with percentages, it doesn't really matter how wrong the numbers are on the probe, as the relative error of the probe will be consistent.

Say you have a bunch of yearling heifers that you want to breed, so you are moving them all over the ranch to "cream" each paddock before other herds move in. You figure if you let them graze the first 25% of the feed that will leave sufficient quality for the follow up herds. Say the probe reads 2000 lbs/acre, 25% of 2000 lbs is (0.25 X 2000) 500 lbs. When the probe reads 1500 lb/acre, 25% will have been removed and it's time to get the heifers out of there. This means of course that you have to go out there and measure again, maybe even a couple of times. But each time you do it, you are developing your eye, pretty soon you can tell when 25% has been removed, just by looking.

In the second and third case, say you want to leave 50% of the green material so you can get a fast regrowth. You have two herds, the follow up herd is twice as large as the lead herd. At what pasture reading do you move the first herd and what reading do you move the second herd and clear the paddock? The reading to clear the paddock of all animals is just one half the reading taken before any animals entered the paddock. If the probe read 2200 lbs/acre on the first reading, then all animals should be out of the paddock when the probe reading is 1100 lbs/acre (you will have left 50%). This means that, as far as the probe is concerned, there is 1100 lbs/acre of available forage. The first herd will eat so many pounds, the second herd will eat twice that amount since they have twice as many animals. So if we give the first herd 1/3 of the available forage and the second herd 2/3's of the available forage we'll be in good shape. There is 1100 lbs/acre available, the first herd should get 1/3 or 363 lbs/acre (0.33 X 1100). So when the probe reads 2200 - 363 = 1837 lbs/acre, shift the first herd out; when the probe reads 1100 lbs/acre, clear the field.

Well that's fine if the two herds are such as one is twice as large, or eats twice as much as the other. How do you figure it when you have, say 52 head that are expected to eat 12 lbs/head/day and the second herd consist of 127 animals that will eat 18 lbs/head/day? Multiply the 52 head times the 12 lb/head/day, that gives 624 lbs/day that the first herd will eat. Do the same for the second herd; 127 head times 18 lbs/head/day = 2286 lbs/day. The total feed demand is 624 + 2286 = 2910 lbs/day. The percent of that feed that the first herd is going to eat is; (624/2910) X 100 = 21.4%. In the above example, if there was 2200 lbs/acre and we were only going to allow them to eat 1100 lbs/acre, the first herd would get 21.4% of 1100 lbs/acre (0.214 X 1100) or 236 lbs/acre. So you move the first herd out when the probe reads 2200 - 236 = 1964 lbs/acre.

A word of caution. It there has been a rain or heavy dew, more water will be on the leaves of the ungrazed paddock than on the grazed one. This will cause the probe to give a higher error reading for the ungrazed paddock than the grazed one. If possible wait until the water has evaporated off before checking.

SCALES:

A scale is to IGM what an inside tip is to the bettor, it gives you an edge.[6] Scaling a group of animals every 3 to 4 weeks, gives you a much better idea of what your pas-

[6] "The race does not always go to the swift, nor the battle to the strong, but that's the way to bet 'em."

tures are doing than a book full of chemical analysis. Used with laboratory analysis, nutritional problems can be quickly identified and corrected. Superior individual performance is also identified and if you're keeping records, their dams and sires.

There are a large number of commercial makes of portable electronic livestock scales appearing on the market, many for around $2,000. Some have the ability to plug directly into a computer, others store information that can be read to a computer later and some are simply scales. Many come with a chute, others with just a platform, load cells and the electronics.

When scaling animals, easy does it, move them gently and handle them calmly and professionally (you lose a half of one percent shrink, for every half hour you chase them around, often more). Remember you are doing all this to obtain information about their performance, not to punish them. If you can, run the group through the scales or scale area a couple of times, a week or so before you start, to get them use to the procedure (patterning). Each time you weigh a particular group of animals, it should be at the same time of day and follow as similar a routine as possible each time. The best time is after their first big graze period in the morning. The idea is to have the same amount of fill in them each time and also not to interfere with grazing patterns too much. Scalings should be no closer than 21 days apart, 28 is better. For best results each animal should be individually identified with an ear tag. If you have a computer, a spread sheet can be set up to handle the input information, either directly from the scale or by hand. Table 13.4 shows one possible set up, an example and the equations used.

TABLE 13.4. SAMPLE FORMAT AND EQUATIONS FOR DETERMINING AVERAGE DAILY GAIN (ADG).

TAG #	DATE 1st weight	DATE 2nd weight	(days) weight change	ADG
	9/15/85	10/12/85	(28)	
105	550	585	35	1.25
106	570	610	40	1.43
107	545	590	45	1.61
108	540	570	30	1.07
AVERAGES	551.25	588.75	37.5	1.34

NOTES:

Averages are determined by summing up a column and dividing by the number of entries, eg. 550+570+545+540 = 2205 (sum); 2205/4 = 551.25.

Weight change is taken by subtracting the first weight from the second weight, eg. 585 - 550 = 35.

Average daily gain (ADG) is the weight change divided by the days between the scalings, eg. 35/28 = 1.25.

When figuring the days between the different dates, don't count the day you first weighed on, but count the second scaling day, eg. don't count 9/15 but count 10/12.

FORAGE FED AND BULL BEEF:

A chapter on Hi-Tech may seem like a strange place to discuss bull and forage or grass fed beef. After all humans have been eating bull and forage fed beef ever since when. Even today forage fed beef is the norm in a great many countries and bull meat commands a premium price in Spain. So what's new about it? Couple of things. The big one is that consumer demand for lean meat is strong, look at the recent "no-roll" and "lite" beef movements. Another is that the perception (the way a person understands or comprehends a situation, which has absolutely nothing to do with facts), of feed additives is bad and that somehow or other feeding all that grain is wasteful. Most "civilians" consider the livestock industry, if they consider it at all, from a knowledge based primarily on TV, pocket book westerns and movies. With that sort of knowledge base, anything is possible.

In the U.S. there has been grain feeding of animals since the early 1800's, as farmers used livestock to increase their incomes by a sort of vertical integration (the farm feed yard). However, just before and after WW II, grain farmers in the U.S. began to produce monstrous amounts of grain. They did such a good job that huge surpluses were built up.[7] All those stockpiles of grain were worrisome, then somebody got the bright idea of feeding grain to cattle on a mass scale but first there had to be a consumer attitude re-adjustment. The public was told that grain fed cattle were best and to prove it, USDA Government labels of "Prime or Choice" were stamped all over it. Now you also have to remember that this was in the days when pigs were of the "root hog, or die" variety and chickens roosted in the barn.

That wasn't the only reason of course, lack of quality control in the bull and forage fed (grass fed) beef was actually responsible. In fact, right now in Australia there is no quality control on the meat sold. The result is that you buy a piece of meat one day and it may not be bad at all, however the next piece, if it came on the bottom of your boots, you'd be stuck. As a result of the lack of quality control the grass fed market will no doubt lose out to the several feed yards that are springing up there. Consumers don't like those kinds of surprises, not at the dinner table.

Right across the Tasman Sea from Australia, the New Zealanders have a quality control program on their grass fed beef (New Zealand feeds no grain). The meat there is very tasty, tender and if you buy a steak one day you can be sure of getting one just as good on the next. Much of the meat that makes up New Zealand "prime" is from intact bulls. The point is that quality beef can be made from either an all forage fed program, or from bulls, or both; anyone that says different, simply doesn't know how.

FORAGE FINISHING:

In the production of beef the feed yard serves as a filter or buffer or an up-grade operation, depending on ones definition and who you're talking to. It takes animals that have been produced under a diverse set of environmental and/or managerial conditions and places them in a carefully controlled environment for a period of time. The entire program is aimed at producing a uniform product that can be marketed anywhere in the U.S. Feed yards are very good at what they do.

The trend in feed yards has been to introduce animals to them at a younger and younger age; some places are taking animals at weaning or shortly thereafter. This is

[7] Farmers and ranchers are the only business groups that, when the price drops, their solution is to produce more of the same.

ridiculous. The simple fact is that feeding grain to young cattle is like feeding grass to chickens, it can be done but what a waste. The argument is, that it's cheaper to feed grain (which must be farmed and harvested with machinery) than it is to let the animal do the harvesting from perennial forages. If you still believe that after reading this far, well, what can we say? The other argument is that young animals make cheap gains and that's true enough, but they don't have to be in a feed yard to do it.[8]

Forage feeding does not mean the absence of grain, it means that grain is used as a supplement and not as the major ingredient in the ration. Forages in the form of high quality hays or silage make up the bulk of the ration. At the very least it means that the animal is carried on forage to 750 to 850 lbs then put in a feed yard for 90 days (short fed). The program doesn't end upon delivery to the slaughter house. As the section on bulls will discuss, handling has a lot to do with the finished product. Also if the carcass is electrically stimulated or not. The advantages are increased apparent grade, improves lean color, increases tenderness and decreases aging time. If handled correctly it is as safe as any saw used, voltages on units are now as low as 35 volts. So why isn't every slaughter house using it? For the same reasons that they don't change their management practices when bulls are slaughtered.

The secret to forage finishing is to maintain a high rate of gain on quality forage and slaughter at a early age (less than 24 months). A goodly portion of this book has been devoted as to just how to get those kind of gains. Correct pasture management, coupled with a forage conservation program, can produce high quality gains on a year round basis and deliver a quality carcass at predetermined times. This type of program can be easily tailored to a specially market or to fit in with a short fed program. However, without some sort of buffering action, whether it be a marketing cooperative or confinement feeding or grading standards aimed at obtaining a consistent quality or a production program geared to deliver a consistent quantity, any program that hopes to capture and retain a significant share of the consumer dollar is doomed to ultimate failure.

There are a number of trends on the mainland that augur well for beef. The National trend towards lean beef (no-roll and lite), the perception that somehow feeding grain to animals instead of humans is wasteful and concerns from health and natural food advocates, means an improved climate for forage finished beef. A very recent trend is the new, gun toting, can-do, "Macho" heroes that are appearing on movie and TV screens, which are replacing the wide eyed, everything is my fault, "Wimps" of the recent past. These new heroes don't eat quiche, but they don't eat fat meat either.

BULL GRAZING MANAGEMENT:

Bulls gain weight better than steers. If you rate bulls at 100%, steers are about 80% and heifers 70%; in efficiency of conversion bulls are about 16% better than un-implanted steers. Most of the work in the U.S. on bull feeding has been in the feed lot, not on pasture. In the U.S. dark cutters are one reason offered as to why bulls should not be used. Research in both the U.S. and Australia on the cause of dark cutters seems pretty clear that it is a primarily a management problem at the time of slaughter and its name is stress. Stress increases the glycogen level in the muscle, this glycogen breaks down to lactic acid upon death. The lactic acid lowers the pH of the muscle and a low muscle pH results in a dark cutter.

---

[8] The grain merchants are like horns on a bull. A point here and a point there, but a lot of bull in between.

The first thing about bull management is to understand that a bull is not a steer; bulls are a lot more emotional than steers. Bulls get up-tight real quick with rough handling. While they may or may not be aggressive towards the handler, they will react towards one another (displaced behavior) in ways that are almost guaranteed to get the handler upset and the vicious little circle goes round and round. Handling bulls is almost an art form and as in art, you have to like what your doing if your expect to obtain anything of substance. This means that bulls need to be kept with their herd mates right up to the time of slaughter; this reduces fighting and emotional stress. They need to be handled differently (read calmer) and fed on a regular basis while in the slaughter pens. They also need to be slaughtered within 24 hours after entering. It has also been found that a little molasses mixed in with their water also helps reduce the incidence of dark cutters.

The management of bulls on pasture is different than in the feed lot. For one thing bulls on pasture spend more time eating, consequently aren't as bored as feed lot animals (don't have as much time to get into trouble). Another major difference is that on pasture the chance of the bulls getting out, or females getting in is much greater. Traditional fencing is not recommended with bulls. Electric fencing is the only type of fencing that will: Hold bulls where they are put; keep other cattle from entering; reduce bull aggressiveness (no more fighting other bulls over the fence); no more tearing up the fence by scratching on it; and in general make them easier to handle. Needless to say, all bulls should be trained to electric fence before placing on pasture.

For the average farmer or rancher the average size of the mob should be kept to 20 to 30 head. There are some farmers that are able to handle over a 100 head to a herd without problems but they know and understand bull psychology and most importantly, apply it. Generally bulls should be grouped according to similar size, age, frame and temperament. Some problems will still occur but the odds of attaining a stable social order are greater. Some operators sort the bulls into these groups as soon as weaned or purchased and keep them as a group until sold; others wait until behavioral problems start occurring. A bull can be removed from a group without disrupting the social order, but if a new bull is added to the group the whole social hierarchy must be re-defined, which means a lot of fighting and not much eating until the order is again stabilized (this can take several weeks). In selling, the whole herd goes, or it is topped off within a period of a few weeks.

Picked on bulls (being ridden a lot) or troublesome bulls should be removed from the herd. Problem bulls are probably best sold or castrated and placed elsewhere. A picked on bull can be returned to it's group without problems, if it has not been away for too long a period of time (a couple of weeks); exactly how long depends on the groups stability and the length of time the bull was a member of that group.

If a new bull or a few bulls must be added to an existing group, it has been found that spraying the entire herd with a solution that contains an odor, such as a mild scented antiseptic, will often trick the group into thinking the new additions are members in good standing. If a new group is being formed of older animals, the above mentioned sizing criteria still apply; spraying or packing the entire group tightly in a corral overnight also helps to give the animals a group identity. Once the group is turned out on pasture the manager must remove problem animals quickly.

Keeping bulls from getting bored also reduces behavioral problems. One excellent way is keep them eating. Frequent shifts, daily or even twice daily, on to good feed (these are growing animals so "Bull" quality feed is not good enough, "Stocker" quality feed is required), also frequent shifts gives them a new area to explore and pasture damage is reduced. Some farmers in New Zealand have reported reduced fighting and hole digging when they added an old bull, a horse or billy goat to the herd.

In shifting bulls, quickly with the least amount of fuss is the rule. To minimize problems in shifting and handling, bulls must be "fooled" into thinking they are being shifted because they want to be shifted. Each bull in the herd is an individual, in addition to being a member of the herd. If he has to much time to think about something he may decide he doesn't want to do it, on the other hand if you rush him, he's liable to balk or get rattled and frighten. Either way fighting, riding or other displaced behavior will result. There is a fine line between shifting quickly and rushing.

Paddock layout is important. Shifting into adjoining paddocks is desirable; if you have to take a trip around the farm every time you shift, you had better do some re-organizing before trying bulls. Bulls should be able to see where they are going; wide alley ways, corridors and gates will help, if too narrow dominant bulls will hold up stock movement. Avoid blind corners in alley systems; gates in the corner or with wings are suggested. In planning a shift to a different paddock, if possible, do not take them by another group of bulls or cattle as territorial and hierarchy dominance behavior will occur.

In order to "fool" them into thinking shifting is their idea, only allocate sufficient forage (grazing area) for one days feeding, this works best on break or strip grazing. Break out the new area, take up a section of the temporary fence, let them move themselves then re-place the fence behind them. Another way to "fool" them is to shift at the same time each day, cattle quickly adapt to routines and form the habit of moving on signal.

The use of growth promoting implants has been used with some success in feedyards, both to increase gains as well as reduce behavioral problems. The advantage in pastures is not that clear cut and in New Zealand mixed results have been reported. Many farmers there prefer to use other management techniques to reduce behavior problems.

CULL COWS:

A major ranching cash flow problem is that dry cows cost money to keep, but don't bring any money in. The cure is simple enough; sell all the open cows. But this can whittle away at herd size, not to mention the cost associated with expanding the replacement program, and/or buying replacements, which can raise hob with the cash flow. If one considers cull animals as part of the total production, rather than something to get off the place of as quick as possible, there may be some money to be made. Cull cows that are thin, but healthy (cutters and canners), if placed on good forage will make remarkable gains (150 pounds or more) for a short period of time, 90 to 120 days. The reason is that all the organs, bones and such are already there, the forage goes to building up muscle and fat. In the U.S. cull cows are about the only class of animal that increase in price the more they weigh; something to do with the hamburger market. However, to work it right requires good forage, especially if they have no front teeth, and most importantly you should have your market secured before you buy or place them on pasture; know the price, time of delivery and who is going to buy them. A 150 pound increase in the weight of a cull cow may add $60 or more net dollars to her final price.

MARKETING:

Selling is getting rid of something you have, marketing is finding out what will sell well, then producing that product. Regardless, there are two things that should be of paramount concern, not only to attract a market, but most importantly, to keep it. You must be able to GUARANTEE QUALITY AND QUANTITY. If you can't deliver a consistent quality and quantity of your product, forget it; the whole thing is doomed before you start, all you'll do is spend a lot of money before you go broke.[9] Many fine programs have failed because somebody got too greedy and started reaching too deep in their coolers. Consumers may be ignorant, but they're not stupid, they catch on real quick (The Baby Beef Program in the South is a case in point). Right now, seems like every retail store has it's own brand of no-roll beef, most of which is selected based on some set of the store's own criteria. If they maintain those standards, well and good; if they don't, the entire industry suffers. The only reason U.S. choice sells so well is that it's a consistent quality, the reason it is consistent is that it is graded by a dis-interested third party (the USDA). Unless "no-roll" and "lite" beef get the same sort of grading by a dis-interested third party, they are doomed to go the way of "Baby Beef". Too many people out there can't resist slipping in sub-standard stuff when the demand gets hot.

Ranchers have been moaning for years about low profits, much of the problem has to do with poor management on the part of the rancher. But a much more significant part of the problem is that the producers have allowed the retail and packing segments of the industry to not only dictate the price they are willing to pay to the producers, but also dictate the type product that they can make the most money with. If the consumer was getting the benefit out of all this, the whole thing might be justified. But he's not; if he was, he was he would be buying more beef and less chicken.

A rancher may spend 2 years producing a slaughter steer that he sells to a packer for around $600. The packer owns it for maybe 2 weeks and sells part of it to a retail store for around $800. The retail store owns it for maybe a week and sells it for around $1200. That never did seem quite right. You ask these people about it, and they come up with all sorts of cost of labor, equipment, buildings, etc., which are all probably true; then they cap off their argument with a statement about how low their margins are, usually less than 5% and often as low as 1%. So you leave their offices with tears in your eyes belittling yourself for not understanding their problems.

If you have stayed with us this far, you remember that in the chapter on Financial Analysis there was a discussion on turnover rates; by increasing the rate that money (or nutrients) move through a system, production is increased. In that case the retail store only made a 1% profit on each dollar spent on meat, but by turning that piece of meat over (selling) each week, it was able to realize a simple interest on its money of 52%. Not bad. Of course if they take a 5% return, profits go right through the roof and might even become a source of embarrassment. This is not to single out packers and retailers as "bad guys", they're no better or worse than any other retail store. It's business, and seems like every other business person knows about it, except livestock producers.

So what can producers do about it? One way is to increase the turnover on your ranch or farm, in methods already discussed. The other approach is to form a marketing cooperative and get into this heady world of fast turnovers, where the big profits are to be made. Most ranching operations are not large enough to do this on their own; remember you have to be able to supply a consistent product all through the year. The only recourse is some sort of marketing cooperative, but one with teeth in the by-laws and arti-

[9] That's kind-a like saying: "I'll die before I'll let you outlive me."

cles of incorporation. A loose organization can be broken very easily and quickly, you need one where the members are totally committed. With this type of set up, the packer and retail ends become service organizations to the producers and not independent businesses. An example of this sort of vertical integration is the chicken industry.

SUMMARY:

Can the U.S. develop a grass fed or bull market? Of course, but probably not on a large scale, the feed lobby is very large and powerful; already one large feed company, Purina, has come out with pamphlets entitled "The Great Grass Myth".[10] There is also considerable resistance to bulls in the packing and retail segments. However, forage fed beef is being sold in up state New York and in Louisiana under special labels, and commanding a better price than U.S. choice (partly as a result of the "organic" or "natural" fad). So far bull beef in the U.S. is mostly confined to specialty markets, however 3 to 5% of all male animals marketed in Alberta are intact bulls, most of which fit into the top grade, "Canadian A". And of course New Zealand, in which intact, forage fed, bulls make up a large share of their "prime" beef marketing. If producers were to ever organize, and the packing and retail end placed in a service position, low profits would cease to be a major topic of conversation at livestock meetings.

SUGGESTED READING:

Ag-Links (Farm Bikes), FPP 637,638,639. MAF, Private Bag, Wellington, New Zealand.

Beef from Young, Intact Males. 1982. Proceedings of the U.S. Beef Symposium, Kansas State Univ. Manhattan, Kansas.

Bull Beef. 1985. Manawahe Seminar Proceedings. Agricultural Training Council, P.O. Box 3109, Wellington, New Zealand.

Bulls vs. Steers - for Beef. 1967. R. Albaugh. Univ. of Calif. Ag. Ext. Ser. report # AXT-240.

Computers and Ranching Operations. 1981. C.T.K. Ching and P.S. Leung. Soil Conservation Service Grazing Systems Workshop, Kona, Hawaii.

Computers on the Farm. 1984. D.T. Smith. USDA Farmers Bulletin No. 2277.

Computer Use in Hawaii Agriculture. 1986. S. Nakamoto, P.S. Leung and H. Marutani. Extension-Research Report, HITAHR Univ. of Hawaii.

Farming and the Computer. 1985. J. Cardiff. Changing Horizon

Forage-Fed Beef: Production and Marketing in the South. 1977. Ed. J.A. Stuedemann. Bulletin # 220, Southern Cooperative Series.

Software. 1984. Moor Data Management Services.

Weaned Calf Decision Analysis. 1983. P.S. Leung and B. Smith. Univ. of Hawaii, Information Text Series 012.

[10] The information contained in these pamphlets is not wrong; it's just sort of slanted, like the information you would expect to get from a used car salesman.

Used with permission of NEA.

White picket fences don't make a ranch successful.

14 INTENSIVE GRAZING MANAGEMENT

Intensive grazing management can be applied regardless of the species, the class of animal or animals raised, the kind of environment (wet, dry, hot or cold), stocking rate (one animal to 300 acres or three animals to one acre). In short there is no biological reason why it can not be applied anywhere forage plants are capable of growing. There are however, economic, social and/or cultural reasons why it might be difficult or downright impossible to apply it to any particular area or country. The reason is that biology is based on the same principals, which are valid throughout the world, regardless of what language is spoken, or what form of government exist; whereas what humans do, or do not do, often begins to make sense when only when their entire cultural and social history is taken into account.

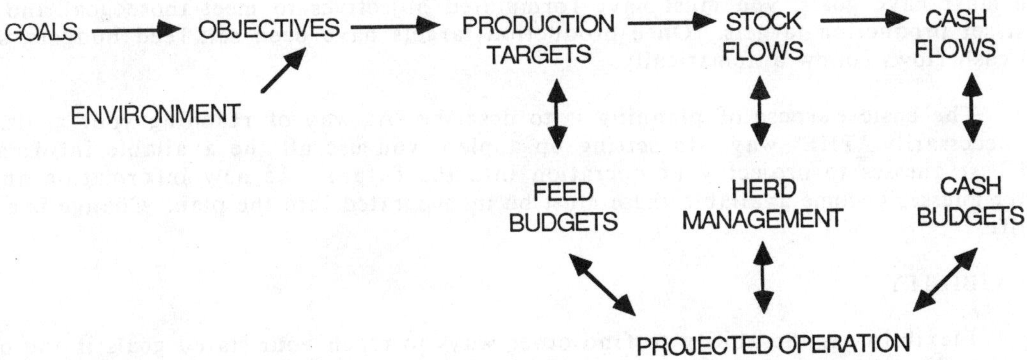

FIGURE 14.1. BASIC FORMAT.

BASICS:

The first step is the formulation of GOALS, the second is the setting up of OBJECTIVES to best meet those goals, the third is the setting of PRODUCTIONS TARGETS. Once this has been done, stock flows, cash flows, feed budgets, herd management and cash budgets follow. The entire projected operation over the specified time has been determined. The next step is to conduct an enterprise analysis of that particular objective and test the economics of actual implementation. At this point an iterative procedures (looping) is employed to maximize the economic returns by adjusting the production targets, either up or down. Figure 14.1 shows a flow chart of this procedure. Depending upon the results, that particular objective is termed economically "valid" or "not valid" and the procedure repeated with the next objective. If this sounds a bit heavy, it is. However, it is essential, if your agri-business is to survive and prosper.

Goals, objectives and production targets have already been covered in Chapter 2. The important point here is to make sure that the objectives and production targets formulated are reasonable ones in view of the environment that you will be operating in. To expect three pounds a day gain per acre, in a range situation of 6 inch rainfall, is a bit unreasonable.

Unfortunately the real world often acts in such a way as to mess up the best laid plans of mice and men. Fortunately, there are ways to deal with the vagaries of the real world; planning, flexibility and knowledge. Planning, so that at least you know where you're trying to go. Flexibility, so if the road is suddenly blocked, a detour can be found. Knowledge, so that you can operate the best vehicle for the conditions of the road (if the detour is through the brush and all you know how to do is drive a car, there will be problems until such time as you either learn to operate an off-road-vehicle, ride a horse or are willing to get out and walk).

PLANNING:

Much has already been said about planning and for good reason: "If you don't know where you're going, any road will get you there". The outline presented in Figure 14.1 is the bare minimum, regardless of your stocking rate or environmental conditions. You must have goals, you must have formulated objectives to meet those goal and you must set production targets. Once production targets have been set, feed budgets, stock and cash flows follow automatically.

The basic purpose of planning is to describe "A" way of reaching your goals, but not necessarily "THE" way. In setting up a plan, you use all the available information and best guesses to project your operation into the future. As new information and/or better guesses become available these must be incorporated into the plan. Change is a fact of life.

FLEXIBILITY:

Flexibility is the ability to find other ways to reach your stated goals if the original plan is suddenly blocked by unforeseen circumstances, or if new opportunities are presented. Blind adherence to a plan can spell disaster. Not trying to single out the military, but down through history there have been some brilliant examples of blind adherence to some prearranged plan; generals that kept wasting their men against impregnable fortresses until there were no more men to waste. A good planning program has built-in flexibility - "Hang Loose".

The owner-operator has a tremendous advantage over the large absentee owner outfits in grabbing the brass ring when it goes by. If you own the outfit you do what you decide. Actually your wife, the kids, the banker, the feed store, the IRS and probably a few others are going to get in on the decision, one way or another. The large outfits, complete with a board of directors, often take several months to make up their collective minds as to what path to follow. The tendency in such collective decisions is to be somewhat conservative, even more so than the average farmer. The owner-operator can react to a favorable or unfavorable situation quicker than can most large outfits. This is not the only advantage a small outfit has, but it's the biggest.

The ability to make a decision and run with it has some inherent risks; the biggest is that you might sail right off a cliff before you know it's even there. This is why the owner-operator must utilize all the expertise available. He may not be able to afford to hire special consultants, he often doesn't need to. There is a tremendous amount of "free" information, just waiting to be picked up. "All's" you have to do is contact your County Agent, SCS or start banging on the doors of your local land grant University. You and everybody else are paying for the knowledge that these service organizations have; if you don't use them, well...

KNOWLEDGE:

Much of what we know, while allowing us to function in society, is of little practical use in the day to day operation of an agri-business. In fact, what we know or think we know, can often work against us.[1] Attitude is the all-important criterion. If you think something is impossible, then you're right. It is impossible for YOU, but not for somebody else. How often have you said: "That's impossible." Only to hear of or see somebody doing it a short while later. Nothing is impossible, difficult perhaps, but not impossible, given sufficient time and incentive.

Knowledge in the context of intensive grazing management is equated to tools and the ability to know how and when to use them. A crescent wrench can be used to pound in a nail, but a hammer does a much better job. Likewise if the only tool you have and know how to use, is a hammer, then you tend to see all problems as nails.[2]

RESOURCE TOOLS:

A tool is defined as "something used". Anything that aids in getting the job done. Since livestock graziers have a lot of different jobs to do, they require a lot of different tools to do them; there are carpenter tools, automotive tools, leather working tools and a whole host of other tools that are used on livestock operations. These are not the kinds of tools that this section is about. What follows are a brief descriptions of the major tools that are available for use by the grazing manager; tools that work on the basic resource. For convenience they are grouped into four general categories; forage, animal, supplemental and sacrifice. Like other tools they are normally used in combination, and like other tools, there is no reason to expect that any particular one will be used each and every time. Which ones, or combinations, apply to your operation, depends on your goals, objectives, production targets and of course the environment.

FORAGE:

1. SPACE: How large an area is allocated for grazing. This will depend on the environment, type of grazing or harvesting program employed and the numbers and kinds of animals to be grazed, and/or the amount of forage to be conserved. The amount of grazing area required per animal per day can vary from less than 100 square feet in very intense environments, to over two acres in some of the high desert areas.

2. TIME: How long the area will be grazed and more importantly, how long the area will be rested after grazing or harvest. The graze period can range from set stocking (long term rotation), to half day graze periods. Set stocking is used only under three conditions; when pasture control is in danger of being lost (fast growth rates), when high individual animal performance is desired or when follower type animals are giving birth (sheep). Half day moves may also be used to maintain pasture control or to maximize per acre performance and is generally restricted to use with high value animals (dairy or stockers). In all but set stocking, graze periods should generally be 5 days or less, but again this may vary depending on production and performance targets.

3. UTILIZATION: How much is taken and how much is left - residual dry matter. There are two terms that describe the extremes in grazing pressure; lax grazing and

[1] Kinda like buying lingerie for your wife's birthday. So you get her something like a one - piece red flannel with a drop seat and then wonder why she gets uptight.
[2] As might be expected, acquiring knowledge is a learning experience.

flogging. LAX GRAZING is the ultimate in allowing animal selection of a high quality diet; creaming the paddock. If the average pasture quality is high, this means allowing at least 44 lbs of dry matter per animal per day; if the quality is poor, closer to 90 lbs per animal. Lax grazing is used in set stocking or in multiple herds where high individual performance of the front herd is desired. To FLOG a paddock is to graze it until bare ground shows, or where there is next to nothing left to eat when the paddock is cleared of animals. Flogging is primarily used to maintain pasture control when a long regrowth period is desired, to remove weeds or to clean up old rank forage. Both extremes have their place in the managers tool box.

4. SPECIES: Different species of forages have different requirements. If you're changing the energy inputs into a grazing area, then very likely the plants that are presently there are not the ones that should, or even will be there at some future point in time. Attempting to keep and maintain plants that did not evolve under herds of hooved animals in a region where such herds have been introduced, is generally counter productive. In most cases (except where the plants were never there originally) the seed sources for the better forage plants are still growing somewhere in un-accessible areas and will gradually make their way back; once conditions have been changed to allow them to do so. If you're the impatient sort, you can always plant strips of different species in some paddocks (range areas) to establish a seed source and hasten the process. As far as tearing the whole thing up and replanting, the use of the plow and other "Heavy Metal" solutions need to be well thought out; the more extensive the area, the more "Heavy Thinking" required.

ANIMAL:

1. SPECIES: Different species have different diet requirements, different grazing patterns and social interactions (Chapter 6). As a result their effect on the plant community will be different. Goats may be used to clean up (flog) a paddock after the main herds have grazed it to clear it of woody and/or weedy plants. Cattle are often routinely used behind sheep, or vice versa; the deciding factor, other than for a specific maintenance job, is which animal on grass has the highest economic value.

2. BREED: Within species, breeds perform differently under various environmental conditions. Holsteins typically require better pasture quality than do Herfords. Brahman cattle are better adapted to hot climates. Dark colored animals generally do better at high elevations than light colored ones. Hairy animals (Scotch Highlanders) perform better in raw (wet, cold, windy) environments than do sleek coated breeds. Some breeds use 75% of all energy consumed just to meet maintenance requirements. One study found that across all seasons, Angus and Brahman cows had the lowest maintenance requirement with Herfords 7% higher and Holstein 23% and Jersey's 59% higher. It's a far, far better (cheaper) thing to match the animal to the environment, rather than attempt to change the environment to meet the animals requirements.

3. CLASS AND WEIGHT: Each class of animal within a breed has different feed requirements; dry or lactating females, intact or castrate males (bulls or steers) and if growing or mature. The weight of the animal also influences the quality and quantity of feed required. Examples of using different classes would be; running dry cows behind stockers, or number 2 milkers behind number 1 milkers.

4. NUMBERS: The number of animals gives control over how fast an area will be grazed. Numbers, more so than length of graze period will determine what will be

grazed. With high density many undesirable plants will be grazed even when sufficient high quality forage is available. High animal densities determine stress on undesirable plants (trampling), it also influences herd effect, nutrient transfer and animal behavior.

5. SELECTION: Perhaps the single most powerful tool that the rancher has for matching animals to the environment that they will be performing in. But there are limits, don't expect animals that have been selected to perform well in one type of environment to out perform animals in a different environment, especially if those other animals have been selected for that different environment. To be effective, selection must follow a plan and must be rigorously applied; "better to lose a few good ones and cull all the lemons". To make beef cattle pay, you need as good a calf crop as any dairy.

6. GENETICS: A scrub animal is a scrub animal, regardless of how much alfalfa you feed it. If you raise the nutritional quality of your pastures, along with your managerial ability, then you may need to raise the genetic potential of your animals to better utilize that superior nutrition and management. But keep your priorities straight. First get the nutritional levels of your operation high enough to meet the genetic potential of the animals at hand; then worry about raising the genetic potential of the animals. The use of genetics is a powerful tool and coupled with AI can quickly bring about desired changes in a herd. In the past genetics was expected to do all things equally well. It is just one tool amongst many.

7. HEALTH: A good preventative health program is a must. Because large numbers of animals are concentrated any sickness must be quickly isolated and treated. However, because the animals are under control, this is no big problem; for example, medication can be placed in the water. All livestock operations should have a hospital paddock or area. Proper immunization for prevalent diseases and sanitary water go a long way towards a good prevention program. In extensive environments, a parasite control program, coupled to your selection should be worked out. On intensive environments, regular treatment is called for. The old dipping vats should be abandoned in favor of alternative methods. After the first 20 or so animals go through, the rest mostly get a swimming exercise, the pesticide is tied up in the dirt and manure washed off the animals, not to mention stress and the many other problems connected with dipping vats. Consult with your local veterinarian in setting up a herd health program.

SUPPLEMENTAL:

"HEAVY METAL" OR MACHINERY: A modern farm or ranch requires some machinery, how much depends on how intensive it is being managed and the environment that the ranch is operating in. The thrill of owning and operating a large tractor or cat, often becomes an end in itself. A great many ranches could sell off half or more of their equipment, contract out what work is required and find themselves in a much better cash flow situation. Most "Heavy Metal" is owned simply for convenience and not because results of an enterprise analysis indicated that such a machine would MAKE, or at lease save the outfit money. The yearly interest alone on a $100,000 hunk of "Heavy Metal" would pay to contract out the clearing of at least 50 acres a year. If you would put that $100,000 in the bank you could clear 50 acres every year on just the interest and still have the original balance in the bank. You also wouldn't have to worry about depreciation, maintenance, repairs and all the other fun things that go with owning "Heavy Metal".

With that said, all of the below require some machinery and/or special equipment, either owned, leased or contracted for. The trick is to know which option to use, and for that see enterprise analysis in Chapter 3.

1. FERTILIZATION: Fertilizers can be used to smooth out dips in forage production curves, correct plant nutrient imbalances, increase over-all carrying capacity and increase forage quality. In an intensive environment, with IGM in place, it has been found on a commercial operation that for every dollar spent on nitrogen, two dollars were returned; the other major elements were in ample supply. This may sound self evident, but fertilization produces an increase in forage, if you can't have the animals there to harvest it when it needs to be harvested, or if you lack the mechanical ability to conserve it, then don't do it.

2. IRRIGATION: Similar to fertilization; forage quality and quantity are increased. Used in conjunction with fertilization and forage conservation, production curves can be predicted with fair accuracy, regardless of how dry it gets.

3. FORAGE CONSERVATION: While usually thought of as making hay, silage or haylage, it is also the planting of a crop to provide feed during a critical period of the year, such as annual rye grass or Leucaena. If used in combination with forage controls, pasture quality can be maintained by harvesting paddocks when the growth rate is very high and paddocks are in danger of getting out of control (becoming rank).

4. ALLOCATION: The most powerful tool for rationing forage. Yet one of the most misunderstood and misused. Continuous graze will utilize between 40 to 50% of the standing crop, twice daily shifts with long strips can get better than 90%. To allocate forage so that the majority gets into the animal, requires many paddocks coupled with high densities and short graze periods. There is no better way, save cutting it and feeding it out bale by bale; and that costs.

5. CHEMICALS: Herbicides and pesticides should only be used as part of an integrated management program. For example, the new treatment for liver flukes in humid regions begins to make economic sense only when used in conjunction with a biological control of the snail host along with chemical treatment of the water for the flukes larval stage. Most applications of chemicals are treatments for effects and do not treat the cause of the problem. Chemicals get the job done and quite often are the most effective tool to use initially. But all applications require equally effective follow-up procedures. If not, it's back to square one, often in a remarkably short period of time.

6. FIRE: You bet, as long as it's a prescribed (controlled) burn.[3] Fire is a natural component of grasslands and was a major factor in keeping the woody plants from taking over under "natural" conditions. Today the "natural" conditions have changed, consequently the use of fire must change if it is to be used as an effective tool. The drier the environment the less need for fire. Occasionally fire may be used to control the invasion of woody plants, decrease tick or other pest problems, increase nutrient flow to overstory crops (tree plantations), or get rid of a forage mat prior to implementing some form of controlled grazing. However, as with chemicals the use of fire must be incorporated into an overall management program. If not, a low producing fire climax species will eventually take over, as has happen in many areas

[3] Your County Agent or SCS office can get you the necessary information on how to conduct a controlled burn.

of Florida, South America and many other areas that are periodically burned. Regardless of the reason for burning it must be integrated into an overall long term program, otherwise, DON'T DO IT. The major goal of any livestock program is to produce forage for livestock, not to see it go up in smoke.

7. SUPPLEMENTS: Livestock supplements mean just that, supplements. All too often a supplement becomes a substitute for forage. The only way to tell if a supplement is working for you is to try it, but at the same time you have to try a part of the herd without it. For this you need a set of scales and at least a 30 day trial period. Then you need to sharpen your pencil and work out the economics. For example molasses will usually put on additional weight gains, but if the value of the weight gain is less than the cost of the molasses and it often is: Why do it?

SACRIFICE:

This does not mean making burnt offerings to the Gods.[4] But it does mean offering up a portion of the ranch in order to save the rest of it. In other words you punish a small portion of the ranch so that the rest may be spared. They are often called sacrifice areas, bare areas around water, stock trails, etc. Often in intensive management a sacrifice decision must be made. Two examples will help illustrate: You are break grazing stockers daily, high density, then it becomes very wet and trampling (pudding) becomes a major problem. If you continue half or more of your paddocks will be severely damaged. "What will you do?" A couple of alternatives: 1. If the growth rate is high, set stocking could be considered, if you can keep them spread out. 2. Decrease the area and increase the rate of movement (half daily breaks on half the area). 3. Select a raceway, alleyway or a small paddock and use this as a bed-ground. The herd is moved to fresh forage once or twice a day but for only 2 to 3 hours each and are then brought back to the sacrifice area.

The second example is in a slow growth situation where you need a few more days rest to get the forage into the linear portion of the growth curve. The situation is delicate, as if you continue at the present pace, you will not have the necessary residual to allow the plants to come away fast for the next cycle and therefore may not have the necessary standing crop at the end of the growing season. "What will you do?" First check soil moisture, if there is sufficient for growth then select the better paddocks or areas within a paddock and flog the hell out of it. Use daily breaks and chew it right into the ground; attempt to get one or two more days grazing out of the same area you were allocating before. This will hurt this part of the pasture and will require a longer rest period to "come away" again. If you offer a bit of green feed every day in each break, this procedure will not seriously bother animal performance. If 7 or so days extra growth will not get the growth rate of the remaining pasture into the linear portion, don't try this approach, you have bigger problems, think of something else; supplement, lower performance targets, etc.

MEETING THE PRODUCTION TARGETS:

It should be crystal clear to the most casual of observers just what should be done and what tools are required to meet any production target that might ever get set. But just in case it's not, all I can say is: "Welcome to the club". There is simply too many variables in a livestock operation to be able to say: "Do this, and that will happen every time". Some researchers are beginning to understand that what happens in the greenhouse

[4] Some wives treat their husbands like Gods; they're always giving them burnt offerings.

or in small plot studies may mean absolutely nothing when placed in the context of the "real world". A plant that has been tramped on and then had 10 lbs of wet manure dumped on it, does not behave the same way as a plant that had the same amount of dried manure lovingly spread around its base.

At this point you might be saying to yourself: "You mean that I have muddied my way through all these bloody pages, just to be told there's nothing I can do?" No, not at all. Just as a mechanic never knows just what part will fail next and what tools will be required to put the machine back in working condition, the grazier also never knows just what will go off course and what means will be required to bring it back on target. The grazier, like the mechanic, must remain flexible, must have a well filled tool box and the knowledge to use each tool or combination of tools in order to make the necessary repairs and get the operation in working order in the shortest possible time.

SOME USES OF RESOURCE TOOLS:

None of the above mentioned tools work independently, all work in combination with others; all are interrelated. One can not just use space, there must be numbers, kinds of animals and time. So how is one to know which combination to use, which one to emphasize? As mentioned before; intensive grazing, regardless of what level, is a learning experience. One learns from what one did before. However, there are a few general principals that will allow one to get started in the general direction of the stated objectives.

Generally speaking, the higher the performance targets, the more tools that must be employed and the higher the overall management applied. If one is content to bring animals to market at four years of age, along with the continued deterioration of his range areas, then continuous grazing or set stocking only need to be used. However, if one is attempting to bring his animals to market at two years of age and at the same time maintain or improve his range, one must use other tools; set stocking may or may not be one of them.

In intensive environments, if plants are green and growing and if there is soil moisture, the greater the residual left behind when the animals leave, the faster the paddock will come away again; up to about 1500 lbs/acre. Conversely, the harder the paddock is grazed (flogged) the slower will be the recovery; when residual levels fall below 500 lbs/acre, damage to the pasture can be expected if grazed at this level for several cycles, especially if soil moisture levels are low. In range areas (extensive environments) leaving 50% or more of the green plant will result in faster recovery.

The greater the density of animals, the more uniform the manure distribution throughout the paddock.

As the grazing period decreases the manager can exert a greater control over allocation and utilization. High density coupled with short stay times, increases the percent pasture utilization; a higher percent of the forage actually gets into the animal.

The greater the density, the more stockmanship is required of the operator. Only one dog should be used in high density grazing, and then only if it is under strict control.

Animal power can be used to plant seed, Figure 14.2. They may be used to prepare a seed bed where a soil cap exist. Feeding hay to animals in the bottom of a earthen dam or pond over the winter period can help seal the bottom (it works a whole lot better if you drain the thing first). Animals can be also be used to smooth out small embankments or ruts, or eliminate un-desirable plants through trampling. This can be done by feeding a supplement at that location; a portable mineral wagon is ideal for this.

FIGURE 14.2. ANIMAL POWER. This New Zealand farmer is using his dogs to run a mob of ewes over an area to tramp seed into the soil. The area had first been flogged, then white clover seed flown on. He continued to run the mob over the area until all parts had been planted (trampled). Fertilizer could also be flown on at the same time, if needed. In warmer climates a pesticide application just after planting would be advised.

All else being equal (similar growth rates), paddocks should be grazed with shorter rest intervals during the summer (hot) than during cooler periods of the year, if similar stock performance is to be realized. When the daily average temperature is above 86 degrees F. the forage quality of grasses decline.

The faster the pasture growth rate, the greater the yield and the higher the forage quality. Consequently, during periods of fast growth either greater numbers of animals may be grazed or higher animal performance may be expected. Because the quality is higher, the rest period may be extended without danger of dropping animal performance, if sufficient numbers are present to graze paddocks to desired levels.

When growth rate stops, for whatever reason, allocation becomes the guiding criterion. Keep the herd moving, graze periods should not be longer than 4 to 5 days, preferably shorter.

In allocation, the more paddocks, the higher the density and the shorter the graze period, the farther you can stretch the available forage. One way to increase the number of paddocks, aside from temporary fencing, is to combine herds and move as a single unit.

The diet the animal selects on the first day in a fresh paddock is higher in quality than that selected on the second day, which is of higher quality than that on the third, etc. This is generally true regardless of the initial average pasture quality. The exception is when the stocking rate is very light with respect to the feed on offer.

Because the quality is higher on day 1, the animal will eat more than it will on day 2 when the average quality is lower, more on day 2 than on day 3, etc. As a result if the paddock contained enough feed for one cow (maintenance ration) for 7 days, the last few days would find the animal going hungry, because it ate most of its feed the first couple of days.[5]

The shorter the stay time in a paddock the better the animal performance. This is particularly true if a bit of green tucker (feed) is present when the animals enter.

Animals entering a paddock with ample feed will tend to graze as a herd and travel all over; whereas if feed is short or they are very hungry they will enter with heads down and graze as individuals.

The ability of animals to select a high quality diet on day 1, allows the manager to use multiple herds and tailor the quality to the requirements of each. Thus, number one milk cows get first graze, number 2 second graze and dry, third graze.

On dairies where two or more grazing units are being used that contain forage of different qualities. You may want to place the number 1's on the best unit, followed by the number 3's; on the lower quality unit, graze number 2's first, followed by number 4's. In general the more herds (if grouped according to production) the better the individual animals feed requirements can be met.

Normally molasses can be fed anytime energy is low. When the plants turn brown a protein may need to be added. If given molasses free choice, the animals will increase their intake as the pasture quality goes down. The greater the molasses consumption the more calcium the animals are eating, phosphorus may need to be fed as mineral, mixed in with the molasses or as mono-sodium phosphate in the drinking water in order to maintain the proper Ca:P ratio. Molasses fed at the rate of 4 lbs per head per day with average forage, will return anywhere from 0.5 to 0.75 lbs of liveweight gain, if nothing else is limiting.

Never flog a paddock if its soil moisture is low. If in doubt, check it out.

Grazing units that have been grazed with only one species of animals will often have an excess amount of feed that is suitable for another species. Grass-clover paddocks that are grazed by only cattle or horses will have a larger amount of clover than those grazed by sheep alone. Range areas grazed only by sheep will have a larger grass component than those grazed by cattle only.

Animals coming from a pasture of low quality and entering a paddock of high quality will generally develop diarrhea and lose weight for the first 10 to 14 days. To

[5] Grazing animals eat, drink and make merry, for tomorrow they move on; or at least their ancestors were able to.

avoid this loss, allow animals in the high quality paddock for only a few hours each day, preferably in the PM (the length of time can gradually be increased each day), for a week or so before placing there permanently. Or, run the new group as "followers" for a week or so before placing in front. Conversely, animals going from a high quality paddock to one of low quality will lose weight with more dignity, unless the poor paddock is really bad, then compaction may develop.

Paddocks that have a problem (poisonous) plant in them for a period of time each year should be taken out of sequence during the critical time. It may be possible to graze these areas earlier or latter depending on when the plant is at its most dangerous. Many problem plants are problems only if the stock are left in for too long a period of time.

Grazing under trees (plantation type) should be of short duration with frequent moves to prevent barking.[6] Most damage to seedlings (trees) occurs by trampling, until the seedlings are tall enough not to be stepped on, grazing should be at low density, lax grazing. Removal of the understory, by grazing, in tree plantations will result in higher growth rates of the trees, require fewer controlled burns and allow for some income from grazing.

Make sure herd management is synchronized with your prevailing weather patterns. If your area has mild winters consider fall calving, especially if heavy rains come in the spring. Fall calving under these conditions will result in less drownings, fewer cases of scours and pneumonia and in general less problems with older calves than with spring newborns. You may not be able to do this with sheep unless in tropical areas (day length effects on estrus).

On growing stock, keep them growing. If the animal stops gaining, it's two to three times as expensive to get them started again.

At weaning, don't change diets on them in addition to the stress of weaning. A great deal of the so called "weaner slump" is due to diet changes at weaning. It takes time for the rumen to adjust to a new diet. The best solution is to leave the calves or lambs where they are and move the cows or sheep away.

Across the fence weaning often gives good results; but the young must be placed on high quality feed and the manager must pay close attention to details. Moving the calves ahead of the cows is one possibility.

If dairy cattle are to be fed grain, feeding should start 9 days before calving to get the rumen "tuned up". This gets the maximum dry matter intake closer to the maximum milk production; 3 days rather than 25 days under regular treatment (first grain at first milking).

Grazing animals, individually, are selective grazers. One study found that cows offered grass silage and corn silage in separate bunks consumed on the average of 55% grass silage and 45% corn silage. However, the range of individual animal selection was from 85% grass silage and 15% corn silage to 18% grass silage and 82% corn silage. This same sort of selection goes on in the pasture. The only way to stop this selection is to mob stock for short graze periods.

When feed gets short tighten up (allow less forage) to the adults and allow the young to creep ahead; wean if necessary.

[6] Barking dogs are often tolerated; barking cows, never.

If you have low quality forages such as Bahia grass, Bermuda grass or any of the digit or star grasses, go for pounds per acre rather than pounds per animal.

Set stocking should not be used: A. After plants have gone to seed. B. In arid or semi-arid (fragile or brittle) environments. C. Where you are attempting to introduce more desirable (palatable) plants to the area. The only exceptions to the above occur when the stocking rate is very low, usually too low to be economical and when the animals can be maintained evenly distributed over the entire area, which is not possible with herd animals.

Set stocking should only be used: A: When high individual animal performance is desired; light stocking rate. B: When the reproductive requirements of the grazing animal require it, especially for follower species like sheep. C: When the pasture is growing fast and you are on the verge of losing control (plants have not yet gone to seed, but there is little doubt that they will if something is not done quickly) and hay or silage equipment is not available. Continuous grazing is a very wasteful method of harvesting forage; Table 14. 1 shows the different efficiencies of various methods of harvesting.

TABLE 14.1. EFFICIENCY OF UTILIZATION WITH VARIOUS HARVESTING METHODS.

Method of harvest	percent lost	percent utilization
Continuous grazing	50	50
Rotational grazing	35	65
Strip grazing	30	70
Hay	18	82
Silage	13	87
Green chop	8	92

This is the result of a study done in Wisconsin in 1957.

STAYING ON TRACK:

Once the objectives and the production targets have been decided on, the problem remains as to how one determines how good or bad, one is doing as time progresses. Of course if one could see into the future with perfect clarity, there would be no problems. Interestingly, one can predict the future one second from now with excellent accuracy; but the farther one attempts to peer into the future, the fuzzier the picture gets. The reason that the one second future is fairly well known and that one year from now is not, is simply because of the amount of information one knows about events that are likely to occur. A great deal of information is known about events that are likely to occur one second form now, whereas to predict events that will occur in one year requires a memory bank greater than any computer will ever have. However, this sort of clairvoyance is exactly what this section is about.

The trick is, of course, to make the future one is trying to predict short enough so that the information required is not overwhelming, yet long enough to be of some practical use. This is precisely why one must continually update budgets. Most of the things that happen do not just suddenly happen, they require time and generally the bigger the event, the greater the build up time required; wars and hurricanes are two examples. Forage growth also does not turn itself on and off like a light switch, unless there is some catastrophic event such as a hard freeze, or flood; even then it may still take a couple days to come to a complete standstill.

STANDING CROP:

Standing crop is the amount of usable forage that is present, or exists on the land at the time the measurement or estimate is made. It is what is there, right now. It is an inventory measurement and does not make allowances for what may or may not happen in the future. By making periodic measurements of standing crop, growth rate can be determined. Once the rate of growth is known, future yield can be estimated. With a fair idea of the amount of forage that is likely to be produced at some future date, intelligent planning becomes a real possibility.

Right now there is only one way to accurately know how much material is present on an area of ground; cut and weigh all of it. But not only is this expensive, it is also impractical. So estimates are used. However, any of the methods that are used to estimate the amount of forage present will introduce an error. The size of this error will vary depending upon the technique used (graziers eye, electronic meters, clipping frame) pasture and the skill of the person doing the estimating. In general, the error is greater the more forage there is and also the more species a pasture contains. It's easy to estimate the amount of forage produced on bare ground; it's a whole lot harder when some plants are 6 feet tall and some only a few inches in height. It's not uncommon to be off 300 or more pounds per acre when the amount of material present is two or three feet deep. Some day it will be possible to accurately measure growing forage with only a couple of pounds per acre error, but until then the emphasis is simply to reduce the size of the error and practice is still the best way to accomplish that.

DROUGHT:

A drought is any time forage growth slows or stops, when it is not normal to do so.[7] There are a couple of ramifications of this definition that make it different than most. First, if the ranch is in a wet - dry area, the dry time is not considered a drought, as it is expected. Second; lack of water or soil moisture is not the only criterion for a drought. Forage growth can slow or stop due to cold temperatures (winter), too much water, not enough sunlight, or because one of the necessary plant nutrients has been leached out. One mis-conception about droughts that many, not in agriculture, seem to have is that all of a sudden, bang, there's a drought. The media is probably responsible for this mis-conception as much as anybody. One day you pick up the paper and there it is, DROUGHT. Droughts don't happen over night. Droughts are cumulative; a good manager must knows the signs.

In a wet - dry environment one plans for the dry time, just as one must plan to get through the winter in snow country. These plans include, but are not limited to the following; short calving season, forage conservation, forage allocation, fertilization, leasing additional land, irrigation and purchase of supplements (keep an eye on those supplement cost). Another "trick" is to combine herds into one large mob and move them through all of the paddocks. Each paddock will end up with a shorter graze period but a longer rest period.

[7] A bad drought is defined as anytime you see two trees fighting over a dog.

While droughts should be anticipated and prepared for in both extensive and intensive environments, the basic emphasis will be different. In both operations drought management is to maintain a certain production level. But on intensive operations, since the animals either can't or aren't able to fend for themselves, the manager must do it for them; the burden is on the manager. Under extensive operations the burden is on the animals. This is true if the ranch has been in IGM for sometime; however, if you have just changed over, then obviously allowances must be made. Don't let your plans interfere with common sense.

Extensive drought management falls into two categories; humane and economic, both of which tend to run to extremes. The extreme humane manager would throw economics to the winds and feed the animals at all cost. The extreme economic manager would keep em standing. He might even buy more animals. As dumb as that might sound there is a sound economic concept in it. Generally during a drought, for any given management program, a certain death loss can be estimated. During a drought the price of animals can vary widely, often more than 50% lower than normal. For example if the drought price is $100 per head (50% below normal price), I buy a 100 head for a total cost of $10,000, if they all live through the drought I could expect to receive $20,000 when I sell them. Not figuring interest or other maintenance cost, I could lose 50% of the animals and still cóme out even. The trick is to know your cost, know the market, understand droughts in your area, plus a little luck.

The totally humane manager will probably not be in business very long; the totally economic manager probably does not exist. Most of us fall somewhere in the middle; it requires a lot of money to be totally humane, and a lack of conscience to be totally economic.

DROUGHT FORECASTING:

First, nobody can predict droughts, the weather service can't do a very good job of it, even with all of their resources. Secondly, there is no such thing as a normal year, every year is different than the ones that went before it. Normal is something we like to think about, but it doesn't exist. However, you can tell when you are moving into a situation in which plants are going to be under stress and therefore the animals, there are guidelines, but they all require keeping records.

Rainfall is one record every one in agriculture should keep. You don't have to read your gage every day. A drop of light machine oil in the gage will float on top of the water and keep it from evaporating. This will allow the gage to be read every week or so. Another important measure, which hardly anybody does is to record soil moisture. This doesn't have to be an elaborate procedure. Use a soil auger, or make one out of a 1 inch wood bit and take a soil sample from three different depths; 6 inches, 1 foot and 3 feet. Take the soil in your hand and squeeze it. If it forms a ball you have ample moisture, if it doesn't form a ball but your hand feels cool when you open it (evaporation), you have some moisture, if the soil blows away, you're dry.

Two other measurements should also be considered, air temperature and soil temperature. Temperature acts like a big forage switch, if it is too cold, plants turn off. Soil temperatures in the 60 deg F range, means slow plant growth for C4 plants; 40 to 50's for C3's. Soil temperatures should be taken from the 3-6 inch depth. Air temperature is a cumulative effect, an overnight low will not affect growth all that much unless it freezes or frost. However, several nights of these conditions, back to back, will.

The major measure of how well plants are doing are the plants themselves. Place a stake, marked off in inches, in an average area of the paddock. The stake can not be very high, 6 to 8 inches above ground and should be place away from the fence, at least 15 feet. Every so often (weekly intervals are sufficient) record the height of growth. Every inch of forage means so many pounds of dry matter that is available for animal consumption. How fast those inches accumulate determines the growth rate.

None of these measurements by themselves means very much but taken together paint a reasonable picture of what exists (right now) and what can be expected in the near future, if the weather doesn't change. For example; it hasn't rained for a month, last weeks growth rate in a freshly grazed paddock was two inches, temperatures are warm, there is wind but there is good soil moisture. From this, one can figure that if it doesn't rain (worse case) good growth will continue for perhaps another month, after which it will slow. The reasons are the good soil moisture, warm temperatures and the fact that there was two inches of growth during one week of rest after grazing, remember the sigmoid growth curve (chapter 5). Records are important, but one does not ranch to collect records, they must be used.

GROWTH RATE:

In multi-paddock grazing, determining the standing crop in each paddock gives the amount of forage present; it also gives, with a little effort, the growth rate. To find the average growth rate of the grazing unit, measure the standing crop in two different paddocks (P1 and P2). Say P1 had 10 days rest and measured 800 lbs/acre and P2 had 20 days rest and measured 1500 lbs/acre. Subtract one from the other and divide by the difference in days rest between the two; this gives growth rate. In this case (1500-800)/(20-10)=(700/10) 70 lbs/acre/day. In equation form: (P2-P1)/(R2-R1) = GR.

Since you are attempting to determine the growth rate of the grazing unit, it is not necessary to sample the same paddocks each time. However, it is important to select the sampling sites to be as near average, or representative of the entire grazing unit as possible. Also, when trying to find the average growth rate of a unit it is best to select one paddock (P1), that has had about one third of the total rest to be allotted to it (R1) and another paddock (P2) that has been rested about two thirds of the overall rest period (R2). If the rest period was 30 days, R1 should be around 10 days and R2 around 20 days.

The growth rate for each individual paddock can also be determined. Say that the rest period is 27 days then on day 9 (R1) take a reading (P1) and on day 18 (R2) take another (P2). Take the difference in weight (P2 - P1) and divide by the difference in rest (R2 - R1), which gives growth rate. It is important that the same general area be monitored each time, so mark it with a stake or something. It is important to select a spot as representative as possible of the entire paddock. If after a while you find it is not, simple, move the stake to where it is. For very intense operations where maximum production is desired, each paddock should be sampled every time the herd is moved.

If one knew how much feed and the quality of that feed, he would not be ranching, he would be running a feedyard. In short, yield and quality determine the number of animals and their performance; which in turn determines the economics. Pasture growth rate (lbs/acre/day) is a measure of two things; yield and quality.

YIELD:

The term yield is used in several different ways and often gets confused with standing crop; as that field yielded a ton of hay to the acre. However, yield is best used as an indicator of future performance, which is how it will be used here. Yield is the amount of forage that the grazing unit will produce over a pre-determined period of time.

Knowing the growth rate and how long that growth will continue at the present rate, determines yield. In other words, knowing growth rate is not enough, you must have some idea as to how long that rate will continue. For this, two other items are required: SOIL MOISTURE and GROWTH POTENTIAL.

SOIL MOISTURE:

An estimate of soil moisture can easily be obtained with a soil auger at the same time forage measurements are made. Generally the best depth to take soil moisture readings is between six inches to one foot. If you are on rock, lava or very thin soil, a subjective estimate must be made. The length of time before soil reserves are depleted will depend on the texture, water holding capacity and depth of the soil, as well as the temperature, wind and the amount of vegetation present. In general the more fine material, clay and silt and organic matter present in the soil, the greater the moisture holding capacity. Three categories are adequate to describe soil moisture; wet, moist and dry. Also the wetter the soil, not soggy, the longer the present growth rate will continue. If soil moisture is good, the growth rate will normally continue for at least one cycle. Again this is a learning experience, the more you do it the better your batting average.

GROWTH POTENTIAL:

Growth potential (GP) is sort of a catch-all grouping and involves such things as temperature, too much rainfall, sunlight, fog, or any climatic factor that limits plant growth. There are three categories; good, poor and zero. If GP is good, forage growth will not be limited. If GP is poor, the growth rate will not be maximum, but will fall in the middle somewhere, this often occurs with cold nights but warm days or when light is restricted due to overcast skies. If the GP is zero, there is no growth. This may be due to complete lack of soil moisture and/or low temperatures.

PREDICTION OF YIELD:

Knowing the present growth rate and the amount of soil moisture and the general potential for growth, some reasonable predictions of yield can be made. Some examples: Let's say that the soil is good, reasonably deep, the wind is moderate, the last measured growth rate was 25 lbs/acre/day, the grazing cycle is 30 days, the soil moisture is wet and GP (growth potential) is good. It would therefore be reasonable to figure that forage growth will continue at the same rate for the next 30 days. However, if growth potential is poor but moisture is wet, then the growth rate can be expected to be about one third, about 8 lbs/Ac/day. On the other hand if GP is good but the soil moisture is moist, then the growth rate can be expected to decline by half over the next 30 days to 12 lbs/Ac/day. To figure the yield for this 30 day cycle, take the sum of the two rates, divide by 2 and multiply by 30 days: [(25 + 12)/2] X 30 = 555 lbs/Ac. Notice that you don't plan on things getting better only what is out there, right now. If the situation improves, great. If you get an unforeseen cold snap, not so great, but you still know what you have on hand.

This procedure is admittedly crude when compared to some of the very sophisticated growth models that have been constructed. There are, however, several things in it's favor. By measuring growth rate, you are measuring a trait that has already incorporated all of the growth variables; something that has happened, a fact. By using soil moisture and growth potential you are extrapolating into the future, but because growth seldom just quits cold turkey there is some rational behind it. Also, soil moisture and growth potential are subjective measurements, which, while allowing for error also means that one can improve with practice. The brain, is in effect acting as the major computer, in that it takes a large number of inputs and decides if they are good or bad. Because it is the brain, it can learn from past experiences. It can also indulge in a lot of wishful thinking which is the major drawback of this approach. This is why, if you go for max production, you must continually take growth rate readings, to keep you on track.

QUALITY:

The faster the rate of growth, the higher the quality. How high? That depends on soil fertility, climate and your management ability. In other words, it's another one of those learning experiences. Two things are required before you can make some use of growth rate as a predictor of animal performance. First you have to have some idea of the quality of the forage and some idea of how that animal is going to respond to that quality. If you know the protein and digestible energy content you can go to the feed and animal tables and estimate animal performance. That's one way, but forage analysis takes time and often two weeks to a month pass before the results get back to you, by then the only value, other than historical, is in case you made hay or silage out of the pasture.

There is a second way, but it only has value if you're very intensive. Along with forage growth rate, animal performance must be monitored and that means weighing the animals on a regular basis. Many ranchers like to brag about their ability to eyeball weights. I ran with a cattle trader for a little while, he was good, plus or minus 25 lbs. He made his living off of ranchers that claimed they had a good eye. In short, unless you make your living by estimating weights, use a scale.

In using animal weights, use the herd average, not one or a few individual animal performances. Average daily gain (ADG) is the one to use. ADG is figured by weighting an animal at two different times and dividing the change in weight by the number of days. On a herd, first determine their average weight (W1), add up the all the weights and divide by the total number in the herd. This is done again at a later date (W2). Subtract the first average weight from the last average weight and divide this number by the difference in days between the two weighings {(W1 - W2)/ (# of days)}. Try to weigh the animals under similar conditions (same time). Now while you are getting the ADG you also need to determine the average forage growth rate during the same time period. This makes one data point; ADG versus forage growth rate. Collect enough of them over time and you have a relationship that you can use to estimate animal performance, given the pasture growth rate. A lot of numbers but with computers not all that difficult to keep track of.

Color can also be used to indicate quality. A rich emerald green indicates high quality, a dull or dark green lower quality, a yellowish tinge generally indicates poor quality and a brown forage has no quality. The shades of color will vary somewhat for different species, but will be along the above. With a little experience you can determine fairly accurately the overall health of your forage. If you take care of your forage, your forage will greatly assist you in taking care of your animals.

HANDLING THE INFORMATION:

All of the above information collected during the course of the year is the beginning of the grazing units "data base". The sampling interval may be as long as a week, however the higher the performance targets the more frequent the sampling. This sort of information is easily handled, with todays computer and spreadsheets. Table 14.2 shows one possible format.

TABLE 14.2. SPREADSHEET FORMAT FOR DETERMINATION OF YIELD AND QUALITY FUNCTIONS.

NAME OF UNIT____________________ YEAR__________

INPUT DATA

	Day of Sampling (1 Jan. = day 1)					
	45	50	54	---	---	---
Paddock #						
P1 (lb/Ac)	1100	1250	1620	-----	-----	-----
P2	1450	1550	-----	-----	-----	-----
P3	1730	820*	-----	-----	-----	-----
etc. ...						
SOIL MOISTURE	WET	WET	MOIST	-----	-----	-----
GROWTH POTENTIAL	GOOD	GOOD	POOR	-----	-----	-----
-	DATE STOCK WEIGHED					
	12	52	---	---	---	---
AVERAGE WEIGHT (lbs)	525	560	-----	-----	----	----
# of ANIMALS	30	30	---	---	---	---

* ANIMALS JUST LEFT THIS PADDOCK

In the above, samples are taken every 4 or 5 days for each paddock. In this case the date has been converted to days beginning from 1 January. Some spread sheets can determine the number of days between two different dates, some can not; in which case you have to convert the date to the number of days in the year.

There is not much to the input. Just the standing crop for each paddock, the day taken, soil moisture, growth potential and any animal data that may have been taken. How this information is used depends upon how you design the spread sheet. The output could be on a paddock by paddock basis or as averages for the entire unit. How the output is presented depends upon your requirements which will determine what calculations need to be made and the equations required. Basically the computer should determine the growth rate for each paddock, calculate the average for the entire unit and, based on soil moisture and growth potential, predict both the paddocks and the units yield at some point in the future, at least one or more grazing cycles ahead. It should also compare the prediction to the feed budget that has been filed (production targets) and let you know if prediction to the feed budget that has been filed (production targets) and let you know if you are short or long compared to what was planned. If animal data is also collected it should give you a comparison with your animal performance targets as well as determine intake per animal and a estimation as to quality of forage. Fine, I'll take two, wrap 'em up.

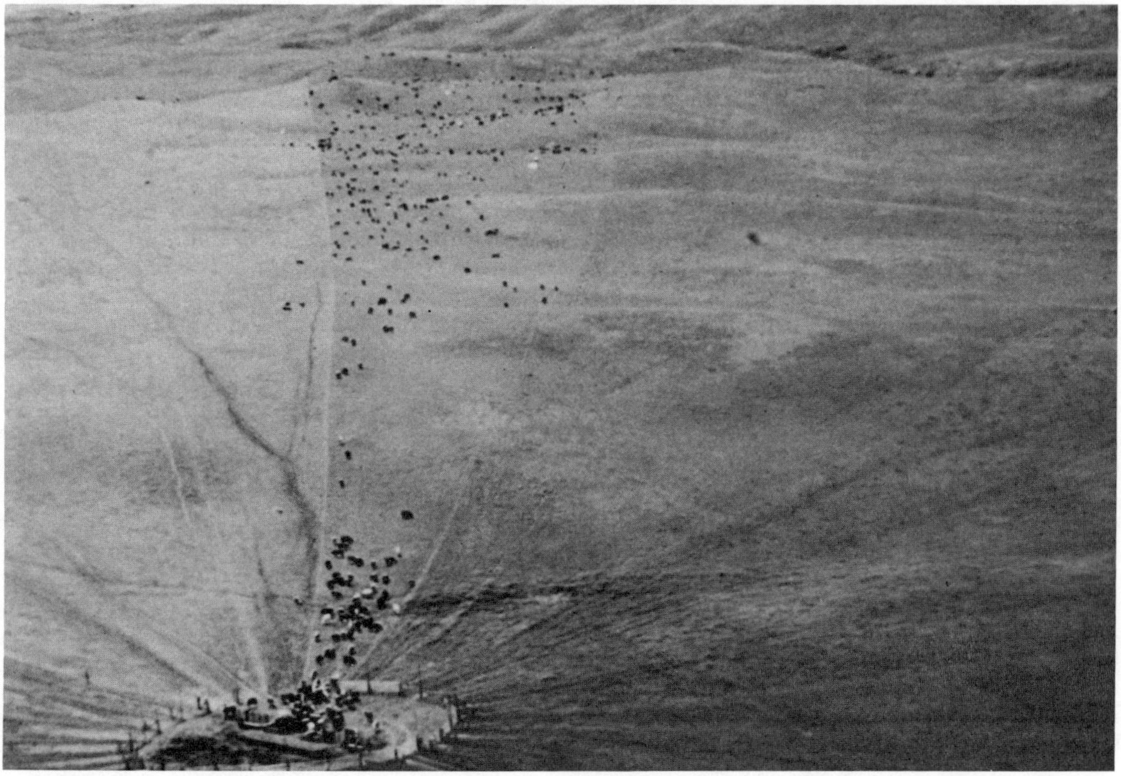

FIGURE 14.3. A well oiled, finely tuned high performance machine does not run unattended. A skilled operator is also required, one that not only knows the tools of his profession, but how to use those tools to obtain the desired production levels.

Unfortunately this sort of program is not an off the shelf item and is best set up by the user. If you know how to use a spreadsheet this is one of the many things that can be done with it. If you don't use a computer, then you must rely on setting up some form of record keeping device in tabular form, keeping track of what was done and when. Certainly not impossible, just takes more time.

AND FINALLY:

The natural world and in particular the world of man, is composed of numerous interactions that cross the boundaries of most scientific disciplines. But conversely, our whole educational process is geared towards specialization. An average Ph.D. is highly qualified within the narrow confines of his discipline, but unless he has, through background, luck, or individual perseverance, gained knowledge or experience in other fields, his ability is limited to his specialty. In agriculture, because it is an applied field, the background of the potential scientist, also plays a major role. Ten to 15 years ago, the typical agricultural scientist, while they did not own, or operate an outfit, at least grew up on a farm or ranch; that is no longer true today. Today the average graduate student was brought up in the city and has had no practical agricultural experience; the curriculum of most Colleges and Universities are so constructed as to almost guarantee that they don't get any, prior to graduation. These are the people that will be doing the research and making agricultural policy decisions in the future, decisions that will be affecting your industry. You need to be aware of this.

The U.S. beef production per cow has risen from 320 lbs to about 505 lbs in 25 years, from 1960 to 1985. Total beef cow numbers in 1985 are lower than they were in 1955, yet total beef production is up. Consumer are buying less beef and the price per liveweight pound are dropping (but not in the retail cases). The new tax laws may well stop the flow of outside money entering agriculture (while this sounds bad, on final analysis it's probably the best thing that could happen, get the speculators out of agriculture.)[8] Farmers and ranchers are going out of business at unprecedented rates. In fact one hardly hears anything good about the livestock industry anymore. There are a lot of naysayers out there whose rallying cry seems to be: "It's always darkest just before it turns black". These sayings were common during the 30's. Somehow or other the industry muddles through. A lot of your neighbors will be different, not only will they be doing different things they will very likely dress differently, look a bit more like business people. Have you been to a dairy meeting lately? Darn few overalls.

Costs, under IGM, can be decreased per unit of production. One Hawaiian ranch has reduced overall cost per pound of liveweight gain, from 60 cents in 1981 to an average of 30 cents in 1986. This is not an isolated case, as other ranches are also bringing their cost down. They have to, weaners are only worth a bit more than half of what they are on the mainland.[9] The point being: Ranching is being forced out of its lifestyle image and now has to run as a business. As in any business, management determines success or failure. Intensive Grazing Management is a viable option, it's your choice.

A word of caution. it's not necessary or even desirable to put your entire operation into an IGM program overnight. As mentioned several times, IGM is a learning experience. Your first grazing unit should be some place that is easy and fast to get to, it should also be where you and others can easily see it; mistakes and all. After you get comfortable with the process the unit can be used to train others. Your first major effort should be placed on your best land, the return will be larger and can finance development of your marginal areas.

Don't worry if you find that you're only using a part of you're land for grazing. Let the rest of it go, either forget it for the time being (hold it as reserve), or rent it out.

Well that's it for now; good luck.

---

[8] As an example of spectators, look at the commodity market. The only time money is being made is when prices are rising or falling. Consequently, there is a great deal of pressure to keep the commodity prices unstable. This works directly opposite toward securing a stable market price for producers.

[9] The Hawaii addendum describes Hawaii and its livestock industry in more detail.

Used with permission of NEA.

Paradise is in the eye of the beholder.

HAWAIIAN ADDENDUM

Why a section on Hawaii in a book on intensive grazing? The reasons are simple. In a period of five years, Hawaii has moved from producing 200 lbs of liveweight beef per acre per year to over 900 lbs, from daily gains per acre of half a pound per day to over 8 lbs per day, from average daily gains of around 1.5 to over 2.8 lbs per head, all on grass. Not everybody's doing it, but it is being done. As you will see as you read through this chapter, Hawaii has a tremendous number of different environments, there are intensive units in most of them. We are almost as close to New Zealand and Australia as we are to the U.S. mainland and there is a lot of information flowing to Hawaii from those places, as well as from the mainland. We are fortunate to be able to learn a great deal about IGM in a very short period of time and sharing is an Hawaiian tradition. Aloha.

FIGURE A.1. WIND IS NOT THE ONLY PROBLEM HAWAIIAN RANCHERS HAVE TO CONTEND WITH.

Contrary to popular belief, Hawaii is not all coconut palms, white sandy beaches, lovely hula girls and endless paradise; we have our share of Mondays. We also have the most diverse environments of any of the 50 states. We have such an ethnic mix of races that no one race is in a majority, even so we get quite a bit done; which should give the rest of the world reason to hope. Located in the middle of the Pacific Ocean and being a stopover for all major flights between East and West, we are truly the gateway to the Pacific Basin for the United States. What happens here does, in no small fashion, affect other parts of the Pacific World.

PHYSICAL:

The State of Hawaii stretches 1,523 miles across the Mid-Pacific ocean spanning 20 degrees North Latitude. If one is not too careful concerning the width of the state, Hawaii actually encompasses a larger area than Texas, however most of it is ocean. There are some 132 islands, reefs and shoals that account for a land area of 6,425 square miles. Eight major islands account for 99% of this area and the Big Island, Hawaii, is larger than all the other islands put together, 4,038 square miles. The other islands, according to size are: Maui, Oahu, Kauai, Molokai, Lanai, Niihau and Kahoolawe.

Hawaii has the widest state border in the nation, some 2,390 miles across to California. While Alaska has the claim to the western most state, Hawaii is the most southern state, even though most residents don't speak with a southern drawl. It is a long ways to anywhere, 4829 miles to the Nations capital, 3,850 to Japan and 4,500 to Australia and New Zealand. Yet, small and remote as it is, Hawaii is perhaps one of the best known places in the world, the original tropical paradise.

Hawaii sits on the north Pacific plate which is moving towards Japan at the rate of 4 inches a year. As the plate moves, it passes over a local "Hot Spot" on the earth's mantel, which is responsible for the volcanic activity that has formed the islands. Kure Island, the northern most island in the state is 25 million years old; of the major islands, Kauai, is 4.5 million and the youngest, Hawaii, is still being formed.

There are three volcanoes that thrust upwards higher than 10,000 feet above sea level, one of which, Mauna Loa (el. 13,677 ft.), is still growing. When you remember that these volcanoes began building on the sea floor, 19,000 feet below the ocean surface, the Hawaiian volcanoes are some of the most massive volcanoes in the world. Volcanism is a fact of life on the Big Island and while the eruptions are very mild, compared to Mt. Saint Helens, they none the less can cause considerable disturbance to the biological community. The recent eruptive phase of Kilauea, through the release of sulfur dioxide, caused massive defoliation of many introduced species on the leeward side; interestingly many of the endemic and native species were not affected. Sulfur dioxide also has a beneficial effect, since most of the soils in Hawaii are deficient in sulfur, consequently after every eruption, clovers and many other plants show a burst of growth not seen during other times.

The lava, which forms the basis for most of the soils in the state, is of two types; pahoehoe and aa. Pahoehoe lava has a smooth surface, often wrinkled and twisted into rope like patterns. Aa lava has a very rough surface of jagged fragments of clinkers, however underneath is a massive dense interior. While aa is more dense than pahoehoe, both are filled with small holes and cracks, which allows rainwater to rapidly seep through.

Basaltic lava is the primary parent material for soils in Hawaii however, in some areas it may be ash or cinders. The more surface area exposed (finer the material) to climate, the more rapid soil formation will take place. All of these materials have the same mineralogical composition. However, due to differences in climate, vegetation, topography and time, there have been more than 190 different soils developed from this parent material. Coral is also a parent material in Hawaii but less than 1% of the soils have developed from this source.

Most of the soils in Hawaii are coarse textured.[1] This allows for rapid infiltration of water and low water holding capacity. This course texture also means that the soils are not as heavy as they are normally thought of. The average soil in Hawaii weighs about one-half a comparable mainland soil; about 1 million pounds per acre to a depth of 6 inches (plow depth), compared to an average of 2 million pounds on the mainland. Generally, coarse textured soils are not as prone to compaction as are finer textured ones.

In general the soils in Hawaii are more highly weathered with increasing rainfall and have a greater accumulation of organic matter with increasing rainfall. However, some soils in Hawaii are highly weathered because the rainfall occurs in sufficient quantity to cause leaching or removal of all the soluble products of weathering although the average annual rainfall is relatively low and may occur in a relatively short rainy season. In addition many soils in Hawaii have positively charged colloids, especially those that are highly weathered and have a very acid pH. Thus sulfates, phosphates and nitrates are tightly held whereas calcium, sodium and other positively charged cations are leached out.

In addition to volcanism, the state regularly experiences earthquakes and tsunamis. There are two types of earthquakes, those caused by volcanic activity and those from faulting and crustal deformation. The entire Hawaiian region is tilting, with Oahu as the approximate fulcrum. Kauai is going up and the Big Island is sinking at a rate of about one foot per century. However most of the major earthquakes are the result of faulting on the ocean bottom, the most severe of which (magnitude of 8.0) occurred in 1868 near the south end of the Big Island. Since 1820 there have been 8 major tsunamis to strike the State, one of which (1946) drove water as high as 55 feet above sea level at some places.

CLIMATE:

While all of the major islands lie within the Tropic of Cancer, the climate is predominantly sub-tropical along the coast. Only along the leeward sides of the islands do the temperatures occasionally reach tropical levels. The ocean acts as a giant thermostat and the northeasterly tradewinds keep the temperatures from soaring to highs that the latitude would suggest. The highest temperature recorded was 100 degrees F (Honolulu) and the lowest was 9 degrees F at the summit of Mauna Kea (el. 13,796 ft.). The tradewinds blow about 90% of the time in the summer and about 50% during the winter. When the trades are absent, the winds are usually from the south and are often light and variable. However, storms originating south of the islands can create severe weather with areas occasionally receiving half of their annual precipitation in a hour or less. Hurricanes are uncommon in Hawaii, but they do occur often enough, there have been two within the last 23 years.

The mountains are the dominating factor in the climate of the major islands. The trades, loaded down with moisture, are forced over or around the various peaks. As the air is pushed up, it cools, and moisture is forced out, resulting in the highest rainfall occurring at elevations between 1,500 and 3,000 feet. After losing its moisture the air passes over the summits and down the other side, however because it is now drier air, the rate of heating up is greater so that the temperatures are higher on the leeward sides of the islands than on the windward.

Rainfall over the open ocean averages 25 inches per year, however over the islands it varies from a low of 7 to 8 inches along the Kawaihae coast to the wettest place in the

[1] Which is a very generous term for lava rock.

world, 463 inches per year at Mt. Waialeale on Kauai. Generally winter is the period of greatest rain on the windward sides, however elevation and topography can vary this greatly. On some leeward sides where convective showers occur during the summer months, summer is the period of greatest precipitation at elevations above 1500 feet, while in other areas rainfall is spread fairly uniformly throughout the year. The general rainfall pattern also appears to be changing due perhaps to the building of new volcanoes or a slight shift in the direction of the tradewinds; Kamuela, located on the northern end of the Big Island, receives almost 50% less rain now than at the turn of century, while Pahoa, on the south end, has noted an increase during the same period.

The tradewinds blow at an average of 15 mph, however due to passage over mountains and saddles and between islands local winds can be much stronger. North Kohala and South Point, on the Big Island, average winds near 20 mph (Figure 15.1), as does most of the extreme north and south ends of all islands. Winds at Hilo are around 10 mph, while along the Kona coast the winds are often light and variable land and sea breezes. During storms, winds on the mountain tops often exceed 100 mph, and in the Kohala mountains, while the average wind speed is 21 mph, gusts of 45-50 mph are not uncommon.

The average change in day length is 2 1/2 hours, this compares to 4 1/2 hours for southern California and 7 hours for Maine. On the windward sides it is cloudy nearly 35% of the time, and partly cloudy 40 to 50%, with clear skies the remainder; however on the lee sides it is a rare day that the sun does not shine at least for part of the day. Relative humidity is high, even in drier regions, seldom falling below 50%. Because of the latitude the sun is north of the zenith for a couple months each year. Because of this and the clarity of the Hawaiian atmosphere nearly three-fourths of the incident solar radiation penetrates to sea level on a clear day. While the ambient temperature is mild, radiation exposure is not. The coolest period is in the latter part of February and early March, and the warmest is in late August and early September; this is due to the seasonal temperature lag in the surrounding ocean.

BIOLOGICAL:

When Hawaii was re-discovered[2] by Captain Cook in 1778, the naturalist aboard the ship recorded 19 species of grass present. Later authorities estimate that there were around 65 species, today there are over 450 species of grasses present. The vegetation of Hawaii has changed dramatically. At elevations below 1,500 feet 90% of the vegetation that is presently there, was not there when Cook first set foot in Hawaii. The primary reason that so many plant species do well in the state is that somewhere there is an environmental niche for them to fit in. According to one climatic classification system (Holdridge), there are 22 different vegetational zones on just the Big Island.

This extreme variety of life zones makes for both interesting and also frustrating problems. Recommendations for one area, often have no meaning for another location only a mile or so distant. Animals that have been raised in one area often are under environmental stress when moved to another. This is particularly true when they are moved from a dry area to a wet one, or a major change in elevation.

[2] I don't know why the British felt that they were the ones to discovered a place, especially when the people that lived there, knew all the time that it existed and even had a name for it. The original people that actually discovered the place, between 400 to 700 AD, were probably from Tahiti. As far as anyone knows, they were the first, although legend has it the Menihuni's were here before that.

The relative short difference in day length often cause animals not to shed their winter hair, or long day plants to flower. However it removes seasonality from breeding herds so that sheep and deer can breed year round. The ever present wind, coupled with mist in the higher elevations, often produces a wind chill more reminiscence of the North Atlantic, rather than the Mid-Pacific. This in turn causes more energy to be expended by animals and forages alike. The result for animals is a lower growth rate, and for plants a higher fiber content and a resultant lower digestibility.

The high humidity over all parts of the state precludes long term storage of hay or other dried forages, unless special drying sheds are used. Mold and mildew are ever present and, unless material is stacked or located where the daily wind can penetrate, will quickly take control. High humidity is just as uncomfortable for cattle as it is for humans. However, artificial shade in hot humid environments, by itself does not appear to be economical, but natural shade does.

The high incidence of solar radiation is a contributing factor in numerous problems of which the most notable is pink eye in Hereford cattle. Dark colored cattle at lower elevations will absorb more radiation than will light colored ones. This will cause an increase in body temperature, which causes more stress in temperate cattle than tropical breeds. However, if the temperature is not hot at night, or if there is a cooling wind, which is usually the case in Hawaii, these animals will re-radiate heat back into the atmospheres faster than light colored ones, thus getting rid of the excess heat load. Dark colored animals at the higher elevations will generally do better than will light ones, especially if there is some shelter they can find at night. The comfort zone for most livestock is between 59 and 77 degrees Fahrenheit; significant departure from this range will cause animal stress and lowered performance.

HAWAIIAN FORAGES:

With the exception of pili grass (tangle head or spear grass), all of the major forage species that are present in Hawaii today are introductions. Most of the 65 or so species that were present when Captain Cook landed are found only in relic areas, the one exception, in addition to pili grass, is mountain lovegrass, which may still be found in some abundance at the higher and wetter elevations.

PASTURES AND ANIMAL FEEDS: by Pete Rotar

With the introduction of cattle by Captain Vancouver in 1793, the search for the perfect forage began. Over the years Hawaii has received over 500 species of grasses representing over 100 genera. Of these, 60 species are either native to Hawaii or were introduced by early Hawaiians. Of the native species Pili grass, or spear grass as it is known in Australia, is the most prominent. As mentioned earlier, the native grasses have not been the best adapted for grazing and as a result many introductions have been made by private individuals and public agencies.

The first recorded grass to be introduced was St. Augustine grass in 1816. Many of the most useful introductions have also turned out to be the worst weed in sugarcane and other cultivated lands. These include guinea grass, elephant or napier grass, California or Para grass, bermuda grass and kikuyu grass. Some of the best established and most persistent grasses introduced are also the least desirable, and in a few cases noxious weeds. Notable examples are fountain grass, broom sedge, sandbur, torpedo grass, yellow foxtail, rat tail, sourgrass and Johnson grass. Table A.1 lists some common grasses found in Hawaii and probable date of introduction.

TABLE A.1. COMMON AND SCIENTIFIC NAMES OF GRASSES AND PROBABLE DATE OF INTRODUCTION TO HAWAII:

COMMON NAME	SCIENTIFIC NAME	PROBABLE DATE OF INTRODUCTION
GRASSES - C3 - TEMPERATE:		
Fescues	Festuca sp.	1911
Orchard	Dactylis glomerata	1906
Perennial rye	Lolium perenne	1911
Rescue	Bromus unioloides	1906
Sweet vernal	Anthoxanthum odoratum	1906
Tall fescue	Festuca arundinacea	1911
Yorkshire fog (velvet)	Holcus lanatus	1906
GRASSES - C4 - TROPICAL:		
Bermuda grass	Cynodon dactylon	1835
Broom sedge	Andropogon virginicus	1932
Buffel	Cenchrus ciliaris	1867
California or Para	Brachiaria mutica	recent
Carpet	Axonopus compressus	1912
Fountain	Pennisetum ruppelii	1926
Green panic	P. maximum v. trichoglume	recent
Guinea	Panicum maximum	1880
Hilo	Paspalum conjugatum	1840
Johnson	Sorghum halepense	1906
Kikuyu	Pennisetum clandestinum	1924
Limpo	Hemarthia altissima	recent
Molasses	Melinis minutiflora	1913
Napier or elephant	Pennisetum purpureum	1912
Pangola	Digitaria decumbens	recent
Phalaris	Phalaris aguatica	recent
Pili	Heteropogon contortus	native
Pitted beard	Andropogon pertusus	1916
Natal redtop	Rhyncheletrum repens	1895
Rat tail	Sporobolus africanus	1906
Rice	Paspalum orbiculare	Hawaiians
Rhodes	Chloris gayana	1906
Sand bur	Cenchrus echinatus	1867
Signal	Brachriaria decumbens	recent
Sour	Trichachne insularis	1906
Star	Cynodon sp.	recent
Sugarcane	Saccharum officinarum	Hawaiians
Torpedo	Panicum repens	1906
Yellow foxtail	Setaria geniculata	1851

Few, if any of the native legumes are useful as pasture plants and many introductions have been made; some with notable success and others failing through lack of understanding of the need for the proper Rhizobium innoculant. It is estimated that several hundred species of forage and browse legumes have been introduced over the years in an attempt to find well adapted forage types. Among the most useful are the white clovers, big trefoil, medics, bur clover, narrow leaf vetch, Kiawe, ekoa or haole koa (Leucaena), greenleaf desmodium, kaimi clover, red clover, perennial glycine, and centro. A number of introductions have turned out to be a pest and are of little value. Among these are gorse,[3] various members of the Crotalaria family (rattlepods), klu, Japanese tea, creeping indigo and coffee senna.

Since most pastures depend upon rainfall, moisture is a major limiting factor on the leeward sides of the islands. Rainfall over much of this area is seasonal, with the majority occurring during the winter months, November to March. The coastal and leeward areas receive the highest solar radiation and with irrigation are potentially the most productive regions in the state. As the elevation increases, temperatures decrease and pasture types change from tropical to subtropical to temperate. In the wetter regions on the windward sides, heavy cloud cover reduces solar radiation and consequently plant growth.

Low temperatures limit forage production during the winter period, when most plant growth slows down, or for some species, virtually stops. For example, when soil temperature at 3 inches depth drop to 68 degrees F, growth ceases for such grasses as sugarcane, kikuyu and pangola. As might be expected, low temperature effects become more pronounced as the elevation increases. The greater number of cloudy days during the rainy season also accounts for seasonal reductions in growth. As an example of this seasonal variation, the growth rate of irrigated pangola, is on the order of 22 lbs dry matter per acre per day during the winter, but increased to 169 lbs/acre/day during the summer. Many ranches are stocked according to the period of lowest forage growth and are often understocked for the rest of the year. Summer feed supplies must be supplemented with stored forages or concentrates in order to effectively utilize the available forage during other seasons.

The fertility of most Hawaiian soils seriously limits crop production. Of the elements essential for plant growth, nitrogen is the most deficient in our soils. The increase in dry matter yields due to nitrogen application is spectacular. Pangola grass yields in unfertilized plots were on the order of 4 tons/acre/year, but when nitrogen was applied increased to almost 20 tons/acre/year. With current shortages and high fertilizer cost, the use of legumes is important. Intense efforts have been made to find suitable legumes for our diverse pasture situations, but as in most tropical areas, has not been very successful. Problems of establishment, nutrition, competition and persistence in association with aggressive grasses and weeds, as well as problems of grazing management, have caused concern about the role of the tropical legume (See Chapter 5).

Phosphorus is a major deficiency in all our highly weathered soils, only the ash soils have anywhere sufficient amounts. These soils tend to "fix" phosphorus in forms that make it largely unavailable to plants. Such deficiencies limit forage production because of poor root development and consequent poor plant nutrition. Phosphorus deficiency also limits legume establishment and growth and is the major limiting factor in legume production.

---

[3] Gorse was introduce as a living fence and has since taken over large areas on Mauna Kea and Haleakala on the Big Island and Maui respectively.

Many soils in the wetter regions are acid in their natural state and may be further acidified by continued use of acid-forming fertilizers without concurrent liming. As marginal pineapple lands are abandoned, many of these areas are converted to pasture. These lands have severe management problems due to aluminum and manganese toxicity, calcium deficiency (pH's often below 4.0), and poor availability of other nutrients such as molybdenum in these very acid soils. Other deficiencies often noted in addition to calcium and molybdenum, include potassium, magnesium and occasionally iron and sulfur. Abandoned sugarcane lands also have their share of problems, the most pronounced of which is their acidity and general run down condition.

Productivity of many Hawaiian ranches is greatly reduced as a result of competition from weedy shrubs and trees. Biological control has been employed with moderate to good success to control several weeds, including prickly pear, lantana, Maui pamakani and emex. Chemicals offer control but extract a stiff application price. Which, unless the ability exist to make use of the resulting increase in forage production, is seldom worth the cost.

Insect pests have severely damaged pastures in recent years. The hunting billbug has caused serious injury to kikuyu grass. The grass webworm has also severely injured kikuyu, pangola and bermuda grass pastures. The army worm occasionally injures kikuyu pastures on the Big Island. Recently a new invader, the psyllid, has caused wide spread losses to ekoa. Predators of this pest have been released, but it is too early to determine the long term effects.

Feed grains can and have been grown in Hawaii, however seldom economically. Hawaii's high relative humidity makes the cost of drying prohibitive. However, forage conservation in the form of hay or silage appears to offer the possibility of economic returns. Several firms are considering producing alfalfa hay on a large scale, and others are looking into silage as a complement to their cattle and dairy operations.

Local by-product feeds or supplements have been used in the past, and will continue to be used on the states ranches. Sugarcane tops, molasses, bagasse and strippings have been utilized with various degrees of success. Pineapple bran is also an important source of feed. Pineapple stubble remaining after the last fruit harvest is also a source of readily digestible starch. However, the future of both of these products for use by the cattle industry is in doubt. There is a slow but steady shift towards marketing fresh fruit rather than canned, thus reducing the prospects of utilizing pineapple bran. The future of harvest stubble is also in doubt due to the possible contamination with various pesticides, most notably PCB's.

RECOMMENDED FORAGES FOR HAWAII: by Sheldon Whitney

Tables A.2 and A.3 list the requirements of pH, temperature and minimum rainfall, overall forage quality and seeding rates, recommendations of various grasses and legumes. To establish a good stand requires proper land preparation and necessary liming and/or fertilization. In most cases, the crop should be mowed a month or so after planting to control weeds. The method pioneered by one ranch to establish rye grass and clover in a kikuyu grass pastures is as follows: On the first pass roto-till with three or four chisel plows mounted behind, this is to bring soil up on top of the mulch. After two or so weeks the second pass is made with a roto-tiller, seeder and culti-packer (all mounted together). Planting by first spreading the seed and followed by using the treading action of animals to actually plant the seed in the soil, requires a higher seeding rate, application of an insecticide and the ability to place a large number (high density) of animals on the area for a short period of time, less than a day. The use of an insecticide is of particular importance in tropical regions.

TABLE A.2. RECOMMENDATIONS OF VARIOUS GRASSES FOR HAWAII.

<table>
<tr><th>GRASS</th><th>AVERAGE TEMPERATURE</th><th>pH</th><th>QUALITY</th><th>SEEDING RATE (lb)</th><th>MINIMUM RAINFALL (inches)</th></tr>
<tr><td>kikuyu</td><td>med</td><td>mod</td><td>med</td><td>veg</td><td>30</td></tr>
<tr><td>pangola</td><td>warm/med</td><td>low</td><td>med</td><td>veg</td><td>30</td></tr>
<tr><td>buffel</td><td></td><td></td><td></td><td></td><td></td></tr>
<tr><td>Biloela</td><td>warm</td><td>mod</td><td>med</td><td>4</td><td>12</td></tr>
<tr><td>Molopo</td><td>warm/med</td><td>mod</td><td>med</td><td>4</td><td>12</td></tr>
<tr><td>Texas A&M</td><td>warm/med</td><td>mod</td><td>med</td><td>4</td><td>12</td></tr>
<tr><td>green panic</td><td>warm/med</td><td>mod</td><td>med/hi</td><td>5</td><td>22</td></tr>
<tr><td>guinea</td><td></td><td></td><td></td><td></td><td></td></tr>
<tr><td>common, |</td><td></td><td></td><td></td><td></td><td></td></tr>
<tr><td>African,|---></td><td>warm</td><td>mod</td><td>med/hi</td><td>5</td><td>30</td></tr>
<tr><td>Coloniao|</td><td></td><td></td><td></td><td></td><td></td></tr>
<tr><td>Makueni|</td><td></td><td></td><td></td><td></td><td></td></tr>
<tr><td>Gatton |---></td><td>warm/med</td><td>mod</td><td>med/hi</td><td>5</td><td>30</td></tr>
<tr><td>limpo</td><td>warm/med</td><td>low</td><td>med</td><td>veg</td><td>30</td></tr>
<tr><td>signal</td><td>warm/med</td><td>low</td><td>med</td><td>5</td><td>30</td></tr>
<tr><td>star</td><td>warm/med</td><td>mod</td><td>med</td><td>veg</td><td>25</td></tr>
<tr><td>rhodes</td><td>warm/med</td><td>mod</td><td>med</td><td>5</td><td>25</td></tr>
<tr><td>yorkshire fog</td><td>med/cool</td><td>mod</td><td>med</td><td>3</td><td>30</td></tr>
<tr><td>phalaris</td><td>med/cool</td><td>mod</td><td>med</td><td>6</td><td>30</td></tr>
<tr><td>orchard</td><td>med/cool</td><td>mod</td><td>high</td><td>6</td><td>30</td></tr>
<tr><td>tall fescue</td><td>med/cool</td><td>mod</td><td>med</td><td>6</td><td>30</td></tr>
<tr><td>rye</td><td>med/cool</td><td>mod</td><td>high</td><td>6</td><td>30</td></tr>
</table>

Temperatures in degrees F: warm = 55 to 104; medium = 46 to 95; cool = 32 to 86.
pH: low = 4.5; moderate = 5.2; high = 6.0.

TABLE A.3. RECOMMENDATIONS OF VARIOUS LEGUMES FOR HAWAII.

LEGUMES	AVERAGE TEMPERATURE	pH	QUALITY	SEEDING RATE (lb)	MINIMUM RAINFALL (inches)
ekoa	warm	high	med/hi	4	30
glycine					
Tinaroo,|					
Malawi |--->	warm/med	high	med/hi	4	30
Cooper, |					
Clarence |-->	warm	high	med/hi	4	30
stylo					
Cook	warm/med	low	med	4	35
Endeavor |					
Schofield|-->	warm	low	med	4	35
Verano	warm	mod	med	4	24
siratro	warm/med	low	med/hi	3	30
centro					
Common	warm/med	mod	med/hi	4	40
CIAT 5065	warm/med	low	med	5	35
greenleaf	med/cool	mod	med/hi	3	35
kaimi	warm/med	mod	med	4	30
white clovers	med/cool	mod	high	4	35
trefoils	med/cool	mod	high	3	35

Temperatures in degrees F: warm = 55 to 104; medium = 46 to 95; cool = 32 to 86.
pH: low = 4.5; moderate = 5.2; high = 6.0.

INTENSIVE GRAZING IN HAWAII:

Hawaii has about 400 ranches with more than 20 head of cattle; also about an equal number with less than 20 head. Over half of the States ranches have adopted some form of intensive grazing management in their operations. Grazing units are being used in various environments, with different animals (cattle, sheep, etc.) and with different goals, (anywhere from red meat and milk production to utilizing forage under macadamia nut trees).

FIGURE A.2. If you raise cattle on one island but your market is on another, a question arises. How do you get your product to the market? Since surfing is so common, swimming seems like a logical choice. However, not all animals were in agreement, some persuasion was usually required.

Over 25,000 acres of land have been placed into intensive programs; this translates into around 3,000 miles of installed electric fence. Approximately 50% of all fencing sales in the State are now of electric fence material. Virtually, every one of the larger ranches that have one grazing unit are planning more. In addition over 300 persons associated with the livestock industry have completed 24 hours, or more, training in intensive grazing management.

FIGURE A.3. Once the animals were driven into the water, they swam with some assistance, to a waiting long boat and securely tied to the gunwale; three to a side.

All this is placed in perspective when it is realized that the first permanent power electric fence was placed in operation at Parker Ranch on January, 1981, and the first grazing cell became operational at Huehue Ranch on January, 1982.

Animal performance data has been collected from several ranches that are into intensive management. All ranches have reported increased carrying capacities of twofold or better over traditional practices. Kahua Ranch, running steers on straight Kikuyu grass pasture, produced 715 lbs (liveweight) in 1982-83; this was increased to nearly 1000 lbs in 1985-86 on a mix of Kikuyu, ryegrass and clover. Kaupo Ranch in 1984-85 produced 726 lbs/acre (also steers) in an guinea/ekoa unit on 22 inches of rainfall; an unofficial world record for a commercial ranch. Both of these ranches were producing 200 to 250 lbs/acre/year under traditional management. An economic survey conducted at Kahua On their first unit (1983) found that net returns to land, management and capital increased 10 fold over traditional management. The cost of development was paid back, based on value added to the animals, within 9 months after start-up.

FIGURE A.4. Upon a signal, a donkey engine on board the transport towed the longboat alongside. A sling was passed under the animal and the halter rope passed to a seaman on board.

The State's sugar and pineapple industries are facing an uncertain future. It has always been assumed that should these lands become available, livestock production would occur by default. With the technologies and methodologies presently available and being demonstrated every day in the State, livestock production may well be the option of choice. What applies to red meat production, applies equally well to milk production. Grass can be used to replace a large portion of high cost grain, in fact, excellent milk yields have been obtained from all grass pastures in other tropical areas of the world. The one commodity that Hawaii grows better than anywhere else in the United States is grass.

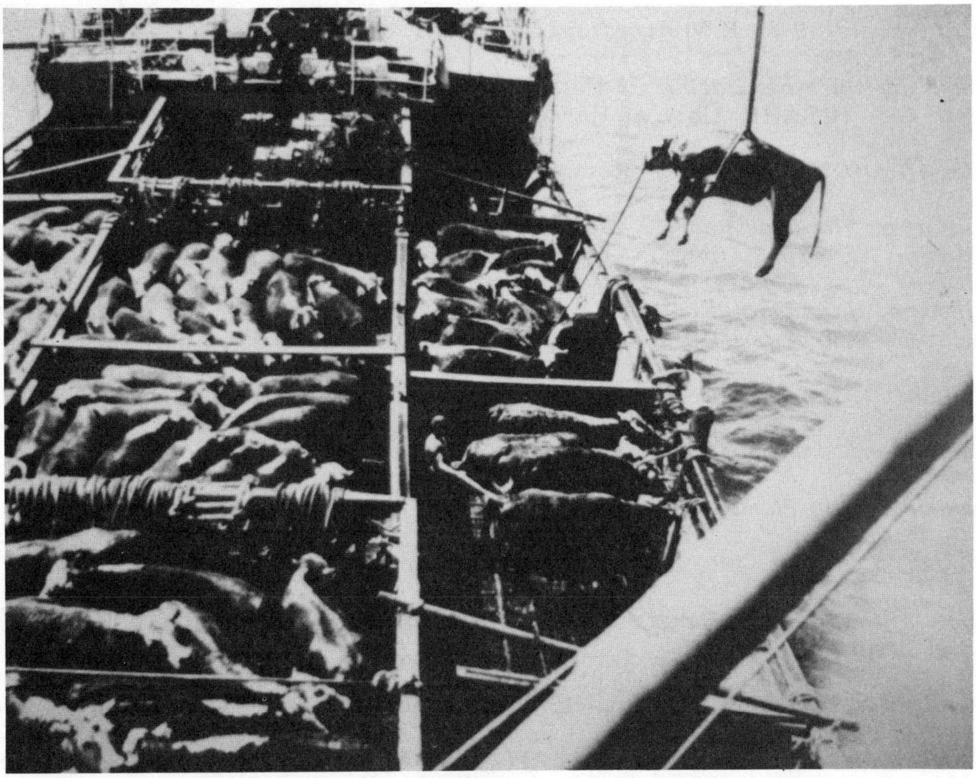

FIGURE A.5. The animal was hoisted aboard, tied up and the salt water washed off. This method of transportation was used until 1949. Today the animals are loaded into a special livestock container at the ranch, hauled to the dock and loaded on the barge. The trip from Big Island to Oahu takes about 24 hours. The animals shrink between 11 to 13% and require 20 to30 days to gain it back. Nobody really knows how much shrinkage there was in the "good old days".

The potential of Hawaii to produce beef on existing grazing land has barely been touched. As the ranchers of this State gain experience in intensive grazing management, and bring more of their operation under this program, the State's beef herd will see a dramatic increase in size and could well double its numbers in a few short years. Today, Hawaii produces only 30% of the beef consumed in the State; tomorrow, there is no biological reason why Hawaii can not become a beef exporter.

SUGGESTED READING:

Alfred W. Carter, Hawaii's Dean of Cattlemen and Notes on Hawaiian Livestock. 1971. L. Brundage. Cathay Press Ltd.

Atlas of Hawaii. 1973. R.W. Armstrong, Ed. Univ. of Hawaii Press.

Economics of Intensive Grazing: A Case in Hawaii. P.S. Leung and B.J. Smith. 1984. Res. Ext. Ser. 045. HITAHR, Univ. of Hawaii.

Grasses of Hawaii. 1968 Peter P. Rotar. Univ. of Hawaii Press.

Grasses of the Hawaiian Ranges. 1939. L.D. Whitney, E.Y. Hosaka and J.C. Ripperton. Bull. 82. Hawaii Ag. Exp. Station.

Handbook of Hawaiian Weeds. 1976. Ed. E.L. Haselwood & G.G. Motter. Lyon Arboretum Association. 3860 Manoa Rd. Honolulu, HI. 96822

Hawaii and Its People. 1933. L.H. Jarrett. Honolulu Star-Bulletin, Ltd.

Hawaiian Land Mammals. 1971. R.J. Kramer. Charles E. Tuttle Co.

Hawaii's Poisonous Plants. 1979. Roger E. Baldwin. Petroglyph Press.

In Gardens of Hawaii. 1965. Marie C. Neal. Bishop Museum Press.

Legumes of the Hawaiian Ranges. 1944. E.Y. Hosaka and J.C. Ripperton. Bull. 93. Hawaii Ag. Exp. Station.

Mammals in Hawaii. 1969. P.Q. Tomich. Bishop Museum Press.

Statistics of Hawaiian Agriculture. 1984. Hawaiian Agricultural Reporting Service. Hawaii Dept. of Agriculture.

Used with permission of NEA.

Some things bear closer scrutiny.

APPENDIXES

APPENDIX 1. FORAGE NOMENCLATURE

TABLE A1.1. COOMMON AND SCIENTIFIC NAMES OF SOME TEMPERATE (C_3 - COOL SEASON) GRASSES:

COMMON NAME	SCIENTIFIC NAME
Barley	Hordeum vulgare
Brome grass	
California	Bromus carinatus
Cheat grass (Downy)	B. tectorum
Chess	B. secalinus
Meadow	B. bilbersteinii
ountain	B. marginatus
Prairie (Rescuegrass)	B. catharticus
Ripgut	B. rigidus
Smooth	B. inermis
Bluegrass	
Annual	Poa annula
Big	P. ampla
Canada	P. compressa
Kentucky	P. pratensis
Nevada	P. nevadensis
Sandberg	P. secunda
Texas	P. arachnifera
Winter	P. bulbosa
Canarygrass	Pharis canariensis
Reed	P. arundinacea
Fescues	Festuca sp.
Chewings (Red)	F. rubra
Idaho (Bluebunch)	F. idahoensis
Meadow	F. elatior
Tall fescue	F. arundinacea
Sheep (Hard)	F. ovina
Sixweeks	F. octoflora
June grass	Koeleria cristata
Needle grass	
California	Stipa californica
Needle and Thread	S. comata
Sleepygrass	S. robusta
Oats	Avena sativa
Orchard	Dactylis glomerata
Ricegrass	
Bloomer	Oryzopis bloomeri
Indian	O. hymenoides

Continue on next page

TABLE A1.1. COMMON AND SCIENTIFIC NAMES OF SOME TEMPERATE (C_3 - COOL SEASON) GRASSES: -- Con.

COMMON NAME	SCIENTIFIC NAME
Rye grass	
Annual (Italian)	Lolium multiflorum
Darnel	L. temulentum
Perennial	L. multiflorum
Hardy	L. remotum
Swiss	L. rifidum
Sweet vernal	Anthoxanthum odoratum
Wheat	Triticum aestivum
Wheatgrass	
Blue bunch	Agropyron spicatum
Crested	A. desertorum
Intermediate	A. intermedium
Quackgrass	A. repens
Siberian	A. sibiricum
Tall	A. elongatum
Western	A. smithii
Wildrye	
Blue	Elymus glaucus
Canada	E. canadensis
Common (dune, beach)	E. mollis
Great Basin	E. cinereus
Giant	E. condensatus
Virginia	E. virginicus
Yorkshire fog (velvet)	Holcus lanatus

TABLE A1.2. COMMON AND SCIENTIFIC NAMES OF SOME TROPICAL (C_4 - WARM SEASON) GRASSES

COMMON NAME	SCIENTIFIC NAME
Alkali Sacaton	Sporobolus wrightii
Arizona cottontop	Trichachne californica
Bahia grass	Paspalum notatum
Bermuda grass	Cynodon dactylon
Bluestems	
Big bluestem	Andropogon gerardi
Broomsedge	A. virginicus
Little	A. scoparius
Bluestems	
Sand	A. hallii
Slender	A. tener
Buffalograss	Buchloe dactyloides
Buffel	Cenchrus ciliaris
Paragrass (California)	Brachiaria mutica
Carpet grass	
Common	Axonopus compressus
Tropical	A. compressus
Congograss	Imperata cylindrica
Cordgrass	
California (Spike)	Spartina leiantha
Prairie (Ripgut)	S. pectinata
Corn (Maize)	Zea mays
Dallasgrass	Paspalum dilatatum
Dropseed	
Blue (Pineywoods)	Sporobolus gracilis
Giant	S. giganteus
Meadow	S. asper
Mesa	S. flexuosus
Prairie	S. heterolepis
Sand	S. cryptandrus
Elephant grass (Napier)	Pennisetum purpureum
Fingergrass	
Rhodes grass	Chloris gayana
Swollen	C. inflata
Fountain	Pennisetum ruppelii
Grama	
Black	Bouteloua eriopoda
Blue	B. gracilis
Sideoats	B. curtipendula

Continue on next page

TABLE A1.2. COMMON AND SCIENTIFIC NAMES OF SOME TROPICAL (C_4 - WARM SEASON) GRASSES -- Con.

COMMON NAME	SCIENTIFIC NAME
Green panic	Panicum maximum var. trichoglume
Guinea	Panicum maximum
Hilo	Paspalum conjugatum
Indian grass	Sorghastrum nutans
Jaragua	Hyparrhenia rufa
Johnson	Sorghum halepense
Kikuyu	Pennisetum clandestinum
Kleingrass	Panicum coloratum
Limpo	Hemarthia altissima
Lovegrass	
Boer	Eragrostis chloromelas
Lehmann	E. lehmanniana
Sand	E. trichodes
Weeping	E. Curvula
Maidencane	Panicum hemitomon
Millet	Setaria faberii
Molasses	Melinis minutiflora
Natal (Redtop)	Tricholaena rosea
Pangola	Digitaria decumbens
Panic grass (blue)	Panicum antidotale
Pearlmillet	Pennisetum typhoides
Pineland three awn	Aristida stricta
Phalaris	Phalaris aguatica
Pitted beard	Andropogon pertusus
Poverty grass	Danthonia spicata
Rat tail	Sporobolus africanus
Rice	Paspalum orbiculare
Saint augustine	Stenotaphrum secundatum
Sand bur	Cenchrus echinatus
Sea oats	Uniola paniculata
Signal	Brachriaria decumbens
Sour grass	Trichachne insularis
Spear (Tangle head, Pili)	Heteropogon contortus
Stargrasses	Cynodon sp.
Sudan grass (Sorghum)	Sorghum bicolor
Sugarcane	Saccharum officinarum
Switchgrass	Panicum virgatum
Torpedo	Panicum repens
Yellow foxtail	Setaria geniculata
Vasseygrass	Paspalum urvillei

TABLE A1.3. COMMON AND SCIENTIFIC NAMES OF SOME TEMPERATE, SUB-TROPICAL AND TROPICAL LEGUMES

COMMON NAME	SCIENTIFIC NAME	ENVIRONMENT*
Alfalfa (Lucerne)	Medicago sativa	TEM
Alyceclover	Alysicarpus vaginalis	TEM
Beans	Phaseolus sp.	ALL
Beggarweeds	Desmodium sp.	SUB
Black medic	Medicago lupulina	SUB
Bur clover	Medicago hispida	TEM
Centro	Centrosema pubescens	TRO
Chickpea (Garbanzo)	Cicer arietinum	SUB
Clovers		
Alsike	Trifolium hybridum	TEM
Arrowleaf	T. vesiculosum	TEM
Crimson	T. incarnatum	TEM
Kenya	T. simipilosum	SUB
Persian	T. arvense	TEM
Rabbitfoot	T. medium	TEM
Red	T. pratense	TEM
Strawberry	T. striatum	TEM
Subterranean	T. subterraneum	SUB
White (Dutch, Ladino)	T. repens	TEM
Creeping indigo	Indigofera endecaphylla	SUB
Leucanena (Ekoa)	Leucaena leucocephala	SUB
Gorse	Ulex europaeus	TEM
Greenleaf desmodium	Desmodium intortum	SUB
Hilahila (sensitive)	Mimosa pudica	TRO
Hetero	Desmodium heterophyllum	TRO
Hyacinth bean (Rongai)	Dolichos lablab	TRO
Indigo	Indigofera suffruticosa	SUB
Kaimi clover	Desmodium canum	SUB
Mesquite		
Algaroba	Prosopis pallida	SUB
Kiawe	P. chilensis	SUB
Velvet	P. velutina	SUB
Koa	Acacia koa	SUB
Lespedeza		
Bicolor	Lespedeza	TEM
Common	L. striata	TEM
Korean	L. stipulacea	TEM
Sericea	L. cuneata	TEM

Continue on next page

TABLE A1.3. COMMON AND SCIENTIFIC NAMES OF SOME TEMPERATE, SUB-TROPICAL AND TROPICAL LEGUMES -- Con.

COMMON NAME	SCIENTIFIC NAME	ENVIRONMENT*
Japanese tea	Cassia leschenaultiana	SUB
Lupine		
Blue	Lupinus angustifolius	TEM
Texas (Bluebonnet)	L. subcarnosus	TEM
White	L. albus	TEM
Peanut	Arachis hypogaea	SUB
Pigeonpea	Cajanus cajan	TRO
Rattle pods	Crotalaria sp.	SUB
Red clover	Trifolium pratense	TEM
Siratro	Macroptilium atropurpureus	TRO
Spanish clover	Desmodium sandwicense	SUB
Stylo's	Stylosanthes sp.	TRO
Sweetclovers		
Annual yellow	Melilotus indica	TEM
White	M. alba	TEM
Yellow	M. officinalis	TEM
Tinaroo glycine	Neonotonia wightii	SUB
Trefoils		
Big	Lotus pedunculatua	TEM
Birdsfoot	L. corniculatus	TEM
Tropical kudzu	Pueraria phaseoloides	TRO
Vetches		
Common (Narrowleaf)	Vica sativa	TEM
Hairy	V. villosa	TEM
Horsebean	V. faba	TEM
Showy	V. grandiflora	TEM

* TEM = TEMPERATE; TRO = TROPICAL; SUB = SUB-TROPICAL.
These environmental designations do not mean that the plant only grows in the environment shown, it simply means that you are more likely to find it there.

APPENDIX 2. SUGGESTED INTAKE LEVELS FOR THE MAJOR MINERALS

Tables adapted from: "The Mineral Requirements of Grazing Ruminants".

TABLE A2.1. CALCIUM (Ca): REQUIREMENTS FOR CATTLE AND SHEEP.

Sheep	weight lb	weight Kg	gain lbs/d	gain g/day	daily g Ca/day	% Ca in diet(DM)
lamb	22	10	0.33	150	1.9	0.38
	44	20	0.33	150	2.9	0.29
	88	40	0.17	75	2.2	0.22
			0.33	150	3.4	0.26
			0.66	300	5.8	0.29
adult ewe 121 lbs (55Kg)						
maintenance					1.3	0.16
late pregnancy			single		3.1	0.23
			twins		4.9	0.28
lactation		2.2 lbs (1 Kg) milk/day			3.7	0.17
		4.4 lbs (2 Kg) milk/day			6.0	0.21

Cattle	weight lb	weight Kg	gain lbs/d	gain g/day	daily g Ca/day	% Ca in diet(DM)
Calf	220	100	1.1	0.5	13	0.50
			2.2	1.0	23	0.66
	440	200	1.1	0.5	15	0.34
			2.2	1.0	25	0.44
	660	300	1.1	0.5	17	0.29
			2.2	1.0	28	0.35
Beef	992	450				
maintenance			-	-	12	0.19
pregnancy			mid third			
" "			last third		15	0.19
lactating (ave.)		11 lbs (0.5Kg) milk/day			26	0.28
"		22 lbs (10Kg) milk/day			45	0.40
Dairy	838	380				
maintenance			-	-	9	0.22
late pregnancy			7 months		14	0.31
" "			last month		20	0.36
lactating		22 lbs (10 Kg) milk/day			27	0.30
"		44 lbs (20 Kg) "			44	0.30
"		66 lbs (30 Kg) "			62	0.35

Calcium requirements are met by pastures continuing 0.29 to 0.44% Ca on a Dry Matter (DM) basis.

TABLE A2.2. PHOSPHORUS (P): REQUIREMENTS FOR CATTLE AND SHEEP

Sheep	weight lb	Kg		gain lbs/d	g/day	daily g Ca/day	% P in diet(DM)
lamp	22	10		0.33	150	1.4	0.16
	44	20		0.33	150	1.6	0.16
	88	40		0.17	75	1.7	0.15
				0.33	150	2.4	0.14
				0.66	300	3.9	0.16
adult ewe 121 lbs (55Kg)							
maintenance						1.3	0.13
late pregnancy			single			2.2	0.16
			twins			3.0	0.18
lactation			2.2 lbs (1 Kg) milk/day			3.5	0.18
			4.4 lbs (2 Kg) milk/day			5.6	0.20

Cattle	weight lb	Kg		gain lbs/d	g/day	daily g Ca/day	% P in diet(DM)
Calf	220	100		1.1	0.5	6.7	0.25
				2.2	1.0	11.8	0.35
	440	200		1.1	0.5	8.2	0.19
				2.2	1.0	13.3	0.23
	660	300		1.1	0.5	13.1	0.23
				2.2	1.0	20.0	0.26
Beef	992	450					
maintenance				-	-	12	0.19
pregnancy			mid third			12	0.19
" "			last third			15	0.19
lactating (ave.)			11 lbs (0.5Kg) milk/day			26	0.28
"			22 lbs (10Kg) milk/day			42	0.37
Dairy	838	380					
maintenance				-	-	7.9	0.18
late pregnancy			7 months			11.0	0.22
" "			last month			16.0	0.30
lactating			22 lbs (10 Kg) milk/day			24.0	0.27
"			44 lbs (20 Kg) "			40.0	0.29
"			66 lbs (30 Kg) "			55.0	0.30

POTASSIUM (K): The requirements for K of ruminant livestock varies from 0.2 to 0.6% of dry matter (DM) diet intake. Most green forage has a concentration of K between 2.0 to 4.0%. Under almost any normal management program a deficiency of K would not be possible. If animals are forced to eat dead standing material that has had excessive weathering, there could be a possibility of deficiency.

High levels of K in the diet does effect the absorption of magnesium (Mg). The greatest effects are when the Mg diet levels are low. Since the Mg requirements of a cow increases 3 to 4 fold at the onset of lactation, high levels of K at this time could lead to grass tetany. High fertilizer levels of K (potash) has been noted to depress the pasture levels of Mg and sodium (Na). The addition of K to pastures that have a low Na level (0.07 to 0.1% DM) could reduce the daily Na intake levels below required levels; lactating cows are at greatest risk. High levels of K (140 lbs/acre) should be avoided unless clear evidence exist that it is necessary for pasture production.

TABLE A2.3. SODIUM (Na): REQUIREMENTS FOR CATTLE AND SHEEP

Sheep	weight lb	weight Kg	gain lbs/d	gain g/day	daily g Ca/day	% Na in diet(DM)
lamp	22	10	0.33	150	0.27	0.029
	44	20	0.33	150	0.34	0.030
	88	40	0.17	75	0.37	0.034
			0.33	150	0.47	0.031
			0.66	300	0.67	0.029
adult ewe 121 lbs (55Kg)						
maintenance					0.37	0.044
late pregnancy		single			0.60	0.050
		twins			0.77	0.055
lactation		2.2 lbs (1 Kg) milk/day			0.85	0.046
		4.4 lbs (2 Kg) milk/day			1.33	0.053

Cattle	weight lb	weight Kg	gain lbs/d	gain g/day	daily g Ca/day	% Na in diet(DM)
Calf	220	100	1.1	0.5	1.6	0.25
			2.2	1.0	2.5	0.35
	440	200	1.1	0.5	2.3	0.19
			2.2	1.0	3.2	0.23
	660	300	1.1	0.5	3.0	0.23
			2.2	1.0	3.9	0.26
Beef	992	450	-	-	6.0	1.103
lactating			22 lbs (10Kg) milk/day		9.5	0.111
Dairy	838	380				
maintenance			-	-	2.7	0.066
late pregnancy			last month		5.1	0.087
lactating			22 lbs (10 Kg) milk/day		9.1	0.112
"			44 lbs (20 Kg) "		15.0	0.114
"			66 lbs (30 Kg) "		21.3	0.125

TABLE A2.4. MAGNESIUM (MG): REQUIREMENTS FOR CATTLE AND SHEEP

MAGNESIUM (MG):

Sheep	weight lb	weight Kg	gain lbs/d	gain g/day	daily g Ca/day	% Mg in diet(DM)
lamp	22	10	0.33	150	0.3	0.06
	44	20	0.33	150	0.7	0.10
	88	40	0.17	75	0.9	0.08
			0.33	150	1.1	0.09
			0.66	300	1.4	0.09
adult ewe 121 lbs (55Kg)						
maintenance					1.0	0.12
late pregnancy			single		1.3	0.11
			twins		1.5	0.12
lactation	2.2 lbs (1 Kg) milk/day				2.0	0.10
	4.4 lbs (2 Kg) milk/day				2.9	0.12

Cattle	weight lb	weight Kg	gain lbs/d	gain g/day	daily g Ca/day	% Mg in diet(DM)
Calf	220	100	1.1	0.5	3.0	0.10
			2.2	1.0	4.2	0.12
	440	200	1.1	0.5	4.9	0.12
			2.2	1.0	6.2	0.12
	660	300	1.1	0.5	6.6	0.12
			2.2	1.0	7.9	0.12
Beef	992	450				
lactating		22 lbs (10Kg) milk/day			16.0	0.19
Dairy	838	380				
maintenance			-	-	6.7	0.15
late pregnancy		7 months			7.4	0.14
" "		last month			8.6	0.15
lactating		22 lbs (10 Kg) milk/day			14.0	0.16
"		44 lbs (20 Kg) "			21.0	0.16
"		66 lbs (30 Kg) "			28.0	0.16

APPENDIX 3. SUGGESTED INTAKE LEVELS FOR THE TRACE MINERALS

There are 15 known trace elements that are essential for animal performance, including fertility. Of these, seven have been observed to occur in nature in such small amounts as to cause reproductive problems, they are: Zinc, Copper, Manganese, Iodine, Selenium, Cobalt and Iron. The following table gives the suggested levels, in parts per million (ppm) for a beef breeding herd; these may need to be adjusted depending on soil and forage conditions and age of animal. Most of the trace minerals are adequately supplied by green growing forages. Some areas are deficient in the trace elements (copper is often deficient) and other areas have near toxic amounts. Some plants accumulate toxic amounts such as species of Astragalus and Stanleya which may contain as much as 3000 ppm of selenium.

TABLE A3.1. SUGGESTED LEVELS OF TRACE MINERALS IN THE DIET OF MATURE BEEF CATTLE.

Element	Level in diet dry matter basis - ppm	Suggested maximum levels - ppm
Zinc	25	1000
Copper	6	115*
Manganese	20	1000
Iodine	0.1	200*
Selenium	0.1	2
Cobalt	0.1	110
Iron	30	1000
Sulfur	0.106 percent	0.4 percent

* Tolerable levels for young animals are less.

For more information check the NRC, Nutrient Requirements of Beef Cattle, 1984, National Academy Press, 2101 Constitution Ave. Washington D.C. 20418; or The Mineral Requirements of Grazing Ruminants, 1983, Ed. N.D. Grace, New Zealand Society of Animal Production; or Minerals for Grazing Ruminants in Tropical Regions, by McDowell, Conrad, Ellis and Loosli, 1983. Send to C.M. Hinton, Publication Distribution Center, IFAS Bldg. 664, Univ. of Florida, Gainesville FL. 32611.

Zinc (Zn): The first effects of a mild deficiency is lower reproductive efficiency. Spermatogenesis and the development and function of the primary and secondary sex organs in the male and all aspects of the reproductive process in the female can be adversely affected. Impaired testicular development and function due to a zinc deficiency has been described in bulls, rams and goats. Effects of a deficiency in females is less documented, but it has been shown to increase conception rates in some studies on dairy cows, sheep and goats. Other symptoms are vision impairment, excessive salivation and a failure of wounds to heal normally. Cadmium, calcium, iron, magnesium, manganese, molybdenum and selenium affect the absorption and utilization of zinc. Dietary levels of of 1000 ppm have not cause lowered production when fed to steers for up to 18 months.

Copper (Cu): The most common symptoms of deficiency are a reduced conception rate and embryonic and fetal mortality. Low fertility of cattle grazing Cu deficient pastures has been associated with delayed or depressed estrus. Cattle showing signs of copper deficiency typically have a yellowish hair coat. Sheep will often give birth to sway backed lambs. Recommendations of Cu levels required to attain optimal fertility depend upon the molybdenum, sulfur, zinc and silver levels in the diet. Mature beef animals require 4-6 ppm Cu (dry matter basis) when the molybdenum level of the diet is less than 1.5 ppm. In the areas where the diet contains high levels of molybdenum (5-6 ppm) and sulfate, the Cu requirement may have to be increased two or threefold. Copper levels of 25 ppm are considered toxic for sheep and levels over 115 ppm for cattle are not recommended. Mature animals can tolerate higher levels than can young growing animals.

Manganese (Mn): The Mn requirements for cattle is substantially higher for reproductive functions than for body growth. Mature females have a higher requirement than do feed yard animals. The main effects of a deficiency on fertility are the suppression of estrus and depressed conception rates. Other effects are lower birth weights and an increased incidence of abortions, stillbirths and fetal deformities. In dairy cows and interaction between the Mn adequacy of the diet and the calcium to phosphorus ratio has been observed. High calcium to phosphorus ratios (over 2:1) appear to increased the Mn requirement. Beef cattle grazing forages high in calcium may benefit from increased levels of dietary Mn, over the suggested level of 20 ppm. Ruminants fed Manganese at 1000 ppm of their diet had near normal growth rates.

Iodine (I): The only function of I in all mammals is as a constituent of the thyroid hormones, tri-iodothyronine and thryoxine. These hormones have important roles in the energy metabolism and protein synthesis of cells; they are closely related to heat production and oxygen consumption. A goitre is an obvious sign of I deficiency, less obvious are the effects on the fetus, reduction of brain size and function and slowing of the development of the lungs. Most I deficiencies are corrected by use of an iodized salt. Calves fed a diet containing 50 mg I/Kg DW, suffered reduced weight gains and lower feed intakes; other signs were coughing and a profuse nasal discharge. However lactating cows fed 200 ppm showed no adverse affects.

Selenium (Se): Up until 1957, selenium was regarded only as a toxic substance, however, it is now known as an essential dietary element. Selenium deficiencies have been associated with reproductive performance, however the effects of Se supplementation have not been consistent. White muscle disease is characterized by white muscles, heart failure and paralysis. The metabolic function of Se is closely tied to vitamin E and together they function to protect membranes from oxidative degeneration. Deficiencies apparently cause a reduction in the fertilization rate and also increase theénumber of retained placentas. Toxic levels are responsible for blind staggers and alkali disease. Symptoms are loss of appetite, loss of tail hair, sloughing of hooves, respiratory failure, starvation, thirst and death. Toxic levels depend on the form that the selenium is in, the duration that the Se is consumed and other components of the diet, for example high protein diets and arsenic appear to offer some protection as does certain sulfur compounds. The recommended levels of intake are 0.1 ppm for feed yard steers and heifers, 0.05 to 0.1 for bulls and cows. The maximum tolerable level is 2 ppm of the diet on a dry matter basis.

Cobalt (Co): Cobalt is synthesised by rumen organisms into vitamin B-12. This vitamin is used in the utilization of propionic acid, which is a volatile free fatty acid and is a major source of energy. Animals with insufficient Co have low growth rates, dull harsh hair coats, and a watery discharge from the eyes. The efficiency of topdressing of Co deficient pasture is low, however the suggested rate is between 0.31 and 1.25 lbs of cobalt sulfate per acre per year, depending on soil type. Oral dosing or vitamin B-12 injections are also used. Cobalt is toxic and in sheep single doses of 44 to 66 mg/Kg body weight were occasionally fatal, a dose of 330 mg/Kg body weight was lethal. In calves a daily dose of 0.9 mg/Kg body weight caused loss of appetite and muscular coordination. In general toxic levels are 300 times the daily recommended rate.

Iron (Fe): While Fe is of considerable importance in animal nutrition the actual content in the body is relatively small, 0.005 to 0.007%. Most of the Fe is found in the red colored pigments of blood and muscle. Young animals need higher rates of iron than do mature animals. Iron deficiency shows up as loss of appetite, reduced growth rates and animals are listless and easily fatigued. Iron requirements of calves is given 100 ppm and that of mature cattle 50 ppm. Iron toxicity shows up as reduced gain, diarrhea, hypothermia and metabolic acidosis. The NRC recommends a maximum tolerable level for cattle as 1000 ppm.

Sulfur (S): The total body content of sulfur is about 0.15 percent and is found in virtually every tissue and organ of the body. The rumen microbes require sulfur and have the ability to convert inorganic sulfur into organic compounds that can be utilized by the animal. The signs of deficiency include reduced appetite, weight loss, weakness, excessive salivation, watery eyes, dullness, emaciation and death. Minimum levels are 0.106 percent of the diet on a dry matter basis. Toxic levels of sulfur result in restlessness, diarrhea, muscular twitching, dyspnea and in prolonged cases inactivity followed by death. The suggested maximum level for sulfur in the diet is 0.4 percent, but some suggest a lower amount for lactating dairy cows, 0.26 percent.

APPENDIX 4. ENERGY AND PROTEIN REQUIREMENTS FOR STEERS

TABLE A4.1. DAILY REQUIREMENTS OF FEED YARD ANIMALS

EIGHT (lbs)	GAIN (lbs)	PROTEIN (lbs)	% con-centrate	MCAL NEM	MCAL NEG	MCAL ME
331	0	0.51	0	3.3	0	5.6
	1.1	0.97	25	3.3	1.2	9.0
	1.5	1.08	45	3.3	1.73	9.6
	2.0	1.19	75	3.3	2.27	10.7
441	0	0.66	0	4.1	0	7.0
	1.1	1.26	15	4.1	1.49	12.1
	1.5	1.35	25	4.1	2.14	13.0
	2.0	1.35	60	4.1	2.82	13.3
551	0	0.77	0	4.84	0	8.2
	1.5	1.37	40	4.84	2.53	14.4
	2.0	1.52	50	4.84	3.33	16.2
	2.4	1.61	75	4.84	4.17	17.0
661	0	0.88	0	5.55	0	9.4
	2.0	1.79	40	5.55	3.82	19.5
	2.4	1.81	75	5.55	4.78	20.4
772	0	1.01	0	6.24	0	10.6
	2.0	1.76	50	6.24	4.29	20.8
	2.4	1.83	75	6.24	5.36	22.4

NOTE: This table is designed for feed yard use and was adapted from the Nutrient Requirements of Beef Cattle, National Research Council, National Academy of Sciences.

If you have an steer that weighs 441 lbs and you want it to gain 1.5 lbs per day, the ME that must be available to do this is 13.0 Mcal. To find out how much digestible energy (DE) is needed divide the ME by 0.81. DE = ME/0.81 or DE = 13/0.81 = 16.05 Mcal. The feed source has a digestibility associated with it. For example assume the source is a high quality kikuyu grass with a digestibility of 68%. A pound of most any grass contains 2 Mcal of gross energy. Therefore a pound of kikuyu will have a DE of (2 times 0.68) 1.36Mcal. Since 16.05 Mcals are required, the animal must eat 11.8 lbs of kikuyu per day (16.05/1.36). The amount of material a grazing animal can be expected to eat will vary between 2.5 and 3.0% of the animals live weight. Thus a 441 lb steer can be expected to consume between 11 and 13 pounds of kikuyu grass. Consequently a gain of 1.5 lbs/day is a reasonable expectation.

The percent protein the kikuyu must contain can also be determined. The amount of protein required, from the above table, is 1.35 lbs, this divided by the minimal amount the animal must eat (11.8 lbs) times 100 gives the percent protein, in this case 11.4%, that the feed must contain.

The next step is determine how much feed should be offered to the animal, in order to obtain these gains. In general, if maximum performance is desired from the forage on offer, 4 to 5 times the actual amount to be consumed must be offered the animal. In the above case if the animal is expected to gain 1.5 lbs/day, and the minimum amount of intake to do that is 11.8 lbs of kikuyu grass, then the animal must be offered between 48 to 60 lbs of forage per day. Heifers generally require slightly more energy and protein than steers.

TABLE A4.2. DAILY FEED REQUIREMENTS, IN Mcals PER DAY, FOR FATTENING STEERS ON PASTURE: Adapted from Sheep and Cattle Nutrition.

liveweight	liveweight gain (lbs/day)						
(lbs)	0*	0.55	1.1	1.65	2.2	2.76	3.3
220	4.5	5.5	6.7	7.6	8.8	9.8	11.0
330	6.0	7.4	8.8	10.3	11.7	13.1	14.6
441	7.6	9.3	11.2	12.9	14.8	16.5	18.4
551	8.8	11.0	13.1	15.3	17.2	19.4	21.5
662	10.3	12.7	15.1	17.4	20.1	22.5	24.9
772	11.5	14.1	17.0	19.6	22.5	25.1	27.7
882	12.7	15.8	18.6	21.7	24.6	27.7	30.8
992	13.9	17.2	20.6	23.7	27.0	30.4	33.7
1103	15.1	18.6	22.2	25.8	29.4	33.0	36.3

* Maintenance requirements under cold conditions are 20-25% higher.

APPENDIX 5. FEED INFORMATION FOR THE NEWBORN: Prepared by Steve Damron.

TABLE A5.1. SUGGESTED WHOLE MILK FEEDING SCHEDULE FOR CALVES

Birth weight (lbs)	1st* week	2nd week	3rd week	4th week	5th week
	------pounds of whole milk per day-----				
50-63	5.0	5.5	6.0	5.0	4.0
64-73	5.5	6.0	7.0	6.0	4.0
74-83	6.0	7.0	8.0	7.0	4.0
84-93	7.0	8.0	9.0	8.0	5.0
94-103	8.0	9.0	10.0	8.0	5.0
103-113	9.0	10.0	11.0	9.0	5.0
113 and above	10.0	11.0	12.0	10.0	5.0

* three to four feedings a day the first week, then twice a day

TABLE A5.2. FEEDING SCHEDULE - OTHER SPECIES

species	age	frequency of feeding	amount per feeding
FOALS	first 7 days	hourly	6-8 ounces (oz)
	2nd week	every 3 hrs	16-24 oz
	3rd & 4th week	every 4 hrs	24-32 oz
	5 week	every 6 hrs	32-48 oz
	weaning	every 6 hrs until weaned	no more than 48 oz per day
PIGS	first 7 days	1.5 to 2.0 hrs	15-20 ml (cc)
	2nd week to weaning	constant choice	free choice
GOATS	first 2 weeks	every 8 hrs	10-12 oz
	2-3 weeks	every 12 hrs	36-48 oz per day
	3 weeks to weaning	every 12 hrs	64 oz per day
LAMBS	first 24 hrs	every 2 hrs	not over 2 oz
	2nd to 4th day	every 3 hrs	3-4 oz
	5th to 7th day	every 4 hrs	4-5 oz
	2nd week	every 4 hrs	5 oz
	3rd to 8th week	every 6 hrs	8-10 oz

TABLE A5.3. COMPOSITION OF DIFFERENT MAMMALS MILK

species	protein (grams)	fat (g)	carbohydrates (g)	energy (Kcal)
Rabbit	10.3	15.2	1.8	208
Cat	10.8	10.8	3.7	174
Lion	9.3	17.5	3.4	228
Dog	7.5	9.5	3.8	145
Wolf	9.2	9.6	3.4	154
Pig	5.6	8.3	5.0	128
Cow	3.2	3.7	4.6	71
Goat	2.9	3.8	4.7	70
Sheep	5.0	7.1	4.9	113
Horse	1.9	1.3	6.9	50
Donkey	1.9	0.6	6.1	40
Human	1.1	4.2	7.0	73

TABLE A5.4. MILK REPLACERS FOR DIFFERENT LIVESTOCK

COW:

A. Processed milk from grocery store (expensive), or commercial milk replacer (available and quite good)

B. Goats milk.

HORSE:

A. 1 pint condensed milk, 1 pint warm water, 2 tablespoons syrup, 4 tablespoons lime water.*

B. 1 pint 2% cows milk, 4 oz. lime water*, 1 teaspoon sugar.

C. 20 oz. whole cows milk, 12 oz. lime water, 2 oz sugar.

D. Commercial prepared formula.

* Limewater is 1 part lime (calcium hydroxide) mixed with 700 parts water. A pharmacist will mix this for you.

PIG:

A. 1 quart of cows milk, 1/2 pint of "half & half", 1 raw egg.

B. Glucose (dextrose) and skim milk in equal proportions.

C. Commercial prepared formula.

GOAT:

A. Whole cows milk, or cow milk replacer.

SHEEP:

A. Add white lard or butterfat to cows milk replacer powder until it is 30% fat. Mix 1 part powder to 4 parts water.

B. Mix 1 part cream with 3 parts cows milk.

C. Commercial ewe milk replacer.

APPENDIX 6. ADDITIONAL FENCE AND UNIT DESIGNS

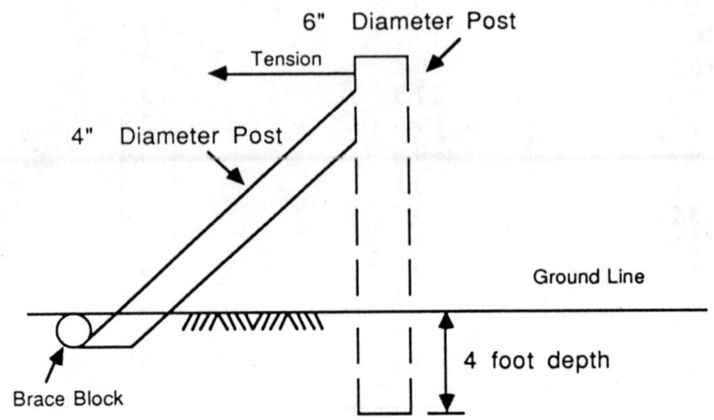

FIGURE A6.1. "N" BRANCE

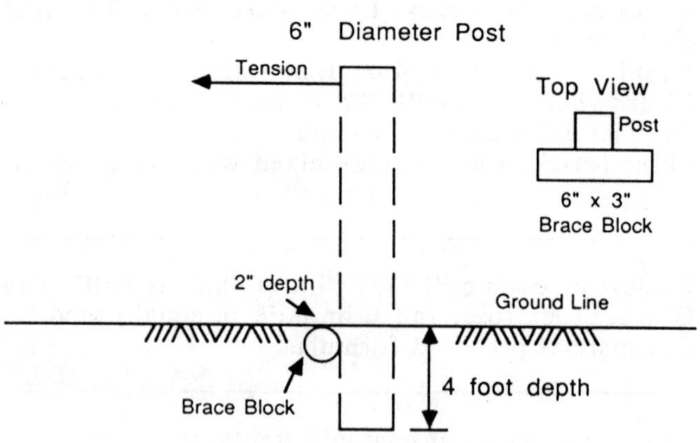

FIGURE A6.2. DEAD MAN POST AND BRACE

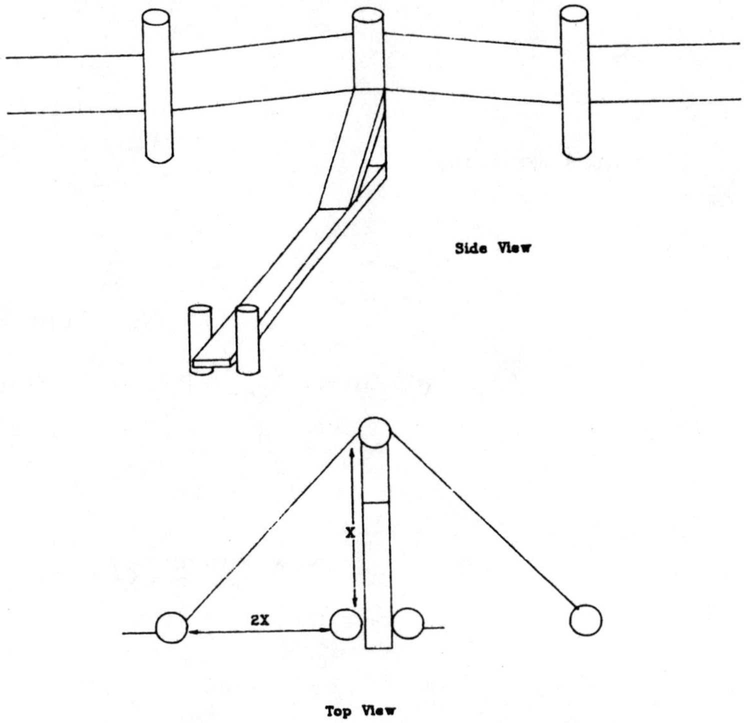

FIGURE A6.3. AUSTRALIAN GATE

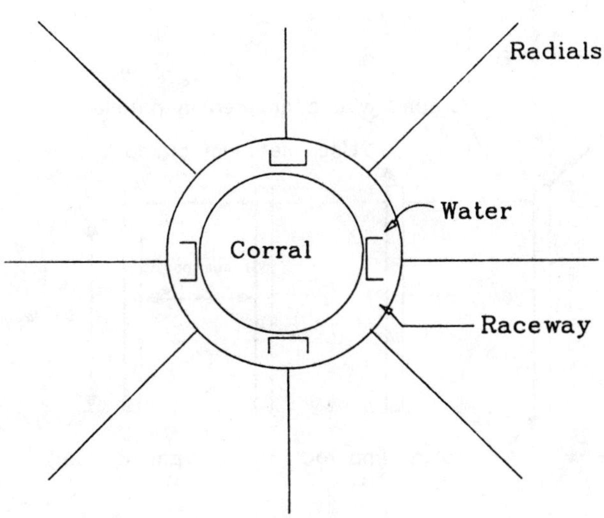

FIGURE A6.4. HUB DESIGN

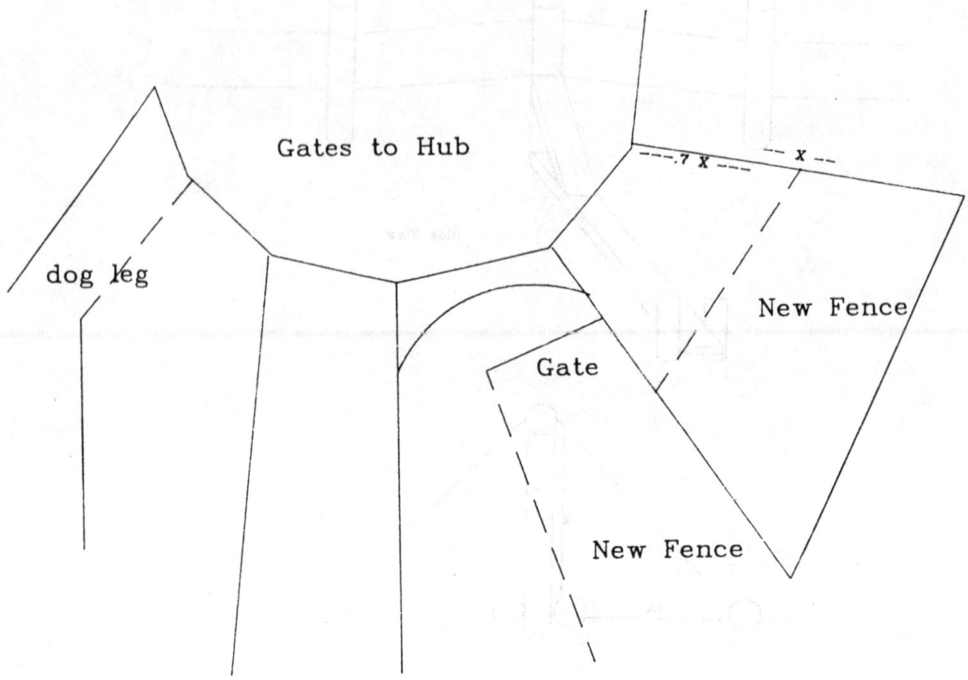

FIGURE A6.5. PADDOCK EXPANSION FOR HUB

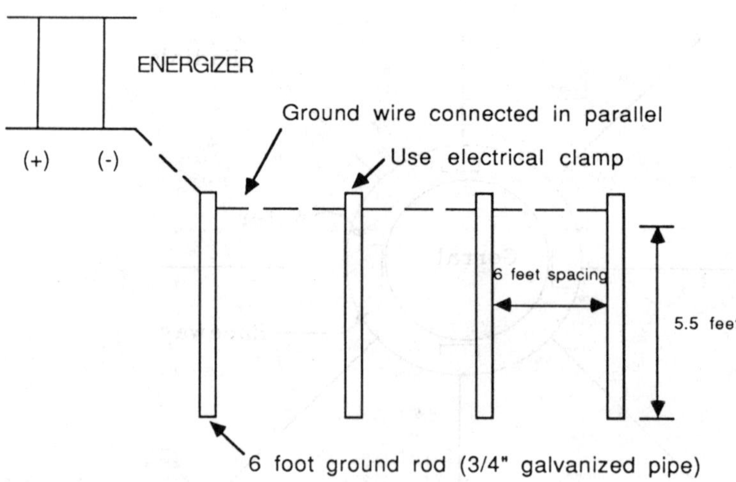

FIGURE A6.6. PROPER GROUNDING

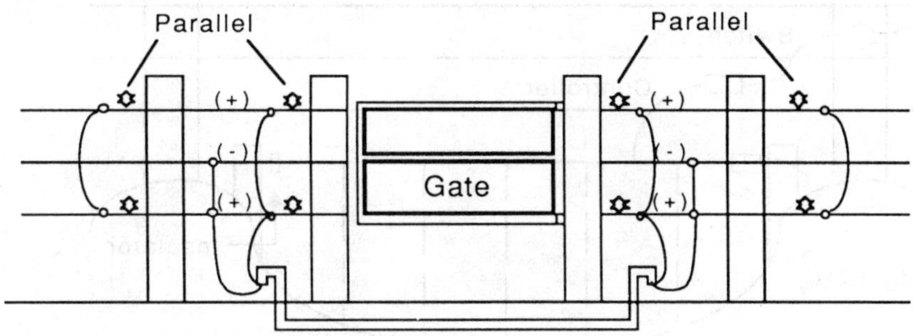

FIGURE A6.7. UNDERGROUND GATE TRANSMISSION

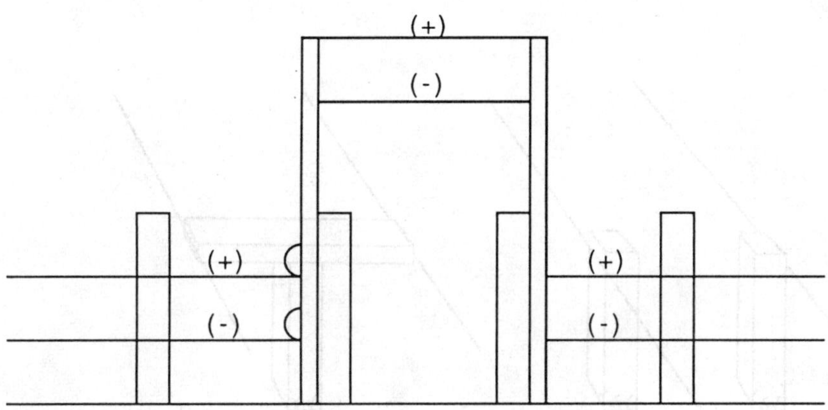

FIGURE A6.8. OVERHEAD GATE TRANSMISSION

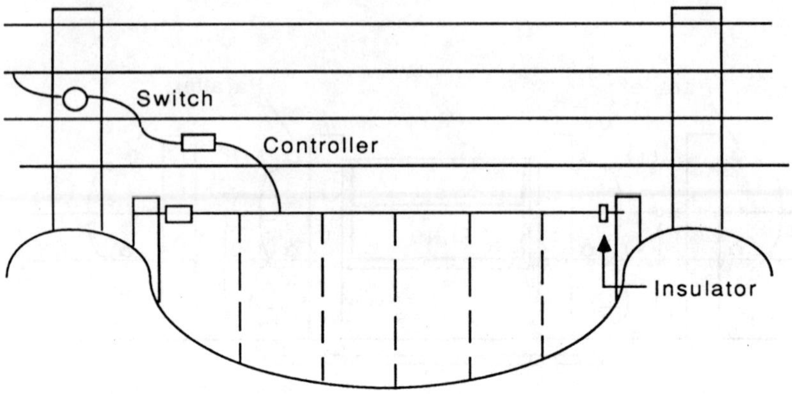

FIGURE A6.9. WATER GAP

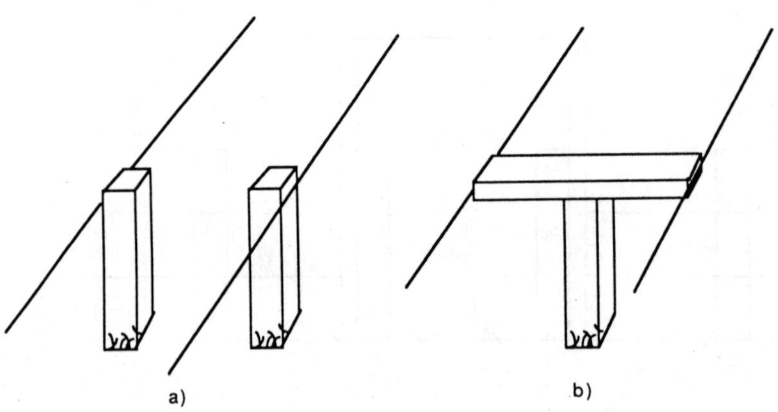

a) Conventional 2-wire visual barrier fence (1/2 No. 3 Post)
b) T-bar 2-wire fence

FIGURE A6.10. TWO WIRE VISUAL BARRIER FENCE

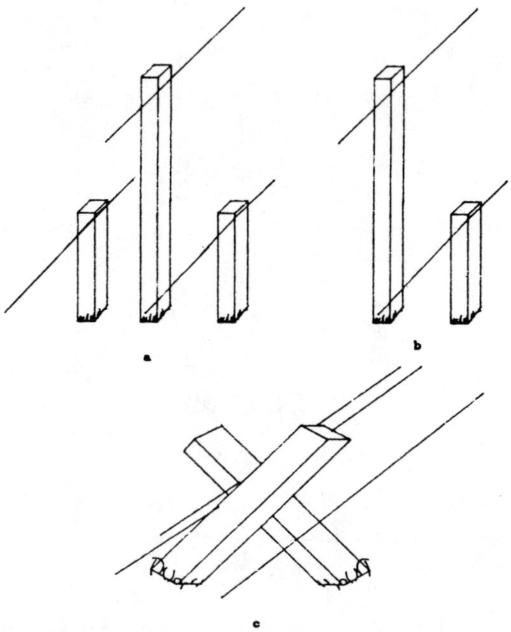

a) Conventional 3–wire visual barrier fence
b) L–configuration 3–wire fence
c) Triangular 3–wire fence

FIGURE A6.11. THREE WIRE VISUAL BARRIER FENCE

Used with permission of NEA.

Rolling along.

INDEX

Alleyway, 215
Allocation, 231
Amine, 271
AMPS, 171
Animal, 93,131
  behavior, 94
  health, 132
  nutrition, 106
  units, 119,232

Block design, 208
Branding, 147
Break grazing, 212
Bull beef, 280
Bull forage, 235,

C_3 Cycle, 65
C_4 Cycle, 65
CAM Cycle, 66
Calf crop, 37
Calibration, 268
Calorie, 3
Calves, 146
Canopy cover, 67
Carrying capacity, 201,292
Cash flow, 42
Castration, 149
Cation, 50
Colostrum, 144
Come away, 293
Computer, 262
  hardware, 264
  software, 265
Continuous grazing, 224
Contract agreement, 27
Control, 6,222
Controlled grazing, 223,292
Corridor, 215
Cost, 32
Cow day, 233
Cow/Calf feed, 235
Cycle, 160,230
  mineral cycle, 161
  water cycle, 160

Days grazed, 226
Days rest, 226
Dehorning, 149
Diseases, 140
Drought, 299
Drugs, 135
Dry matter, 76

Ecology, 152
Ecosystem, 154
Economics, 31,245,
Electronics, 276
Employees, 19
Energizer, 173
Energy, 153
  flow, 2,117
  pulse, 156
  requirements, 337
Enterprise analysis, 31
Environment, 242
  extensive, 242
  intensive, 244
Ester, 271

Feed on offer, 82
Feed, 234
  banking, 255
  budgets, 234
  requirements, 338-341
Fencing,188
  designs, 342-347
Fertilization, 58
Financial analysis, 31
Fire, 166,167
Flip-Flop, 226
Flogging, 77
Floodgates, 195
Foliar application, 267
Follow through, 206
Forage, 231
  allocation, 231
  conservation, 254
  legumes, 84
  nomenclature, 324-329
  quality, 79,234
  quantity, 79
  toxic, 88

Game fences, 190
Gates, 193
Genetics, 159,123
Goals, 8,10
Gomer, 98
Grazing, 221,286
  continuous, 224
  density, 226
  formulas, 226,232
  intensive, 287
  pressure, 223,226
  rotational, 228
  tether, 223
  unit, 226

Grounding, 175
Growth rate, 301

Hay, 252
Health, animal, 132
Heavy metal, 11
Herbicides, 267
Herd, 38
  replacement, 40
  structure, 38
Herding, 100
Holistic, 37
Hub, 204

Intensive grazing, 287
  Hawaii, 306,308,318
Income, 31
Intercropping, 256
Inventory, 11
Irrigation, 55

Japanese quarter horses, 275

Lamb, 82
Lax grazing, 85
Leaf area index, 67
Leaseholder, 26
legumes, 84
Livestock fences, 190
Living hay, 255

Management, 20,23,241
  extensive, 241
  intensive, 242, 287
Manure, 60
Marketing,284
Minerals, 330-336
Mob, 72
Motivation, 23

New Zealand block, 211
Nitrogen, 48,58

Organic matters, 50
Organization, 17
Objectives, 11
OHM, 171

Paddock, 226
Parasites, 140
Pasture meters, 277
Pasture, 81
People, 18
Performance targets, 16
PH, 53
Planning, 288
Production targets, 14,293
Protein requirements, 337
Put and take, 225

Range, 81,166
Rational grazing, 229
Reproduction, 121
Residual dry matter, 76
Rotational grazing, 228

Savory grazing method
  (SGM), 229
Scales, 278
Selection, 124
Set stocking, 224
Shifting, 238
Sigmoid growth curve, 72
Silage, 248
Soil, 47
  erosion, 47
  fertility, 55
  moisture, 302
  organism, 52
  profile, 51
Standing crop, 299
Stock flows, 37
Stocker feed, 234
Stocking rate, 225,226
Stress, 102
Strip grazing, 212
Subdivision, 226
Supplementation,257
Surfactants, 272

Tether grazing, 223
Texas block, 210
Tools, 289
Toxic, 88
Training,185
Tucker, 296
Turnover rate, 37

Vitamins, 109-110
Volts, 171

Wagon wheel, 202
Watts, 171
Wildlife, 11,158
Wire, 176
Worming, 139

Yield, 33-37,301-302